ISBN 978-1-330-13985-1
PIBN 10035511

1 MONTH OF
FREE
READING

at

www.ForgottenBooks.com

By purchasing this book you are eligible for one month membership to ForgottenBooks.com, giving you unlimited access to our entire collection of over 700,000 titles via our web site and mobile apps.

To claim your free month visit:
www.forgottenbooks.com/free35511

English
Français
Deutsche
Italiano
Español
Português

www.forgottenbooks.com

Mythology Photography **Fiction**
Fishing Christianity **Art** Cooking
Essays Buddhism Freemasonry
Medicine **Biology** Music **Ancient
Egypt** Evolution Carpentry Physics
Dance Geology **Mathematics** Fitness
Shakespeare **Folklore** Yoga Marketing
Confidence Immortality Biographies
Poetry **Psychology** Witchcraft
Electronics Chemistry History **Law**
Accounting **Philosophy** Anthropology
Alchemy Drama Quantum Mechanics
Atheism Sexual Health **Ancient History**
Entrepreneurship Languages Sport
Paleontology Needlework Islam
Metaphysics Investment Archaeology
Parenting Statistics Criminology
Motivational

THE TROPICS

THEIR RESOURCES, PEOPLE AND FUTURE

A DESCRIPTION OF THE TROPICAL LANDS OF
AFRICA, ASIA, CENTRAL AND SOUTH AMERICA,
AUSTRALASIA AND THE PACIFIC ; THEIR
NATURAL PRODUCTS, SCENERY, INHABITANTS
AND INDUSTRIES, AND THE POSSIBILITIES OF
THEIR FUTURE DEVELOPMENT

BY

C. R. ENOCK, C.E., F.R.G.S.

AUTHOR OF "THE ANDES AND THE AMAZON"
"THE GREAT PACIFIC COAST," ETC.

WITH 64 ILLUSTRATIONS AND MAPS

LONDON
GRANT RICHARDS LTD.
ST MARTIN'S STREET
LEICESTER SQUARE
MDCCCCXV

PRINTED BY THE RIVERSIDE PRESS LIMITED
EDINBURGH

CONTENTS

v

CONTENTS

LIST OF ILLUSTRATIONS

vii

For some of the illustrations in this book the author is indebted to the following sources:—Royal Mail Steam Packet Company; Soudan Government Railway Agency in London; The Church Missionary Society; *The Journal of the Royal Geographical Society*; *National Geographic Magazine*, Washington; British South Africa Company, and Elder Dempster Steamship Company.

INTRODUCTION

THE purpose of this book is, first, to provide the general reader and those specially interested in the natural resources, people and topography of the lands within the tropics with a broad description, in the compass of a single volume, of these lands; and, second, to afford a basis for the consideration, as far as it concerns the tropical races, of what the author (here and elsewhere [1]) has ventured to term a new science of "constructive human geography," or constructive economics. In his study of the world's peoples and their surroundings, in the endeavour to educe and establish this proposed science, the author found there was no existing book dealing with the tropics as a whole. The field is, of course, an enormous one, covering, as it does, about one-third of the land surface of the globe, and embracing lands and peoples so widespread and varied as those within the torrid zone: and it is with some diffidence that the result of the task is put forward. Yet it is believed that the work—apart from its second and special purpose—will fill a vacant place in the literature of topography, and it is to be recollected that the social, economic and climatic conditions of the people and lands of the tropics may be regarded as an entity from the point of view of their relation with the people of the temperate zone, who are called upon to administrate or develop what are, in effect, the world's backward and coloured communities.

What, it will be asked, is this proposed science of constructive human geography? The term is used for lack of a better one. Anthropogeography, or human geography, is already an existing branch of geography, well known to students of that science. The main difference—and it is a wide one—is in the element denoted by the word constructive. The human geography of the geographer is observatory, not constructive. It examines and records the effects of human agency on the face of the globe; modifications or features brought about—generally

[1] In lectures before the British Association and Royal Society of Arts, etc.

i

undesignedly—by his actions, his habitation, agricultural, forestal, engineering and other operations, in times ancient or modern. Just as geography is a record of what nature has done on the face of the earth, so anthropogeography is a record of what man has done on the face of nature. Constructive human geography transcends these. It aims, or rather will aim to set forth in addition what man *should* do on the face of the earth, and thus it enters upon the field of economics, and even of ethics. But economics itself as a science is not yet constructive in the sense here meant. There is, in brief, no real " science of living on the earth," so far, and this is somewhat remarkable, in view of the fact that all other sciences have developed, embracing every phenomena and natural circumstance. Economics is little more than an observatory science so far, and scarcely emerges from its academic beginnings.

Thus " constructive human geography " will bear somewhat the relation to geography that medicine bears to anatomy. Geography, like anatomy, is concerned with examining and depicting an existing framework. Medicine, on the other hand, is constructive. The business of the physician is to build up, to cure. So on the bones of the world's physical structure will constructive human geography seek to build up the life of humanity, to cure it of those ills which are resultant upon the misunderstanding of its togographical and economic environment (for, as will be shown, topography and economics are in reality inseparably bound up together).

Constructive human geography, or the science of corporate living on the earth, thus will seek the way by which the reaction or adjustment of mankind to its topographical and economic environment may be brought about ; the permanent adaptation of man's economic life to his surroundings, and the ultimate attainment of " economic equilibrium." It finds the economic and industrial life of the world at variance with its surroundings, governed by haphazard, the prey of unexplained or unstudied social phenomena ; conditions which have existed throughout the ages and must inevitably exist—however advanced our civilisation—until corporate life be taken hold of as a science and brought into harmony with its topographical environment.

The proposed science embodies two main lines of action. First is a comprehensive and impartial survey of the conditions of life ; of the world's people and how they live ; a sort of stock-taking of natural resources, topographical potentialities, and human resources and capabilities: What is the natural wealth of the world ? What is the waste ? What is the amount and distribution of the things by which mankind lives, and the possibilities of the places he inhabits ?—such are the matters which will form the foundation of the science, and which show how greatly it must rest upon geography, especially upon what has come to be termed "regional" geography. This is the observatory part of the science. The second is the constructive part. We require to find out how a balance may be brought about between these natural resources and topographical potentialities of the globe and the human element which inhabits it.

Here we plunge into what is practically a new field. We have never entered upon a diagnosis of life of this character with a constructive purpose in view. From the times of Pharaoh to the present age man has never attempted a comprehensive grasp of the materials by which he lives, nor striven to work towards that economic world-order which is undoubtedly the goal of evolution.[1] Empires and civilisations have risen and fallen, and will continue to rise and fall until man strives to lay the basis of a science of corporate living on the earth. We have, of course, the enormous work that has been carried out by geographers, economists, sociologists and others, who have acquired great store of scientific detail about countries and communities, such as will form the material for constructive effort. But that broad and intensive science of corporate living which we now require, if society is to consolidate and advance—nay, even to exist in security—calls for far-reaching constructive methods, combined with ethical considerations : methods such as geographers and economists have not conceived, and which it was scarcely their business to conceive.

We cannot here go into the details of this proposed science except to accentuate the concrete manner in which topography bears upon it. It is hardly necessary to say that such a science does not only affect tropical lands, whose description forms the subject of this volume. It is a matter which even more vitally affects the most advanced industrial and manufacturing nations, who themselves have to seek the method of setting their houses

---

[1] And the purpose of ethics and of Providence.

in order, or suffer from perennial strife, disorder and waste. But the lands of the tropics are equally concerned therein.

The relation of topography to the subject is a fundamental one. Geography shows us that mankind is made up of races and countries. But mankind is not only composed of races and countries but of " localities," or neighbourhoods. What is the natural unit of corporate life ? It is the locality or neighbourhood. The locality is the natural " crystallisation of humanity," so to speak: the native subdivision of man-earth. Thus a nation or country consists of numerous social groups or " economic tribes." The extent of these tribes or localities (whether it be in Great Britain or Darkest Africa) is governed by what may be termed their natural " radius of action "— that is, the radius through which their daily activities may ordinarily range. In civilised communities it is the radius of the baker's cart and the morning's walk to work, and the fact varies little even in savage or primitive communities. We can never escape from this economic " tribal " grouping. And the endeavour to disarrange or destroy it is responsible for many of the economic and social ills of civilised life, although the fact is not yet recognised. For example, we take our workers, in manufacturing towns, considerable distances from their home to their work, and back at night, using up both mechanical energy and human vitality in transport ; we bring into a locality from the outside articles and commodities which ought to be made on the spot, which involves various forms of waste, commercialises industry, leads to unemployment, congestion of industries in certain spots, and other illogical and prejudicial matters.

For, in the study of constructive human geography we have very early to note or accentuate this fact : that human life depends upon industry. All economic facts and theories come down sooner or later to what men do every day with their hands or their brains—principally the former. If, therefore, as we have said, people's lives are inevitably lived in their localities, and ought to be governed by their natural radius of action, and if industry and manufacture is the only basis of livelihood, should it not naturally follow that their industries and occupations should be of the locality ? But this is not the case, in great degree. The commodities used in a locality—food-stuffs, clothing, furniture and many other matters—in the majority of cases are not produced in each locality, but are brought in from the outside. The business and profit of a locality should consist in making—as far as is physically possible—the things it has to consume. We see, however, that manufacture and many agricultural and other industries are centralised in the places where they happen to have grown up. The people of many places are deprived of the right of manufacture or production ; their local markets are neglected, or invaded by outsiders. We see that agriculture is generally divorced in any given locality from manufacture, instead of the two things being supplemental to each other. We see that countries import and export a host of articles regardless of whether they could or could not be produced upon the spot where they are to be consumed : a system which follows in great part the dictates of a sheer commercialism and has its basis in profit-making by those who carry it on, neglectful of scientific considerations and careless of the waste of energy, human and mechanical, which it involves. We carry over seas and mountains goods of all description : goods which under a more intelligent system could be made where they are required, yielding employment to the consumers, economising effort and tending towards economic equilibrium. We have not yet discovered, apparently, the law of nature that organisms must be, *as far as is physically possible*, self-supplying and self-supporting. The only excuse that can be advanced for the reverse process is that it is the present way of the world ; that it is the stage at which human industry has arrived. Oversea trade has conferred enormous benefits upon the world, and has been one of the most powerful agents in opening up the darker places of the earth to civilisation, but it must now be accompanied by more scientific methods. Moreover, foreign trade has now evolved to such an extent that the acute rivalry for the possession of foreign " markets " which it has induced has become a cause of friction between nations. This rivalry is, in large part, one of the causes of the Great War. The advanced nations are eagerly bent upon commercial dominance. They appear to regard trade, the making, buying and selling of manufactured goods, or of raw material, as synonymous with civilisation. This is a phase which cannot endure. Trade, moreover, cannot expand indefinitely. Rivalry will therefore become more and more bitter, with possible wars in the future between those nations engaged in such commercial throat-cutting. Will it be possible to curtail such rivalry ; to define and agree upon commercial spheres of influence ; to establish a quota of export for each exporting nation ? It could be argued that such a course is necessary if peace is to be maintained in the world. But unfortunately the " pacifists " (who bitterly condemn rivalry in armaments) are generally acute traders and have not realised the element of commercial rivalry as a war factor. Herein lies some hypocrisy.

Thus we see how strongly topographical and industrial considerations enter into the proposed science of constructive human geography, and how necessary such a science is. To develop the industries of each country and each locality, to the end that it may become as far as possible self-supplying, becomes a positive duty. Under such a development trade rivalry would perforce diminish and economic equilibrium be approached. This is the direct antithesis of that recent school of thought which proclaimed that the growth of international trade and finance was a safeguard against war ; that war " would not pay," because of the dependence of the nations upon each other's commercial and banking operations. Constructive human geography will rather teach that it is in independence, not in interdependence, that security will be maintained.

How strongly these considerations affect the tropical and coloured peoples is shown by the fact that such communities are regarded specially as available " markets " for oversea trade. The teeming coloured people of the torrid and sub-tropical zones have not arrived at the modern stage of manufacturing what they require—their textiles, hardware and so forth—and the mills of the manufacturing nations work eagerly and unceasingly to supply them. But—and the fact has been strongly accentuated throughout these pages—it is not necessarily that these backward people cannot or could not manufacture for themselves, but that they are not encouraged to do so. Their white masters prefer that they should remain producers of raw material and purchasers of manufactured goods. One of the principal motives of this book is to set forth the capabilities of the tropical people in arts, crafts and industries. We shall see, on a perusal of its pages, that often from time immemorial the black and brown races of the torrid zone have shown extreme dexterity in such, whether in textile, stone, metal or other material. It is a severe reproach to civilisation that the development of these arts should be kept back, or swamped by the trade of the white nations who hold them in economic thrall. It is bad, not only for the tropical people but for the whole of society, because it is part of an unnatural system ; a contravention of the law of the self-supplying organism, and, as already remarked, a field fertile with rivalry. It is the duty of a science of constructive human geography to combat this system ; to encourage the development of native arts and the spirit of self-support. Such a science, moreover, would tend to conserve the earth's resources (which are being wasted), to induce a simpler life, to elevate and ennoble labour, to re-create the association of art with manufacture.

It may here be exclaimed that the tropical lands must always produce great quantities of raw material and food-stuffs for export : articles which for climatic reasons cannot be produced in the temperate zones ; and this is, of course, true. But it bears out the argument that communities should become self-supporting and independent as far as is physically possible. The system of what in this book has been termed " monoculture "— that is, the production of one kind of raw material in a given place—must to some degree be maintained, but it must be shorn of its abuses and supplemented by that varied production which is the right of any community and the true basis of economic equilibrium, and consequently of civilisation.

It is a remarkable fact, and one which goes to prove the need for a scientific study of economic affairs such as that here advocated, that the political party in Britain (the greatest tropical empire in the world), which has, or professes to have, at heart in greatest measure the welfare of the poorer and labouring classes—that is, the Liberal and Radical party—should, never- theless, be the upholders of that doctrine and prevailing economic system known as Free Trade, which has as its basis the keeping open of national markets for the admission of foreign commodities. It is remarkable that this political party has never recognised that the true basis of prosperity and order would be in the utmost encouragement of home industry, whether for white communities, whether for black

b

or brown, and that such a system would promise the best development for democracy. They do not, or will not, see the injustice meted out to the coloured tropical races in making of them hewers of wood and drawers of water, and forbidding them to exercise their own ingenuity in manufacture. To produce cocoa, coffee, rubber and other raw material for the white man and to buy his cheap cotton fabrics—such seems to be the " Free Trade " ideal. The reasons for this attitude are several. Scientific ignorance is the foremost. Next is the fact that Liberals are generally traders and manufacturers, whose traditional maxim has been to buy in the cheapest and sell in the dearest markets—a maxim surely very remote from true civilisation. They have, however, certain righteous and logical reasons in their support. The opposite policy, that of Preference and Protection, under present conditions would not be likely to serve the ends of economic equilibrium or social advancement, for the reason that the benefits of the increase of such domestic business as it would produce would tend to accrue to those few who would control it, and lead to the growth of trusts and monopolies, which would offset any advantages gained. The truth of this assertion can be shown by comparing the industrial life of those nations which respectively exist under Free Trade or Protection—for example, Great Britain and the United States—when it will be seen that the one is no better nor worse off than the other.[1] It is here that the proposed science of constructive human geography would operate, with its doctrine that not only nations but localities should be, as far as physically possible, self-supporting and independent. The localities of a country, the towns and districts, require economic protection from each other as much as nations require it, and the development of their own industries within their own radius of action.[2]

The absence of the constructive or human element in geography is noteworthy. An example will illustrate this. Let us open any of the geographical text-books, such as are produced in their scores for the use of schools, and turn to the generally very brief paragraph dealing with native industries in this or that land. We are told, perhaps in one breath, that the natives

[1] The author has examined both systems on the spot.
[2] Not, of course, in a mediæval, but a scientific way.

are dexterous in their textile industries and that the chief articles of import are Manchester (or other) cotton goods. If we consult a work of a more extensive character we may perhaps be told that such native industries have been swamped and destroyed by such imported goods.[1] But in none of these books shall we find any suggestion of the immorality of this system of commercialism. It does not seem to occur to their writers that it is a grave economic error to destroy native industries. The suggestion is never made that they should be built up.

Geography as a physical science has made great strides of late. Those who follow it will note how intensively land forms, sea forms, mountain structure, wind action, river work, ice sculpture, climatic conditions, distribution of human races, animals and plants, and almost all natural phenomena and circumstance are closely studied by an ever-increasing band of scientific (men and women) geographers. It is, of course, no reproach to geography to say that the constructive human element is absent from it. But the comparison is curious—the enormous and painstaking effort expended upon these natural phenomena and the entire absence of a science which should deal constructively with human phenomena and conditions. We cannot say that geography is not constructive in its way, but it has not yet learned the real connection between the life of man on the earth and the inanimate things of the earth. Geography, for example, will point to the destruction of forests, but the destruction of human life by famine, or the destruction of native industry by commercialism, leave it quite cold. Indeed it is part of the work of geography—commercial geography— to encourage commercialism, to lay bare the conditions of weakness in native economic life, showing where imports may be taken in, or the natives set to work producing raw material for export, without any particular regard for those almost indefinable rights of local markets, or of local economic convenience, and the needs of the future.

We have illustrated this geographical connivance in economic outrage—to use, perhaps, rather a severe term—in the case of

[1] One of the worst examples is that of Dacca, on the Ganges, early a famous centre of textile production, including the most beautiful muslins. But Dacca was ruined and declined mainly " because it could not compete with Manchester ! "

the Burma and India textile industries. Many other examples occur throughout this book. We may here cite another. In the dense forests of tropical America, those of Yucatan, are found exceedingly beautiful ancient buildings, the work of the old Mayas. The stage of culture reached by these people must have been a very high one. Indeed a recent observer of high standing has proclaimed that the Mayas " were the equal of the Greeks, and their brilliant ideas as numerous." This is possibly true. But it does not occur to any geographer to ask if the descendants of these people, who now are peons, or practically serfs and semi-slaves on the plantations of Mexican millionaire hemp planters, under the lowest conditions of life, are not capable of being built up again to a high civilisation. There are numerous geographers and archæologists who spend praiseworthy energy in the examination of ancient ruins, in any part of the world, or in expeditions to the most remote places, but at present we have not arrived at the idea of expenditure upon constructive human geography. We may examine the rocky structure, climate and natural history of the most dismal valley, but we do not ask : Can these dry bones live ?

It is possible that there is something of this feeling underlying the popular view, in some cases, of polar and other exploration. We have heard it asked, both by the educated and the uneducated, when the heroic exploits—and the heavy expenditure —of expeditions to the North or South Pole have been under discussion : What is the use of it ? There is little defence of such a question ; it is shortsighted, to say the least. The value of such expeditions and explorations, whether in a material or moral sense, is very great. But the inference is, although it is not expressed or expressible, that such things are done when constructive human geographical exploration is left undone.

It may be that the same idea underlay the words of the late Dr Alfred Russel Wallace, who, a little before his death, was looking over the author's manuscript of a hitherto unpublished work[1] with the view of writing an introduction thereto. " Your book should create a new science," he said, " but, if I may advise, do not call it a geography. The word will put

[1] *The Need for a Constructive World-Culture : a New Science of Human Geography and Industry-Planning.*

people off, and they will think it is only another exposition of what they regard as a *capitalistic* science." [1]

Whatever may underlie such a view, it can scarcely be said that it is the business of physical geography to enter into matters connected with economics and sociology. It is a thing apart. The argument here adduced is rather that the world now requires a new science, whose business shall be to teach how corporate human life on the earth may best be lived; to correlate natural and topographical conditions with human requirements; to study the real condition underlying labour and industry, without fear or favour; to strive to do for the body politic what the physician does for the body. Such a science will be based upon geographical and topographical conditions to a large extent. Its first axiom will be: That the world is capable of maintaining all its people in plenty, and that this primary and rational condition can be carried out if the earth's resources are scientifically organised. Such a science has never yet arisen. Why? Largely because its teachings would clash with what are termed selfish interests. Those early geographers who proclaimed that the earth was round, and that it moved, were burnt or tortured, or held up to ridicule. But if the world no longer directly threatens or opposes the advance of science, nevertheless prejudice, custom and selfishness have still to be overcome. If it be possible to inculcate a greater spirit of detachment, a much greater spirit of progress might be aroused. The earth and its kindly fruits form, in reality, an enormous property, heritage or estate for the human race, but if we are now and hereafter to profit thereby in full degree an organising science will have to set up its machinery and control to direct us. Furthermore, a science of constructive human geography is one which might enlist the sympathy and even the active work of a very large class of people. The poorest peasant in his village could work towards putting into practice its principles and the rich man and the unoccupied would find wide scope in its opportunities. It would, moreover, make for patriotism, for the sense of nationality, for the feeling of local pride, as against the teachings of that mistaken internationalism which of late years has grown to being. The love of the locality and the native land is a thing which cannot be killed, which should be

[1] The italics are the author's.

fostered. It is part of that natural crystallisation of humanity of which we have earlier spoken. *Coelum non animum mutant.* That spot of mother earth on which we live is the spot which should claim our greatest thought and effort.

The new science might open to the mind of the traveller a new point of view ; might provide a new outlook and occupation for him. He would go about not merely in an observing but in a constructing spirit. What can we do to build up the life of this place, we the intellectuals of the world ?—he would ask himself. Is the geographer, the traveller, content merely to look on at the miserable state of life of the teeming backward people of the earth ?[1] Is he content that age after age should roll on whilst they reproduce their kind, generation after generation, into poverty, ignorance and hunger ? Is there no vision before him of a better earth, such as a constructive economic science would assuredly build up ? There are possibilities in every hamlet. Every feature and fold of a landscape contains latent potentialities. The material, the human and topographical material, ready to hand, is asking mutely for organisation. The traveller or observer who would awaken to this new spirit would find a new charm in the map. He might develop what we may here term a new sense of " place-possibility " : a sense which will learn to regard a place as an organism, capable of being brought to a flourishing and permanent state of life, just as we try to bring an individual, a son or daughter, to the best development of life by our expenditure of money, example and precept. In this spirit we should not come to the place merely to see what possibilities it offers to trading and profit, or whether we can teach it a new religion, or what are its archæological remains, or its curious customs, or the shape of its people's crania ; matters such as may form the subject of a book of travel or entertain a London lecture audience. We should come animated by a spirit of order, to show it a gospel of ordered life : an ordered life not for a small part of its population, which lives more or less at the expense of the balance or neglectful of them (as even in our modern cities). Viewing a locality in this spirit, we might well receive a shock, and feel the blush of shame upon our cheeks. Is it possible, we should ask ourselves, that we have been so long

---

[1] The same reflection applies to our London or other slums.

blind ; that these miserable people are of our own kind ; that we have not before remarked the shameful conditions of life of the teeming poor, nor the almost ludicrous disorganisation of the natural resources around them ?   And lest we should raise up too proud a feeling of our own intellectuality and our own country's advancement, let us turn our eyes equally upon the savage conditions of life in the poor streets of our own towns and remark the absurd lack of scientific system that governs our own country-side.   Such consideration might well arise in him or her who studies the map in this awakened sense of place possibility, which the proposed science of constructive human geography would inculcate.

Whether in peace or war Society will be forced, early or late, to consider what are those true and positive economic principles by which the life and business of a world may be securely carried on. We are far from them to-day, although the pressure of war and of industrial conflict—both—are turning men's minds to those truths which are never acknowledged except under suffering.

The tropical lands and their people, upon whose description we here enter, afford an amost limitless scope for observation and constructive work in the field here advocated.   But whether or not such considerations may enter into the mind of the reader, these lands offer much that is of interest and attraction. Further, this book happens to be published during the Great War, and some interesting considerations arise therefrom.   We have generally regarded many of the tropical peoples as barbarous.   But barbarities have been committed, during the War, by one of the most advanced and Christian nations—the Germans—such as the world has probably never witnessed among Western people, or west of Suez.   On the other hand, the coloured troops or adherents of Britain and France have comported themselves with dignity and restraint.   More, from India, from Fiji, from Zanzibar, from the remote Moslem dependency of Bornu, in the heart of Darkest Africa, have come voices spontaneously raised in indignation of Teutonic barbarity and proffering men, money and support to the righteous cause of Britain, the character of whose rule has impressed even the savage potentate who rules under the equatorial sun.

THE AUTHOR.

LONDON, *October* 1915.

CHAPTER I

WHEN we speak familiarly of " The Tropics " there may arise before us scenes and pictures more or less attractive, visions perhaps of blue skies and feathery, gracile palms, under which the dark-skinned native lives his simple life in a world far removed from that prosaic one of the white man. Memory, or imagination, may revert to fruitful lands beneath a perennial sun, bounded by mountain ranges soft in distance ; or to the humid, tangled forests wherein Nature has housed her most varied animal creation and displayed her most gorgeous hues in flower and winged creature ; lands where the torrent flows unhindered over sands of gold, or where the summer wavelet beats upon the coral shores of islands beautiful and remote.

Or again we seem to march with that host of intrepid explorers who have pressed onwards through burning deserts and tiger-haunted jungles, or amid fever-stricken swamps and snow-capped precipices, braving all Nature's moods and the treachery of savage man. From Chimborazo to Timbuctoo, from the Nile to the Amazon, from the dark forests of Malay to the scorching deserts of Sahara and Arabia we follow where these pioneers have mapped their courses, to whom Cancer and Capricorn offered no bar, and who gave perhaps their lives and left their bones beside the trail that the round world might be better known to they that dwell therein, and that men might enjoy those kindly fruits of the earth which for so long had hung beyond their reach or knowledge.

> " Antres vast and deserts idle,
> The Anthropophagi, and men
> Whose heads do grow beneath their shoulders."

From that nobler picture we turn to another, painted in tropical colours of a darker hue, whose contemplation conduces to no soft adjectives but rather to speech " disrespectful of the Equator " (as the humorist has put it), or to condemnation of what has been done and still to some degree is carried on beneath the equatorial sun. Here the silence of the gloomy forest has been broken by the sound of the lash and the groan of the tortured Indian or negro, staggering beneath the burden piled upon his naked loins by the white man. Here the slave was torn from his native village, driven to his death in field or mine, or the poor peon passes his life in debt-bondage to create wealth which he himself will never taste. We contemplate a picture which began thousands of years ago in the making of bricks without straw,[1] and closes with the gathering of rubber without pay—if indeed it be closed yet.

To this last picture we shall have continually to revert in these pages. It gives place, we may say for the credit of our civilisation, to one painted by a wiser governance and by the efforts of the earnest missionary, as well as to the growth of the social conscience in matters economic, and to some realisation now of what might have been supposed to be a primitive wisdom in commercialism, which at least should have bid its votaries refrain from the destruction of the means which gave it wealth —the coloured man of the tropics. The unspeakable savageries of earlier trafficking between Cancer and Capricorn remain here and there, but amid the milder forms of selfish exploitation of the torrid zone we obtain broad glimpses, like cultivated patches in the sombre jungle, of contented and growingly independent natives, and see broad plantations where formerly dark barbarism and ignorance lived. What has been done here serves to accentuate how much remains to be done.

Nor do our eyes range necessarily only over wild man and wild nature in the tropics. We contemplate handsome cities, centres of wealth and civilisation, and communities where science, banishing the dreadful fevers and malarias of those delusive spots which have formed the " White Man's Grave," has created healthful and habitable homes. We have, more-over, the proofs of ancient cultures to consider within the equatorial zone : old cities and monuments, and history and traditions alluring ; the work of bygone tropical peoples who have left, buried in the jungle, or upon the mountain plateau,

---

[1] The land of the Pharaohs lies, of course, outside the tropics, but the simile holds good.

examples of their skill and knowledge such as excite our admiration even to-day.

A picture of the tropics which, however, we are here largely concerned to depict is that of regions only half exploited, yielding up but a relatively small portion of their resources and capabilities, whether to the outside world, whether to their native inhabitants. Viewed as a whole, the tropical lands may be regarded as a natural storehouse for the benefit of the world at large, yet one which to a large extent we are still without the knowledge and power—or the will—to organise and enjoy. The time is upon us when we shall have to take stock, and in a much kindlier and more intelligent way, of the earth's resources and people, in whatever land and climate, employing the methods of what we shall here term a "constructive human geography," a science whose development as regards the tropics it is among the main purposes of this book to advocate. Than this there is perhaps no more insistent lesson before the world, and, as concerns tropical lands, the intelligent nations into whose hands the control of the tropical and backward coloured races has been given will have to awaken to a greater sense of responsibility and justice and more intelligent methods in their dealings therewith.

The widespread lands of the tropics, which with their manifold resources and varied peoples we have here to consider, are becoming of increasing value and interest to the world. They abound with elements which we urgently require, whether of food-stuffs or raw material of commerce and manufacture, whether in matters of travel, scenery, sport and change. But it is scarcely an unfounded charge to lay against an intelligent public to say that knowledge of the life and topography of these regions is in general scanty. "The tropics" are but a vague and indefinite term; their lands are scattered broadcast around the world, and an entire library of travel and topography must be studied to grasp them as a unit; if indeed such a grasp be possible. It is in the hopes of affording such a grasp, partial it may be, that we shall here enter upon this survey and conspectus, acknowledging with gratitude the numerous works of travellers and students which have rendered possible the attempt.

The tropical zone, defined on the north by the tropic of

Cancer and on the south by that of Capricorn, intersected by the Equator, over 3000 miles wide and extending round the world, has very much less land than water within its boundaries. Nature has disposed the greatest land areas of the globe in the temperate and cold zones. A plethora of tropical lands would have presented an unbalanced scheme. The land within the tropical belt covers an area of about 18,000,000 square miles, or only one-third the land area of the globe.

Moreover, considerable portions of the tropics possess a cool climate, by reason of the high elevation of the land above sea-level, and some of the highest and bleakest inhabited places in the world are found in parts of the equatorial zone.[1] Conversely, some of the hottest places are found near or beyond the edge of the tropics; and we see that temperature and climate thus depend, to some extent, on geographical conditions other than latitude. In the elevated and some other regions of the torrid zone the diurnal range of temperature is often very marked, and it has been well said that " night is the winter of the tropics."

The tropical regions, as a glance at the map shows, cut through great portions of all the continents except Europe. From the Greenwich meridian the zone sweeps through Africa, covering by far the greater area of that continent, including half the vast Sahara, a small portion of Egypt, the whole of the Soudan, Abyssinia, the enormous territories forming the British, French, German,[2] Portuguese and other European colonies, the Congo, Rhodesia, a fragment of the Transvaal and nearly the whole of Madagascar.

Thence it crosses the Red Sea and the Indian Ocean, cuts Arabia and India each in half, enclosing Ceylon, covers the southern extremity of China, the whole of the Philippine Islands, Indo-China, Malaysia, the Straits Settlements, Borneo, Sumatra, Java and adjacent islands, and New Guinea—dividing the island continent of Australia near its waist.

Crossing the Pacific, it covers innumerable islands and archipelagos in that ocean, and reaches America. Mexico is cut midway, and the whole of Central America and the West Indies are included. It covers the great bulk of South America

[1] Such as those of Bolivia and Peru, for example.
[2] German before the Great War of 1914.

A COCO-NUT GATHERER

in the thickest part of that continent, including nearly the whole of the enormous republic of Brazil, the whole of Bolivia, Peru, Ecuador, Colombia, Venezuela and the Guianas, and portions of Paraguay, with the northern extremities of Chile and Argentina.

It is perhaps a point of geographical interest to note how the northern tropic, that of Cancer, cuts some of the great peninsulas of the globe, which seem to form a fringe hanging down from the continental bulk : projections or mighty larders whose function in part is to produce those peculiar foods which the temperate and colder lands do not contain, but which their people require for their nourishment.

The influence of the tropics upon mankind has furnished interesting matters of history. It might have been supposed that the cradle of civilisation would have first lain in those lands where climate and nature are least inclement or most beneficent ; where fruits hang ripe to the hand of man without effort, such as the luxuriance of tropical climates affords. Indeed it is probable that man originated in the tropics. But the high civilisations of the world, or at least the enduring civilisations, did not have their origin under the equatorial sun, but were born of more temperate skies, where the struggle against Nature was keener and aroused man's ingenuity and energy. To-day it is the cool, temperate belt of the earth's surface in which mankind advances in civilisation and social development, and even the modern or alien inhabitants of tropical lands are obliged constantly to refer their standards of progress and conduct thereto, lest they should degenerate.

On the other hand, certain tropical and sub-tropical lands produced early civilisations and culture, some of which assisted in laying the foundation of human progress ; and for such elements the world is indebted to China, India and Arabia, whilst the monuments still existing, scattered in many tropical lands, of mural, sculptural and literary arts and sciences, show how considerably mankind had advanced in certain regions. To the old cultures of tropical Asia must be added those of tropical America, the monuments and laws of the Inca, the Aztec, the Toltec and others, which, if they have not been of lasting benefit to the world (due in part to their ruthless destruction by the white man), at least bear out the scheme of

Nature in autochthonous progress, and the exemplifying of the reaction of man to his environment. We find many beautiful examples of old native tropical art here, in some cases thousands of years old, whether in stone, metal, or textile fabrics.

Within the tropics we encounter some of the most stupendous natural features of the earth. The world's greatest desert, the unconquerable Sahara, occupies the most extensive homogeneous land belt therein, covering the continent from the Atlantic to the Red Sea. One of the mightiest mountain chains, the cordillera of the Andes and its continuations, traverses the entire width of the torrid zone; and the largest and least-known river on the globe, the Amazon, parallels and crosses the equator. The Nile has its rise amid the very snows of the equator and crosses the (northern) torrid zone, and the Congo, the Niger, the Orinoco and other great rivers have their courses entirely therein. The forests of the Amazon are perhaps the most extensive area of virgin forest in the world, and like those of the Congo form in places the densest and most impenetrable barrier of vegetation that Nature has reared against man's advance.

The gradual discovery and development of the tropical regions under the people of more northern climes is of interesting study. We remark how Arabian culture—often mainly predatory—extended along the African coasts and penetrated among the equatorial negroes of that continent, and how powerful Moslem-African kingdoms arose in the dark continent, grafting the religion of the Prophet upon the barbaric fetishism of the natives. We trace the wide-spreading field, covering almost every continent, of Portuguese exploration, where the vessels of the early explorers pushed in upon every African and South American coast, dominating, trading with or persecuting the natives. We see how enormous were the activities of adventurous Spain, how she implanted her culture throughout thousands of miles of savage cordillera and forest, and stamped her language and customs permanently upon half the area of the New World.

Finally we see the arrival of the British, bent less upon mere conquest than upon trade, but establishing, with their peculiar genius, the best governing conditions that the poor and outraged coloured people of the torrid zone have ever enjoyed;

FOUR AFRICAN KINGS AND THEIR PRIME MINISTERS: BRITISH EAST AFRICA AND UGANDA

and we take note also of the Dutch and French activities in remote tropical lands, and of the endeavour of the German, late in the day, to secure for the Fatherland some share of prestige and dominion within the tropical belt.[1]

Many of the commodities most valuable to manufacture, commerce and the food supply of the civilised world are products of the tropics. There is the coffee of Arabia, Africa, Mexico, Costa Rica, Guatemala or Brazil, and indeed of almost all lands in the torrid zone. Sugar comes from an equally wide source. The potato was a development of the cold Andine uplands. Quinine, to which mankind owes a debt of gratitude, was a gift of Peruvian forests, as was cocaine. Nature herself placed the antidote to fever in the malarious jungle of the Amazon here. Cocoa or chocolate, an original gift of Mexico, is grown in every tropical land in increasing quantities. Tea from India and Ceylon is given bountifully, especially to the people of the British Empire. Rubber, another gift of Mexico and the New World, is found or propagated in almost all lands which approach the Equator, and its planting forms a remunerative industry. The coco-nut, but a short time ago regarded merely as a curious refreshment, is now cultivated on nearly all tropical shores and exported in growing quantities for its vegetable fats ; one of the most important articles of manufacture. Indeed, the numerous oil-bearing palms and plants are becoming of prime importance. Cotton is indigenous to or cultivated in nearly all tropical lands, and steps are being taken to sow vast areas with this prime necessity of life. Rice, the food of one-third of the human race, is enormously grown in the tropics. For the gift of tobacco the smoker owes a debt of gratitude to the whole tropic zone—tropical America first.

The host of products important to civilised or refined life, and the luscious fruits which the lands between Cancer and Capricorn provide, will occur to the least initiated. He who does but enter an ordinary grocer's shop will be greeted with sufficient evidence in sight and odour that here is an emporium —a veritable place of geographical romance—in which are

[1] The German colonies fell one by one, in the Great War, to their opponents. It is not our purpose here to judge of the merits or demerits of German rule therein, but, as regards the African possessions, Germany has freely been accused of avarice and sharp practice in obtaining them, and of doubtful methods of governance of the native.

congregated innumerable articles from lands adjacent to the
Equator. Further—and it is a lesson we shall here strive to
convey—if he or his housewife be blessed with somewhat more
than the ordinary imagination, he or she might be tempted to
speculate how much more for their sustenance or comfort
might not be derived from the same oversea source ; to
inquire why they pay so heavily in coin for plants, fruits,
leaves, and berries which in their native environment are
almost weeds, and for whose further cultivation there remains
untold vacant space.

Especially is this true of the British citizen, whose empire
embodies some of the most extensive and prolific parts of the
tropical zone, but about which he, it is not too much to say, is
often profoundly ignorant or apathetic. The tropical Crown
colonies and protectorates are, in one sense, extremely important
parts of the Empire for the people of the British Isles, for the
reason that they are under the direct control of Britain, and
not, as is the case with Canada, Australia and South Africa,
practically separate states. Their importance is in the fact
that they are vast areas of food supply and material of manufacture available for British consumption, whilst the food and
material of the self-governing states will, sooner or later, be
required for their own white population. It is Britain's
enormous good fortune to hold these possessions, a fact that
growing land-hunger and the demand for supplies in the future
among the nations is likely to accentuate. The total tropical
area of the Empire covers between 5,000,000 and 6,000,000
square miles.[1]

The widely scattered regions which here come under our
survey embody many places of exceeding interest, although
often difficult of access. We penetrate the equatorial forests
of Africa, ascending the Niger and the Congo, traverse the vast
Soudan and descend the Nile. We enter the fruitful if sometimes fever-scourged European colonies on the African coasts,
and stand upon the borders of the great equatorial lakes of
that continent, overlooked by the once mysterious mountains
of the Nile, and ascend to Ethiopia's rocky heights. Southward we traverse the fertile land of Rhodesia, and eastward we
sojourn in ancient Zanzibar.

[1] Before the Great War.

Crossing from Africa to Asia by the "Gate of Tears" we traverse Arabia Felix and Arabia Deserta, and make a pilgrimage to Mecca ; passing thence to India and the old East Indies, with Malaysia, Siam and Cambodia, and southwards to the famous isles of Borneo, Java, and its companions, and to tropical Australia.   We voyage thence among the innumerable tropic isles of the Pacific, and beyond them reach the shores of tropical America and the beautiful West India Islands.

As elsewhere remarked, if any excuse be needed for describing all these scattered lands in a single volume it is that the torrid zone in general becomes an entity when we regard it as the home of the coloured and backward races of the world; lands which, due to their climate and general environment, and the problems of their future, are naturally and inevitably marked out for special treatment by the world's governing people.   With our constructive economic purpose in view we shall take special note of the capabilities of the native peoples here, and of their arts, crafts, industries and cultures, as also their capacity for education and powers of self-government.   These elements may in some cases be far more important than we had previously pictured them.

Far more general discussion needed for proper introduction

CHAPTER II

TROPICAL AFRICA

THE name of Africa bears with it a sense of vastness and diversity which perhaps that of no other continent conveys. From the times of Pharaoh and the early civilisations which flourished and fell upon Africa's northern shores, through the great explorations of last century which laid bare the heart of the unknown continent, down to the modern colonising activity which has resulted upon its partition by European people, we are carried over a period covering all known history.

> " Geographers, in Afric maps
> With savage pictures filled their gaps,
> And o'er unhabitable downs
> Placed elephants for want of towns."

If we contemplate upon its chart the huge bulk of Africa and compare it with a map of fifty or sixty years ago we shall realise the changes which geographical exploration and colonising and trading avidity have brought about in relatively so short a time. The savagery, superstitions, fetishes and juju of Africa's dark races, although by no means things of the past, are becoming more and more circumscribed, and the abomination of slavery remains only in its milder but still inadmissible forms of domestic slavery, peonage, and forced and contract labour. The last-named is often little less than slavery.

The blank spaces on the earlier map have given way under the heroic exploits of traveller, missionary and hunter. Incidentally the elephant and his kindred have also, in some cases, given way to the efforts of the last-named agent of civilisation. The lust of commerce and gain, of gold, rubber, ivory, palm-oil and a host of other natural resources here so lavishly planted has its own peculiar history : part of the ruthless process of development from savagery towards civilisation, itself but a relative term so far, in any climate.

NATIVE CANOES, UGANDA

We remark the comparatively regular outline of Africa, and the absence of very high or very low lands such as we encounter in other continents. South Africa discloses a great plateau, falling away at its edges like an inverted saucer ; East Africa is made up of table-lands, mountain ranges and great depressions or " rift-valleys " where the earth's crust has sunk, affording place for the equatorial lakes and the source of the Nile. The West African highlands give birth to other great rivers, the Niger and others, whilst through North Africa stretches the dreaded Sahara.

Africa, lying balanced, as it were, across the Equator, is tropical throughout, with the exception of its northern and southern portions, and due to this circumstance we do not encounter very marked extremes of heat and cold over its enormous extent. But the heat is modified by various agencies, and even in the Sahara, whose torrid climate results partly upon the great width of the continent and distance from the sea, there is occasional frost at night ; whilst the moisture brought from the ocean renders the greater elevations of Africa cooler.

The tropical zone of Africa, considered as a whole, is the most unhealthy portion of the earth's surface for the white man, its insalubrity exceeding that of tropical America, especially in the low, moist region of the West Coast, with its malarial scourge. Yet there are high and healthy plateau regions, such as in parts of British East Africa, Abyssinia, Rhodesia, Somaliland and elsewhere. The intensely hot Sahara, due to its exceedingly dry climate, is not unhealthy, although portions of it lie below the level of the sea. Among the natural agencies in tempering or disturbing the climate of Africa the monsoon and the harmattan winds are those we constantly remark.

A scanty plant life follows upon a scanty rainfall over a large portion of Africa, but the heavy rains of the equatorial region give rise to such a profusion of tree, plant and flower as it is almost impossible to depict in words, and even in the desert the date-palm flourishes where nothing else will grow. The valuable nut and oil-bearing palms and plants of the primeval forests are becoming a source of wealth which may almost equal in value the gold of Africa's mines. The great interior of Africa is by no means the desert land which popular imagination has depicted it to be. It has also been well described as more greatly provided with "footpaths" than any other land: the hard-beaten negro trails which form a veritable network upon the continent.[1]

The animal life, which is among the natural glories of Africa, is, as before remarked, becoming diminished with the advent of civilisation. The great herds of game, the uncouth, the ferocious, the beautiful, the interesting of the animal world here have been seriously menaced. But wiser counsels are now tending to conserve, in their native state, many of these irreplaceable creatures of the equatorial continent. Also, the hunter is happily learning to substitute the camera for the rifle.

With the exception of the Lower Nile Valley and "Roman" Africa—which lie outside the tropics—the continent contains no ancient history and very few archæological remains, such as are so noteworthy in tropical America. The Black man's "culture" was inferior to that of the Brown. Yet the absence of great natural barriers might seem to have facilitated the spread or intercommunication of native tribes, and indeed sharp division of people and races did not prevail here.

The chief races of Africa are the Negro, the Bushman, the Hamite (including the Egyptians, Abyssinians, Berbers, Fulas, etc.), the Semite or Arab, and the Hottentot. Negroes and their Arab admixtures form well-organised states, professing Mohammedanism, throughout vast areas of Northern Africa; a negroid people inhabiting savanna and forest, to where, below a line drawn across the continent from the great Cameroon peak in the western angle of Africa to Lake Albert in the east, the Bantu negro prevails and Bantu languages are spoken.

We have now briefly to regard Africa politically. It is an example of a continent absolutely carved up by what might be termed absentee proprietors, and it forms a veritable chequer-board of colonies of one or the other European Powers. With the exception of Morocco in the north, Abyssinia in the east, and the little negro republic of Liberia in the west, there is no independent area, state, kingdom or republic on the continent.

The "scramble for Africa," as the division has been termed, which has resulted in its partition by Europe—largely since the Franco-German War of 1870 and after the lethargy of Great Britain and other nations in the development of the dark

[1] Well described by Drummond in *Tropical Africa*.

continent had begun to give way—became very marked after 1885, and was practically completed at the end of the century now past, and it has formed one of the political and colonising romances of the world. The struggle was precipitated, however, less by the action of the Great Powers than by the ambitious pretensions of the King of the Belgians in the Congo and by the interest following on the discoveries of Livingstone and Stanley.

On the one hand traders and manufacturers saw the vast possibilities for commercial activity here, and on the other the missionary and philanthropic element contemplated millions of savages plunged in heathenism. The one condition offered wealth, the other a field for proselytising, and although the two elements have inevitably their points of difference, together they gave birth to a new movement, and have resulted in painting the map of Africa in many colours, an operation which was performed in the incredibly short space of about a quarter of a century.

In this metamorphosis Great Britain, with that singular good fortune which, in conjunction with good management, has attended the expansion of her Empire, has remained with vast areas of the best and most useful portions of the continent. But by whoever they be owned or influenced, vast areas of Africa still remain to be brought under proper control and a true economic administration.

At the time of the great European War of 1914 the partition of Africa had resulted in the following political divisions :—

French Africa controlled 3,866,950 square miles ; Britain, 2,101,400 ; Egyptian Africa, 1,600,000 ; Germany, 910,000 ; Belgium, 900,000 ; Portugal, 787,000 ; Independent Africa, 613,000 ; Turkey, 400,000 ; Italy, 200,000, and Spain, 79,800 square miles.

Leaving outside our field the Mediterranean Coast and Morocco, which are beyond the tropics, we first reach the Saharan Coast, belonging to France, out of which is carved the small Spanish colony of Rio de Oro. Then, in their order, come the French possession of Senegal, the British colony of Gambia, Portuguese Guinea, French Guinea, the British colony of Sierra Leone, the black republic of Liberia, the French Ivory Coast, the British Gold Coast and Ashanti, German Togoland, French Dahomey, the British colony and protectorate of Nigeria, the German colony of Cameroon, the Spanish Muni River Settlement (with the Spanish island of Fernando Po to the north), Gaboon and French Congo or Equatorial Africa, cut by the Equator (with the Portuguese islands of San Thomé and Principe off the coast), the narrow edge of the Belgian Congo, where that vast river debouches into the Atlantic, and Angola or Portuguese West Africa—the whole of these possessions forming in a wide acceptance the Guinea Coast and old Slave Coast.

Thence follow German West Africa and the British possessions, rounding off the continent beyond the tropic of Capricorn. Turning north we reach—again in the torrid zone—Mozambique or Portuguese East Africa (with Rhodesia inland), and Madagascar, controlled by France, off its coast; German East Africa, British East Africa—the latter crossed by the Equator— and Italian Somaliland, taking us to the sharp eastern horn of Africa; British Somaliland, French Somaliland, Italian Eritrea —the four last cutting off Abyssinia from the Indian Ocean and Red Sea—and finally, the coast of the Anglo-Egyptian Soudan, whence, in Egypt, we again cross the tropic of Cancer.

We have now to consider these various territories in some detail.

NEGRO BVELA CHILDREN OF RHODESIA

CHAPTER III

WE have here to enter upon a vast territory covering about 2,000,000 square miles, extending from beyond the northern tropic, through the Soudan and Sahara, over portions of the vaguely named Guinea Coast and Slave Coast, through many a hundred leagues of tropic jungle, fertile savanna, burning desert and tortuous, alligator-haunted stream ; the home of ancient Mohammedan states, of vast, fetish-ridden negro hordes, and the old, blood-stained kingdom of Dahomey ; a region fringed by the surf of interminable Atlantic rollers beating against the great rounded shoulder of Africa, which, at Cape de Verde, protrudes farthest west into the ocean. Such, in general terms, is the vast empire of French West Africa.

This region is divided, as regards its western side, by the colonies of British and other Powers, which we shall enter in their places ; but it unites in the interior, forming a huge bulk of territory of which more than half lies in the Sahara.

The first and northernmost of these French colonies is Senegal, with Upper Senegal and Niger extending thence far inland. Next in order comes French Guinea, separated from the above by the British colony of Gambia, and surrounding the colony of Portuguese Guinea, with Sierra Leone, belonging to Britain, on the south. Third is the French Ivory Coast, with Liberia adjacent on the west and the British Gold Coast Colony on the east. Finally we reach Dahomey, between German Togoland and British Nigeria, and still to the south is French Congo.

This wide-spreading territory is mainly administered by a French governor-general at Dakar, capital of Senegal, upon Cape Verde, assisted by the lieutenant-governors of the various units, forming a Council which rules over a population of 13,000,000, of which about 12,000 are Europeans.

15

As we skirt the shores of Mauretania, the Atlantic front of the Great Sahara, to the colony of Senegal, we approach the mouth of the Senegal river, flowing to the sea from the south and east.   For a thousand miles along the Sahara Coast, from beyond Cape Blanco, we shall encounter no river until we reach the ancient Senegal.   The Portuguese navigator, Diaz, who entered its mouth in 1455, thought the river was an arm of the Nile.   A flat, desolate, arid dune-lined shore unfolds, whose monotony is only relieved from time to time by an occasional cliff, and this aspect is maintained until Cape Verde, Africa's westernmost point, is passed.

Cape Verde does not bear out the significance of its designation.   It is no longer verdant.   Due to climatic changes which have brought increasing drought—such as influence also the Cape Verde Islands, 300 miles out at sea—that promontory has now relatively no clothing of vegetation.   But after rounding the Cape, when the stretch of littoral which conducts us towards the Gambia unfolds, we remark a changed topographical character, green marshes appearing, and the soil is clothed with luxuriant vegetation ; whilst the unbroken coast-line gives place to the gulf and island of Goree, and farther on to the archipelago formed by the Salum estuary.

The larger ocean steamers, except when lightened of their cargo, do not cross the bar of the Senegal river to ascend to St Louis, which lies about twelve miles up-stream ; for, except at flood-time, the depth of water is insufficient.   However, a ten-foot draught is always available for 275 miles up-stream.   The capital of the Senegal colony is built upon an island connected with the river-bank by long bridges, one of which gives access to the railway to Dakar, 160 miles to the south.

We experience extreme heat at St Louis, and the low-lying position of the town renders the place unhealthy, but the suburb of N'dar Toute is a pleasing summer watering-place, with villas, gardens and a fine park, approached by a magnificent avenue of palm-trees.   The governor's house at St Louis, the great mosque, the cathedral, barracks and court house are among the principal buildings, turning from which we observe the flat roofs, balconies and terraces of the Europeans' houses, in contrast with the round beehive huts of the native fishermen, which cluster nearer the coast.

The population of St Louis, with its suburbs, is about 30,000, and the town is the oldest foothold of France in Africa, having been founded by merchants of Dieppe in 1626. The forts, which face the landward side of St Louis, were built to protect the place from the savage tribes of Cayor.

Dakar, the general capital of French West Africa, is strategically in a commanding position on the sea route between Western Europe and South Africa and Brazil. It nestles on the shore of the Gulf of Goree, under the shelter of Cape Verde, and is the only port of Senegal in which the largest vessels may lie with security. Its broad and regular streets, fine palace of the governor-general—administrator of the French West African possessions—and other handsome public buildings attract the traveller's eye, and the position is fairly healthy. Extensive naval and commercial docks have been established : Dakar being a regular port of call, coaling and provisioning for the large steamship lines. The population of this important town is about 25,000.

Politically, or rather as regards its administration, the Senegal colony proper consists of the four towns of St Louis, Dakar, Goree and Rufisque—whose combined white population is somewhat over 5000—and a narrow strip of territory adjacent to the coast ; the rest of the territory under this heading including native states under French protection, the whole having an area of 74,000 square miles, with about 1,800,000 inhabitants. One of the principal native states is Bondu, inhabited by the Mohammedan Fula people.

The town of Goree lies upon its island, in the great natural harbour which Cape Verde forms, and though the place has suffered in importance from the rise of Dakar, the healthy climate has preserved its value as a sanatorium. Above the narrow streets, whose flat-roofed houses are built of dark red stone, arises the castle of St Michael, which, with the governor's residence and other administrative buildings, attest its former importance. The island has changed hands many times since the Dutch took it in the seventeenth century, having had English and French masters successively. Rufisque merits our attention as the chief point of export of ground-nuts,

B

Senegal's chief commercial product, amounting to many thousands of tons yearly. The place lies on the railway from Dakar.

The climate of the Senegal Coast, though hot and unhealthy generally, is cool and agreeable in the dry season when the harmattan sweeps seaward from the Sahara. The flooded low lands of the Senegal in the wet season are very unhealthy, and none of the country can be said to be suited for permanent European settlement, although the malaria and yellow fever are being successfully combated by sanitary methods. We may penetrate this stretch of coast south of Cape Verde by various navigable rivers and estuaries, among them the Casamance and the Salum, and here we shall encounter the teeming reptile, bird and animal life which is so marked upon the Guinea Coast.

Proceeding inland we cross a barren plain, stretching east wards for perhaps 250 miles, giving place to a fertile mountain ous region, whose ridges rise to 4000 feet above the level of the sea, their steep slopes sinking rapidly to the valley of the Niger and, to the south, merging into the highlands of Futa Jallon the birthplace of the Black Senegal. The White Senegal also flows from the south, both rising in French Guinea, sweeping through that part of Africa where the mighty Niger and the Gambia have their sources.

The Senegal in some respects is a remarkable river, the curious regimen of its waters embodying a series of natural "reservoirs," closed by dams ; reservoirs partly protected by "curtains of verdure" which prevent the too rapid evapora- tion of the water so marked in the lower reaches of the river across the arid littoral. Were it not for these natural locks and dams which cross the channel at intervals in the high reaches, the Senegal would be but an insignificant stream at its mouth, after its thousand-mile course. But in the rainy season the water accumulates in the great receptacles thus provided, the barriers being submerged one after the other at the flood, and thus the waters are first stored up and subse- quently restored to the river in the dry season.

From Bafulabé, which in the native tongue means "the meeting of the waters," where the White and Black affluents mingle, the Senegal flows through swift narrows and over high cascades, receiving the waters of other rivers. At the town of Kayes the traveller may embark for St. Louis, the capital of Senegal, the river being navigable by steamer from August to November for 500 miles to its mouth.

Otherwise, we may travel by the railway, which runs eastward to the navigable waters of the Niger at Kulikoro. The railway from Dakar and St Louis also reaches Kayes, and thus we may traverse a vast stretch of territory by railway and river, to Timbuctoo and beyond, thence descending the Niger into British Nigeria.

The railway linking the Senegal and the Niger, from Kayes to Kulikoro, is 347 miles long, and twenty-five years were occupied in completing the work, due to bad management at Paris. The total cost was £3,500,000, and the line was not opened until 1905. The first portion of the works were found to be hopelessly defective in their construction, an unfortunate example of incapacity in French colonial management, whose rectification was not entered upon until the closing year of last century.

The Lower Senegal forms the boundary between the Western Soudan and the Sahara, and here the wandering Berber, or " Moor," comes into contact with the settled negro, both peoples forming the population of the territory, although a mixed negroid race constitutes its bulk. Arabic, Fula, Mandingo and other languages are spoken. The slave-raiding which formerly prevailed has been abolished by the French, who no longer recognise even domestic slavery.

The natives carry on their primitive weaving, pottery and brick-making, also the fashioning of their trinkets. They cultivate maize, rice and millet—the latter their principal food ; and cattle, sheep, camels and horses are raised in large numbers. Gum, collected by the Arabs, and gold and rubber in their respective districts, are produced, whilst iron and quicksilver are among the mineral resources. The most important articles of import are British cotton goods.

Our way lies into the vast Sahara and those territories of the Upper Senegal and Niger which are tributary to such ancient and formerly mysterious places as Timbuctoo, Jenné and other towns of the semi-barbaric people with whom the early explorers

contended, or by whose malice they were murdered, and where
they rest perhaps beside desert trails or in unnamed graves
The French West African possessions are enormous, and, upon
the map, might seem to cause the smaller adjoining colonies
of other European Powers to sink into insignificance, although
this would not be true in an economic sense.

The Upper Senegal and Niger territory covers an area o
210,000 square miles, extending far to the east, with French
Guinea, the Gold Coast, Togoland, and French Dahòmey upon
its southern side.   Largely upland desert, there is much fertile
country on the borders of the Niger; west of that rive
agriculture and stock-raising being carried on; and there are
forested areas.

The people we encounter in this great territory are Berbers
and in the Niger land, which forms a large part of the adminis
trative unit, negroid and Fula races.   The climate is not un
healthy, except in the districts inundated by the Niger floods
This is a gold-bearing land, deposits of auriferous quart
and gravel being found, and possibly these gave rise to th
early stories concerning " hills of gold " which lured on th
adventurers of the seventeenth and eighteenth centuries.   Bu
the gold is secondary to the more permanent wealth which th
fertile soil displays, in its grains and fruits, and the abundan
pasturage which the wild guinea-grass yields, such as afford
rich sustenance to the many herds of cattle whose master
are the Mandingos, or to the forests abounding in valuabl
trees.

The wild animal life includes the lion, elephant, hippopotamus
wild boar, antelope and other great fauna, whose abundanc
attests the wealth of nature in this region.   Game, large an
small, is plentiful.

By far the greater part of the Niger's course is throug
French territory.   We have already spoken of the Senegal-Nige
railway, and towards its Niger terminus stands Bamako, th
seat of Government of the colony, at the head of the navigabl
waterway, and beyond it, at Kulikoro, the railway end
Descending the river we pass Segu and Sansandig, both in
portant river ports, the first-named place being that wher
Mungo Park reached the Niger.

Lying upon a natural canal connecting the Niger with it

A VIEW IN DJENNE

Lander photo.

affluent the Bani stands the ancient and one-time famous town of Jenné, which merits our attention. Jenné—which gave, perhaps, its name to the Guinea Coast—is believed to have been founded in the eighth century, on the site of the ancient palace of the Longpoi negro " kings," and there is some resemblance to Egyptian architecture in the pylonic buttresses upon the façades of its homes, though it is remarked that there is little trace of Arabian or Moorish art. Long flat bricks of sun-dried clay form the material, and massive walls of the same surround the city. The ancient citadel was converted by the French, after 1893, into a fort.

. When Jenné was at the height of its power, from the twelfth to the sixteenth centuries, the merchandise of its people—who are to-day extremely active traders—was found in every port upon the West African Coast, and from a corruption of its name it is conjectured the nomenclature of the "Guinea" Coast may have arisen, as above remarked. Under the Moors, Tuaregs and Fulas, who invaded the place, Jenné declined, however. Its position is singular, in the wet season the place becoming an island, surrounded by *marigots*, as the natural channels which dry up during a portion of the year are termed in this part of Africa. The people of Jenné build a class of boats which are famed throughout the whole region of the Western Soudan.

Some 250 miles to the north-east as we descend the Niger lies Timbuctoo "the mysterious," as it was earlier termed, the " meeting point of camel and canoe," known also as " the port of the Soudan in the Sahara." The town lies little more than two degrees west of the Greenwich meridian, and at an elevation of about 800 feet above sea-level, on a terrace of the Sahara's southern edge, overlooking the periodically flooded lands of the Niger, and surrounded by sandy wastes, where the dunes of the mighty desert press forward. It has been left, in a sense, high and dry in the lapse of time during which the drying up of the Sahara region has been going on.

Like Jenné, with which in its history Timbuctoo has been closely associated, the ancient town has been frequently captured or devastated by negro, Tuareg and Moorish princes and bandits or Niger pirates, ever since its foundation by the Sanghai, and until its occupation by the French. So beset

was it by marauders that the district around became known a
Ur-immauders, meaning "God hears not," and when in 189
the French flotilla arrived, the inhabitants, wearied of thei
insecurity, readily opened their gates to the white man's adven
and protection.

Then little more than a vast ruin, its ancient trade revivec
Its fame as a mart for gold and salt had, long before, reache
Europe, and great caravans of thousands of camels, laden wit
salt, still cross the desert therefrom. Goods to the value c
£50,000 annually reach the town across the Sahara, and a
equal quantity via Senegal. Gold, ivory, wax, cereals and nativ
cotton goods are brought to its port of Kabara by the Nige
steamers, and it has been proposed to dig a canal into the city
thereby stimulating the transit trade, whose value is alread
over £800,000 per annum. Traffic is the mainstay of Tim
buctoo ; the people are traders above all, and its local industrie
of weaving, pottery, leather work and embroidery are sub
ordinate thereto. The rich countries south and west of th
Niger, and the produce of North Africa, make of Timbuctoo
vast exchange.

Three prominent towers stand up in the town ; one the ugl
earthen tower of the great mosque, forming a landmark fa
around, and, beneath, the French forts protect the roads to th
desert and the river. The extensive ruined areas attest a
earlier importance, dating partly from that time when, in 159c
the Sultan of Morocco, whose cupidity had been aroused b
stories of its wealth, dispatched an army across the Sahar
and occupied Timbuctoo, making it his capital city.

We cannot here follow in much detail the exploits of th
brave French or British explorers who entered Timbuctoo fron
time to time, some of whom were murdered, some of whos
names have been commemorated in mural tablets placed ther
by France. The British might have controlled Timbuctoo an
its district, but they failed to avail themselves of circumstances

The gallant French expedition which first occupied the town
acted against the orders of their Home Government. A handfu
of marines left the gunboats at Kabara and entered the town
but the Tuaregs attacked the boat party and killed the office
and some of the crew. The French colonel, Bonnier, who wa
stationed 200 miles away to the south-west, and who marche

Landor photo

A VIEW OF TIMBUCTOO

in relief and entered Timbuctoo, left the town and with 100 men pursued the enemy far into the desert. While camping at night an overwhelming attack was made upon the white men, and Bonnier and nearly all his men were slain.

Timbuctoo and the Niger were the scene of the exploits of brave Mungo Park. In the service of the African Association, Park crossed the Upper Senegal basin, and, traversing the desert, was laid captive by a Moorish chief; but, escaping with nothing save his horse and compass, he reached Segu, thus being the first European to gaze upon the long-sought Niger. Exhausted and without means, he turned back from a point somewhat beyond, and reached Bamako, afterwards falling ill, his life only being saved by the kindness of a negro, who nursed him. Returning home, Mungo Park's account of the discovery of the Niger aroused great public interest. Studying Arabic and developing a friendship with Sir Walter Scott filled in the time until, in 1805, with a captain's commission, he again reached the Niger; most of his party dying of dysentery or fever. Nevertheless Park, with the only soldiers left who were capable of work, embarked at Sansandig to explore the unknown river, in a craft fashioned out of two canoes.

The explorer had written home to the effect that he would " discover the termination of the Niger or perish in the attempt." This, and a letter to his wife, were the last messages received from Mungo Park, and afterwards the account of his disasters and death came through. Attacked at various points on the river in the downward journey, the craft in which he floated at length stuck in the rapids at Bussa, below Timbuctoo, and the hostile natives on the bank attacked the voyagers with bow, arrow and spear, rendering the situation desperate. Park, his companion Mostyn, and the two soldiers who had survived, were obliged to leap into the waters, which closed over them, the sole survivor being a slave, from whom the account of the gallant Scotsman's death was afterwards obtained.

The Sahara, spoken of as a whole, belongs principally to France, portions of it pertaining to Morocco, Tripoli and Egypt. The eastern boundary, under an Anglo-French Agreement, is partly formed by a line running south-east from the intersection of the tropic of Cancer with the sixteenth meridian east to where that line cuts the twenty-fourth meridian, which it then follows south to Darfur. East of this boundary lies the Libyan Desert, and the Anglo-Egyptian Soudan. Thus the French dominion extends from the Atlantic Coast eastward towards the Nile, over some forty degrees of longitude.

It would be impossible here to enter upon any close description of this mighty tropical and sub-tropical belt. The northern side, beyond the torrid zone, is typically marked by the Atlas range, where the observer looks out as over an illimitable sea of sand, with shores and promontories and towns upon its " coast." Towards the south the desert also comes suddenly to an end, in some places, as if it had been " cut off with a knife," but elsewhere it merges into the fertile Soudan. However, the Sahara is not necessarily such as the popular mind has pictured it or, as observed at certain points, merely an ocean of sand, as it has a very irregular surface in general, ranging from 100 feet below sea-level to as much as 6000 or 8000 feet above. Sand dunes, oases, rocky plateaux and vast tracts of loose pebbles, and valleys, where at one time abundant water-courses must have made their way, diversify its surface, and it still possesses the skeleton of a river system, with defined water-partings. These numerous " wadis," or sometimes dry river-beds, are associated with evidences of vast denudation in the past, the surface of the desert having been worn down by the elements.

This latter consideration destroys also a further popular and even scientific misconception formerly held—namely, that the Sahara was once covered by the sea—although portions of it appear to have been so submerged. The sand is mainly derived from the erosion of the rocks, throughout the ages, and due to desiccation caused partly by the diurnal range of temperature, although the causes are not yet fully explained.

Again, although the sand dunes shift under the wind, the popular stories as to armies and caravans being engulfed therein are, in reality, regarded as unfounded, except possibly in the Libyan Desert. Through this last-named dreadful wilderness there is but one route, that from north to south, passing midway through the Kufra oasis, near the tropic of Capricorn.

The most famous tree of the Sahara is the date-palm, in the

oases; and without it human life in the desert would be impossible. There are, of course, many other products. Plans for railways have at various times been brought forward to traverse the Sahara, and the idea of " forming an inland sea " of the desert has also been advanced, by the method of admitting the waters of the ocean, but the area it would be possible so to cover is calculated as only about fourteen times the size of the Lake of Geneva.

The Sahara and Soudan, as before remarked, link all the possessions of France in West Africa, as well as in Central and North Africa, and the next French colony' which comes under our survey is French Guinea, a territory covering 100,000 square miles, with a seaboard upon the Atlantic 170 miles long, extending eastwardly into the interior for 450 miles, with a population of 2,000,000 to 2,500,000.

A coast low and sandy, much broken up by deep estuaries and the mouths of rivers confronts us on approaching this land—formerly named Rivieres du Sud; earlier a difficult and little-known shore, which the slavers, driven from other parts of the Guinea Coast, made their last stand, and here the ruins of their strongholds, where the slave traders defended themselves with cannon, still exist.

Beyond, the country slopes upwards towards the Futa Jallon highlands. This hydrographic centre of West Africa rises in places to 5000 feet above sea-level. The dense forest and park-like savanna lands of the coastal slopes give place, on the plateau, to a short herbage, upon which large flocks and herds of sheep and cattle are raised by the Fula inhabitants.

The capital is Konakry, and as the steamer approaches the little island of Tombo, sheltered by its promontory, the town comes into view, with its docks and long jetty, the port being among the principal on the West African Coast. Its Government House and large public garden and good water supply are evidences of European progress in what was formerly a place of little importance. An iron bridge connects it with the mainland, and the railway, of which it is the terminus, runs inland for 342 miles to join the Niger at Kunissa, which river gives access by boat to Bamako in Upper Senegal, and so to St Louis and to Timbuctoo. A good road, 300 miles long, also runs to the Niger, and the French have built other highways. Konakry has a population of about 20,000.

The few hundred or more white inhabitants of the colony are mainly French. Over the greater part of the country the native princes, as in Senegal, retain their sovereignty under French administration, and the French have shown marked solicitude in the spread of agriculture and education, whilst the natives, in their turn, are in general friendly to their white masters.

The climate, whilst fairly healthy for Europeans in the highlands, is hot, moist and unhealthy on the coast, and in the rainy season the calm periods are followed by tornadoes. The harmattan prevails in the dry months.

The dense forests of this region contain those varieties of trees, lianas and flowers we have seen elsewhere on this coast ; the beautiful oil-palm, and the huge bombax, and, in the Niger regions, the baobab and the shea-butter tree. The soil of the colony is generally fertile, yielding the rich products of the tropics, among them, from the coastal districts, the palm-oil and kernels. From the interior, rubber is brought, worth half-a-million pounds annually ; coffee, wax, and ivory and hides, as well as ground-nuts and other products. British cotton goods are the main article of import.

The elephant is rare, and the lion only haunts the highlands, though panthers and leopards are common. Crocodiles infest the rivers, and fierce combats have been witnessed, it is recorded, between these reptiles and the great pythons which share the river-banks therewith. Large oysters, turtles and fish are abundant, and brilliant-plumaged birds flash through the woods, where huge chimpanzees and many venomous serpents also have their home.

Returning now to the Guinea Coast we pass the British colony of Sierra Leone and the black republic of Liberia, both described in their places, and, rounding the shoulder of Africa, the French Ivory Coast unfolds to the view. Here the great Atlantic billows thunder unceasingly upon the sandy verge and break over the submerged sandbanks which parallel the shore. No bay or promontory breaks the monotonous regularity of the seaboard, though in the east lagoons fringe it. Landing is difficult from these causes.

The Dieppe merchants and the Portuguese discoverers were the first arrivals here, followed by the traders who sought ivory, and the slavers who trafficked in human flesh and blood. Now slavery is a thing of the past, and the ivory is almost exhausted, the elephants having been exterminated or hunted into regions still more remote.

But the primeval forests, lying beyond the low littoral, are rich in palm-tree products, rubber and mahogany, which form the principal articles of export, the rubber being shipped mainly from Assini, largely from Ashanti, and goes to England. Races of cannibals dwell on the northern confines of the great forest belt, and the bulk of the inhabitants are fetish worshippers, although the coast tribes in general are placably disposed, but are not reputedly industrious or intelligent. The torrential rivers, descending from the highlands, are navigable only for a few miles from their mouths.

Built upon the sandy spit of its lagoon, forming a good harbour once the bar is passed, Grand Bassam, the commercial centre of the colony stands, its inhabitants numbering somewhat over 5000, with perhaps 100 Europeans. The climate of this place is unhealthy, the dreaded yellow fever being endemic. However, the capital, Bingerville, nineteen miles distant, is more healthily situated upon a hill.

Among the interior towns is Bontuku, with 3000 Hausa and other Moslem inhabitants—a place surrounded by walls, its mosques and other characteristic structures giving the appearance of an Eastern city. The town lies on the cavaran route to Sokoto. Kong, Koroko and Bona are centres of importance for trade with the Middle Niger.

From Little Bassam, adjacent to Grand Bassam, its seaway free from a bar, a railway runs for 250 miles to Katiola. Canals for navigation have been cut between Assini and the lagoon.

The area of the Ivory Coast colony is about 120,000 square miles, and its seaboard extends for 380 miles along the Guinea Coast. The natives—numbering 1,000,000 or 2,000,000, no near estimate being forthcoming—live in comparative ease on the product of their small cultivated lands, where they grow maize, bananas, cotton and fruits, fishing and hunting adding to their supplies. They also smelt iron and weave cloth, after the manner of many African tribes. Coffee culture was introduced by the white man, and gives good results. The white population is less than 1000.

The ancient gold mines in this part of Africa, abandoned since the tenth century, prove the existence of the yellow metal, and some native villages have their own placer mines. British capitalists have taken up concessions and spent a good deal of money in mining here since 1901. Britain is also represented in the customary import, cotton goods, the main article of import; and half the trade of the colony is done with Britain. The colony is, as to its revenue, self-supporting.

Leaving the Ivory Coast, we pass the shores of the British Gold Coast Colony and of Togoland, elsewhere depicted, to reach the next portion of the vast West African dominion of France; the ancient negro kingdom of Dahomey, whose name, even in the popular imagination, carries with it a savour of barbaric cruelty and bloodshed.

Dahomey, before the French occupation, was one of those pagan negro states which had attained to some considerable organisation in its governance and a certain amount of native "culture." In this respect it was not unlike Ashanti and Benin, to the west. Hidden in its dark interior, beyond the swamps and forests of the coast, lived a people and their kings who practised the most repugnant rites, periodical saturnalia of bloodshed perhaps only equalled by those of the ancient Mexicans in another continent, also under the sun of the northern tropics. Here, moreover, existed that remarkable army of female warriors, or "Amazons," which at one time formed the flower of the Dahomeyan forces.

The colony of Dahomey is a relatively narrow strip of territory 75 miles wide on the Guinea Coast, running northwards for 430 miles to the Niger, where it merges into the great Upper Senegal and Niger region. Near its middle portion it becomes 200 miles wide, and the area covered is about 40,000 square miles, sheltering a population of over 1,000,000.

Here we encounter that type of scenery common to parts of the Guinea Coast, low and sandy, a line of lagoons, the flat country extending inland for 50 miles and covered with dense vegetation. A vast swamp, the Lama Marsh, covers a considerable area of the littoral, the land then rising to the high plateau which forms the water-parting of the Niger with the Dahomeyan Coast streams.

Like that of the neighbouring territories, the climate is hot and moist upon the coast, subject to heavy rains and tornadoes in their seasons, with the dry harmattan alternating, when the vegetation droops and the skin of the European cracks. However, for a large part of the year the sea breeze blows, tempering the climate, its influence reaching inland to Abomey, the ancient capital of the kings of Dahomey, to which we shall presently ascend.

This ocean wind generally begins in the morning, and at sunset has increased to a gale, and its cool temperature dowers the chief port and centre of administrative authority, the town of Kotonu, with a climate which, in comparison with that of other towns of the Guinea Coast, is a healthy one for the white man. An iron pier, extending out beyond the surf, affords some facilities for landing.

Kotonu is a town which has arisen from its own ashes, the original place having been burnt out in 1890 ; and due to this circumstance is its modern town-plan, laid out upon the beach between the sea and the lagoon of Porto Novo.

Twenty miles across this lagoon, upon which small steamers ply, we reach the town of the same name, Porto Novo, so designated by the Portuguese arrivals in the seventeenth century. This, the earlier French headquarters and the present main commercial centre, has a population of over 50,000, and contains numerous public buildings, merchants' houses, mosques and churches.

To the west, some twenty-three miles from Kotonu, standing also upon its lagoon, is Whydah, an old and at one time thickly populated town, famous in slave times, when it was divided into English, French, Portuguese, Brazilian and native quarters ; the first three of which were protected by powerful forts, of which the French alone remains. To-day we may picture the scene when cargoes of unfortunate negroes were dashed through the heavy surf in surf-boats to the anchored slave ships. This type of craft is specially constructed for landing on the beach, for there is no harbour, and the rollers break with unmitigated force. Within and about this old slave pirate centre thousands of orange and citron trees yield their grateful shade and luscious fruit, giving to Whydah the name of the " Garden of Dahomey." The population, large in earlier times, has dwindled to about 15,000.

We shall not linger on the coast but, taking the train at Kotonu, on the railway which runs northwards through the entire length of the colony to the Niger, reach Abomey. The Dahomey railway is of a metre gauge, 270 miles long from the port to Paraku, near the British frontier of Nigeria. Completed in 1907, it gives access to that vast line of communication elsewhere described.

Abomey lies upon this railway 70 miles from the coast, built upon a rolling plain 800 feet above sea-level, and it has a population of over 15,000. When this old barbaric stronghold was burnt by King Behanzin, its slave-raiding potentate, before the oncoming French expedition which had set out to punish him, Abomey was surrounded by a wall, with six gates, around which a moat, thickly grown with prickly shrubs extended, a protection common to such African strongholds, the circumference of these defences being six miles. Herein were villages separated by fields, whose cultivation—after the system we shall observe in Nigerian towns—yielded food for a beleaguered population, and contained royal palaces, market-place, and native barrack square. The French rebuilt the place and dowered it with a water supply, by means of artesian wells.

The kings of Dahomey had their burial-places and country residences at Allada and Kana, nearer the coast, both now served by the railway. The country we traverse in the interior of the colony consists of undulating land, with occasional forest, and barren in places. However, the soil is generally fertile and capable of high cultivation, but agriculture is backward. The Toffo district is famed for its orchards of oil-palms, and indeed the palm-oil is the principal product of the country, as it is of others on the Old Slave Coast. The making of palm wine, formerly important, has not been encouraged, as it leads to the destruction of the trees.

Around the villages and in the cleared forest spaces maize, guinea-corn, cassava, sweet potatoes, yams, plantains, oranges and others of those fruits and foods common to these savage regions are raised. The African apple grows wild ; coco-nuts yield well, and ground-nuts, kola nuts, sugar-cane, ginger and other cultivated and forestal products are obtained, and some plantations of rubber-trees and vines have been made.

The export of palm-oil and kernels is considerable, and the railway serves a region rich in the palms. Copra, rubber and dried fish—the latter a product of the considerable fishing industry of the lagoons—are other articles of foreign trade, in exchange for which cotton goods are obtained, largely from Great Britain, with the customary arms, ammunition and alcohol. Much of the trade is carried on through the adjoining British port of Lagos, and its volume fluctuates around £1,000,000 per annum. Various tramway lines and roads give access to certain districts not reached by the railway.

The people of Dahomey are of well-formed stature, tall, and in their demeanour proud and reserved, warlike and keen in trade, but polite in their intercourse with strangers. They are in general fetish-worshippers, but Mohammedanism has made some progress in the north. On the coast the skilled surfmen have acquired the name of the Krumen of Dahomey, and there are some communities of Fula, Hausa and Yoruba in the country. Native laws, where the character of such permits it, have been maintained, under French supervision.

Let us lift the veil a moment upon the past history of the savage potentates who flourished before the French dominion was established. Their history began early in the seventeenth century, with the extensive kingdom of Allada and Whydah, which was invaded by the Dahomeyans, under King Gezo, who also attached Abeokuta, in the Yoruba country, now part of Nigeria. In 1852 England blockaded the coast to prevent the slave trade. France, Germany and Portugal also had or desired to obtain " rights " on the coast in view of their Whydah " forts," and they protested against the action of Britain. In 1861 the King of Dahomey, Gezo, quarrelled with Britain, which annexed Lagos, followed by France, which took Porto Novo, but the independent existence of Dahomey did not come to an end until 1894, when the French entered Abomey.

The reign of the kings was marked by the horrible ceremonies known as the " customs," hundreds of men, women and children being butchered whenever a monarch died, in order that the departed sovereign might be supplied with attendant spirits in the shadowy world to which he was supposed to have gone. It is recorded that in the ancient rites of 1791, 500 victims perished thus, dancing, feasting and ceremonials attending the slaughter. Dressed in calico shirts and gaily decorated, the victims were confined in wicker baskets, with alligators, hawks and cats in similar receptacles, taken up to high platforms and hurled thence among the crowd, which, mad with frenzy, vied in their swift butchery of the human sacrifice. Within the palace, simultaneously, eunuchs and women were privately done to death, the walls being afterwards adorned with the skulls. The king's chamber, it is recorded, was paved with the heads of victims, taken as prisoners of war, and the skulls of conquered kings were used as drinking-cups. It is also related that a canoe was floated in human blood by the King of Dahomey, and that the bodies of the slain were cooked and devoured, although some of these horrors have been denied.[1]

The most curious institution of the Dahomeyan was the training of women as soldiers, and it is stated that a quarter of the female population, after the most favoured in appearance had been apportioned to the king and chief men, were drilled, armed, formed into distinctive regiments, and always took the battle front, their arms being flint muskets and bows and arrows. In the " manœuvres " which the king occasionally held for the edification of foreign visitors a test of endurance was given by the Amazons. One behind the other barriers of thorny acacia were piled up, and at a given signal the women, barefooted and half naked, charged these terrible obstructions, and disappeared underneath them, presently emerging torn and bleeding at the other side, having made their way bodily through the thorny lines, seemingly insensible to pain, and thence passing calmly in review before the monarch.

Leaving Dahomey we enter another vast tropical territory belonging to the French republic, facing upon the Guinea Coast ; that of the French Congo, or French Equatorial Africa.

Although not part of French West Africa in an administrative sense, for the colony exists as a separate unit, the French Congo territory joins the Sahara on the north, and thus is linked with

[1] Sir Richard Burton, *A Mission to the King of Dahomé.*

the other French possessions which we have examined in this chapter. It is divided from Dahomey by Cameroon.

The French Congo is an enormous territory, covering four separate " colonies " under· lieutenant-governors, these four divisions being the Gaboon, the Middle Congo, and the Ubangi and Chad divisions. The whole area is estimated at 700,000 square miles, with a population variously numbered as between 6,000,000 and 10,000,000, of which less than 1500 are white.

The seaboard of this great, irregular-shaped region lies nearly north and south, and is cut by the Equator near the upper part of its 600-mile length. Above, to the north and west, lies the little Spanish foothold of the Muni River Settlements and German Cameroon, the westerly boundary terminating at Lake Chad, whence the frontier is formed by the Sahara, the line running in an easterly direction to the Anglo-Egyptian Soudan, which territory forms the eastern boundary. South and east lie the Belgian Congo State and River Congo, and the little Kabinda territory of Portugal, near the mouth of the Congo.

The Gaboon, which forms the coast-wise division of this colony, was early visited by Portuguese explorers, who appeared here in the fifteenth century. The first object of French occupation, in 1842, was to secure a revictualling port for men-of-war, and Libreville, the capital of Gaboon, founded in 1849, owed its establishment to a settlement of negroes taken from a slave ship. To-day it has a small population of a few thousands. The town lies near the mouth of the Gaboon estuary, which deeply indents the coast, almost upon the Equator.

Southwards the sharp promontory of Cape Lopez projects, and the Ogowe, the principal river of the colony, forms its delta here. We may ascend this river by boat for a distance of over 230 miles to N'Jole, a busy trading centre. The other rivers flowing across the coastal zone are navigable only for short distances, but the inland waterways formed by the Congo and its affluents are very extensive. The Shari river waters the north-west part of the colony, and flows past the Cameroon border into Lake Chad, affording a navigable waterway between the Gulf of Guinea and that lake by means of its affluent, the Logone, connecting with the Benue and the Niger.

As we ascend from the Guinea Coast the Crystal Mountains unfold, rising in places to 4500 feet, sloping thence to the plateau,

c

itself cleft by deep river valleys, whose walls in places rise sheer for nearly 800 feet above the stream bed. These higher rivers, on their way to the coast, traverse a widely changing landscape, from terraced regions floored with bare sand, to grassy plains and virgin forest, and thence to the dreary coastal flats. The climate is almost everywhere hot and dangerous to European life.

The dense, primeval forest areas of the coast alternate with open lagoon, mangrove swamps and park-like stretches of country, which in their turn give place to prairies covered with high grass, and to cultivated patches. Along the river banks dense walls of vegetation arise, and here the hippopotamus and the crocodile have their home. The curious climbing fish inhabits the mangroves, and fireflies light up these weird places at night. Enormous trees, sometimes 200 feet in height, abound in the forests, and the wine-palm, the oil-palm, the baobab, the silk-cotton tree and other great or valuable trees familiar to the West African traveller are encountered.

The products of the forests, in conjunction with ivory, are the chief sources of the colony's wealth. Elephants are numerous. Apes and chimpanzees, with beautiful parrots and other tropical birds, inhabit the untravelled forests, which at times are firmly bound together by an inextricable mass of lianas, whilst fragrant jasmines are interwoven along the margins of the rivers, where the papyrus—often 20 feet high—and the cottony hibiscus grow. In the lagoons numberless oysters are found. The insect life becomes very apparent to the traveller by reason of the abundance of sand-flies and mosquitoes, and there are strange and poisonous fish, reptiles and serpents, which urge him to untiring vigilance as he penetrates their haunts.

The most formidable animal of these African wilds, however, is the gorilla. This terrible beast ranges this part of the continent upon a zone extending through to German East Africa, and it is found nowhere else in the world. None of the wild denizens of the continent is feared by the natives more than the gorilla, although it is rarely seen, and a full-grown specimen has never been brought away in captivity.

Ascending the River Congo we reach the town of Brazzaville, on the north side of Stanley Pool; and opposite is Leopoldsville, one of the district capitals of the Belgian Congo State. Brazza_ ville is the seat of the governor-general of the French Congo.

From Stanley Pool the steamer takes us 680 miles into the heart of Africa to the north-east, ascending the Congo and its great affluent, the Ubangi, the river forming the dividing line between the French and Belgian Congo. The rapids of Zongo then bar the course, but the river again becomes navigable beyond, and small vessels may even gain access to the Nile, through the Bahr-el-Ghazel tributaries of that river. Upon the Sanga, a Congo tributary entering 270 miles above Brazzaville, steamers ply for 350 miles, crossing the frontier of Cameroon. Railways have also been surveyed, and some portions built, with the purpose of connecting the coastal rivers with Brazzaville, and the districts served are rich in deposits of copper and other minerals.

In the northern part of this French territory the people are chiefly Mohammedans, Fula, Hausa and Arabs, and they formed powerful, organised states. The negro peoples are often of a very low type of civilisation, fetish-worshippers, and in some cases cannibals. Tribal law is recognised over their lands, but unoccupied territory belongs to the state.

These last-named areas, embodying the principal part of the colony, have been carved up since 1899, when the " concessions " regime grew up, into vast estates belonging to limited companies, with monopolistic rights. In some cases areas of 54,000 square miles are held by a single company. These concessions caused grave prejudice to the older trading companies of the Gaboon, some of the most important of which were formed of Liverpool merchants. The system also came into collision with the Berlin Act, which had established equal rights in the Congo basin, and worked against the economic interests of the natives, and the monopoly of trade and of land thus set up was an undoubted outrage. However, a settlement was reached in 1908, and French reorganising decrees were issued.

The value of the foreign trade of the French Congo has risen to about a million pounds annually. Palm nuts and oil, ivory, ebony, cocoa, coffee, vanilla, copal, copper ores are all exported, and gold and iron are found. The natives live largely on manioc, which they cultivate, with other customary products.

We have now to describe the colonies of other European Powers on this coast, which intervene between the foregoing.

CHAPTER IV

In the partition of Africa, Spain, the wonderful colonising nation of the New World, had neither sought nor obtained any considerable foothold in the tropical regions. The total area owned by Spain in tropical Africa does not exceed 79,800 square miles, and is made up of two or three colonies—namely, Rio de Oro in the north, forming part of the Sahara Coast, and the Muni River Settlements upon the Guinea Coast, with the islands of Fernando Po, in the Gulf of Guinea. There is also the small island of Annobon.

A nearly waterless land, dotted with a few oases, sparsely peopled by wandering Arabs and Berbers, with an area of 70,000 square miles, this Saharan land, the protectorate of Rio de Oro, unfolds before us. It is cut midway by the tropic of Capricorn, and thus lies in the same latitude as Mexico, that extensive and varied " New Spain " of America.

The principal town lies upon the peninsula of Rio de Oro, united by a sandy isthmus to the mainland, its adjacent bay affording good anchorage. To the south lies a hilly district, and the climate throughout is generally temperate and not unhealthy, except at certain seasons. If the land is sterile the sea is prolific in fish life, and cod and other fishing is an important industry. The country is administered from the Canary Islands.

The next Spanish possession in Africa lies over 3500 miles away, south and east along the vast African coast : the island of Fernando Po.

As the traveller enters that great right-angled corner of the African coast washed by the waters of the Gulf of Guinea, his attention is suddenly arrested by a beautiful island, seemingly like a single mountain rising from the ocean, its sides clothed with vegetation, whose verdant hue contrasts with the blue of the tropic sea and sky, its summit culminating in a magnificent cone, penetrating the atmosphere to 10,000 feet. It is Fernando Po, and the mountain is Clarence Peak, or Santa Isabel.

Struck by the beauty of the scene, more than four hundred years ago, the Portuguese navigator whose name it bears called this island-gem Formosa, "the beautiful," and a colony was established by Portugal on its shores. But in 1778 the island was ceded to the Spaniards, who, in their turn, their attempts to develop it proving disastrous, made it over to the administration of Great Britain, which country made of Port Clarence—named by them after William IV. when Duke of Clarence—a naval centre for the men-of-war which sailed those wide seas of the Old Slave Coast, in their work of suppressing the slave trade. From hence the white-winged cruisers of Britain put forth to intercept the slave ships which strove to pursue their nefarious traffic from Africa to America, with their human freight of negroes, torn from their villages on the Guinea Coast.

The island of Fernando Po is the largest in the Gulf of Guinea, over forty miles long and half as broad, and is of volcanic origin. Steep and rocky shores give access to a narrow coastal plain, whence the slopes of the mountains arise, covered with forests to the summit, scarred with craters and gemmed with crater-lakes. From these high sources torrential rivers rush downwards through the deep valleys they have carved out in the mountain-sides, traversing forests rich in oil-palms, tree ferns, ebony, mahogany and oak, the home of monkeys, tree snakes and pythons, passing thence to the plains and to quieter reaches, where the crocodile and the turtle have their homes, and where malarial fevers menace the human inhabitants. Antelopes roam over the uplands, and the civet cat is found.

In the densest parts of the forests live a singular people, their villages fashioned amid the brushwood, cunningly to conceal the whereabouts of their dwellings. These are the Bubis, a finely developed race of Bantu stock, who, jealous of their independent life, love not in general to enter the service of the white man. They go naked, pursuing their life of sportsman and fisherman. A mixed negro people, descended from slave ancestors, with some admixture of the Portuguese and Spanish blood of their one-time masters, occupy the coast, and the whole population numbers about 25,000, of which perhaps 500 are Spaniards and Cubans.

The steamer casts anchor at Port Clarence or Santa Isabel, as the Spaniards term the principal settlement, whose safe and commodious harbour lies on the north coast. The climate is unhealthy, and in the town's graveyard lie the remains of several African explorers. But the seat of Government has been removed to a little town in the hills high above sea-level, a few miles away.

The palm-oil industry, formerly the commercial mainstay of the community, takes a place second now to the cultivation of such tropical produce as coffee, cocoa, sugar, tobacco and vanilla upon the plantations developed by the Spaniard, and the kola nut is also grown. The planting of cocoa has assumed considerable importance, yielding perhaps 2000 tons yearly of this commodity for export. This development was largely brought about as a result of Spain's loss of her colonies in the war with the United States, when the wealth of cocoa in the Portuguese plantations of the West African mainland urged the Spaniards to cultivate this important article. Cuban prisoners were sent here at this time.

English is the common speech of the coast people of Fernando Po, a legacy of the British occupation, during which West Indians, freed slaves and Sierra Leoneans settled in the island. The British control terminated in 1844, the Spaniards refusing to sell out their rights, although they appointed an Englishman as governor. The Baptist missionaries who laboured to reform the natives were dismissed, however, and Jesuits appeared, but Methodists are also at work now, in conjunction with the Roman Catholic religious element.

A hundred miles away from Fernando Po, on the African mainland, a small rectangular territory carved out of the French Congo and bordered on the north by the German colony of Cameroon, lies the colony of Spanish Guinea, or the Muni River Settlement. It covers about 9800 square miles of territory, with a coast-line 75 miles long, stretching between the rivers Compo and Muni, and running inland about 125 miles.

The bold bluffs of Cape San Juan arise as we approach,

Corsica Bay developing thence, with the mouth of the Muni river.   Access to this channel, which the depth of water on the bar affords, has led to the name of the Muni River being given to the small protectorate, all the other streams being obstructed and so unfitted for navigation.

Making our way inland we traverse the Crystal Mountains, whose ridges we shall also pass in entering the Congo territories, for the range parallels the West African coast in this part of its course.   Beyond, we reach the tableland, 2500 feet above the sea, and above us still rise the granite peaks for 1000 feet.   Dense primeval forests cover the face of the country over large areas, a growth which is brought about both by the great fertility of the soil and the heavy rainfall, which latter, with that of the neighbouring territory of Cameroon, is among the heaviest in the world.

In the little-trodden wilds of the interior the elephant has an undisturbed home, and is still numerous.   Innumerable birds and insects, whose colours are of the most striking and brilliant, inhabit this remote land, and the forest life is further animated by the numerous tribes of monkeys.

A trying climate, consequent upon the great humidity and excessive heat, characterises the region, lying as it does but a few miles north of the Equator.   The dense forest gives way to more open country in the eastern districts, but the European settlements lie upon the coast, where trading stations exist at the mouths of the rivers, and the customary West African commodities of palm oil and kernels form the mainstay of the people, with rubber and timber.   But cocoa and coffee plantations flourish, and the natives cultivate those foods common to the maintenance of their kind in these regions ; and they are, in addition, expert in fishing and hunting.   The whole population of the colony numbers little more than 150,000, formed mainly of Bantu negroes.   The colony is governed from Fernando Po, with sub-governors in the principal centres.

We have now to continue our survey from the north.

CHAPTER V

GAMBIA

THE singular ribbon-strip of British territory upon Africa's western coast, a little below Cape Verde (the westernmost point of the continent) ; a strip only 27 miles wide on its sea front and less in the interior, but 250 miles long, forms the colony and protectorate of Gambia. Through this narrow land a silver thread, the great Gambia river, winds and doubles, falling into the Atlantic Ocean ; an important stream, the only river in Africa navigable at all seasons of the year for ocean-going vessels for 200 miles. This small, narrow possession of Britain, covering under 4000 square miles of territory, or about equal to the combined size of Devon and Cornwall, presents on the map the appearance of having been crushed by the surrounding French territory, which we have described in a previous chapter. This in effect is what has taken place, for, due to lack of fore-sight and to negligence at the Colonial Office in earlier times, France by degrees became possessed of piece after piece of territory which once was British. However, the river, almost in its entirety, is British, a port at the head of its navigable waters having been ceded to France in order that she might enjoy an outlet to the stream.

The climate of the Gambia during the dry season is the best in British West Africa, but in the rainy season it is so un-healthy, due to the great change from excessive dryness to excessive damp with its debilitating effect, that business at that time almost comes to a standstill, and those Europeans who are in a position to do so go away on furlough.

Bathurst, the capital, is generally fanned by a refreshing breeze from November to May, and in the pleasant month of April the soft sea wind prevails. The mean temperature here is about 80° F., with, however, considerable variations, and up-stream the average is much more. The highest recorded rainfall is about 67 inches, and the average about 51.

The town has a population of about 8000, and is one of the cleanest places in West Africa. It stands on St Mary's Island, near the mouth of the river, and a bridge connects it with the mainland. Red sandstone forms the principal material of its houses; the main buildings, facing the sea, including the Government House, hospital, barracks and churches; whilst a fine avenue of bombax and other wide-armed shade trees forms the market-place. Bathurst was founded in 1816 and named after the then Secretary of State for the Colonies, and M'Carthy's Island, upon which is situated the trading station of Georgetown, the only other place of importance in the colony, took its name from the Governor of Sierra Leone, who was beheaded by the natives in the Ashanti War.

Still in good preservation on the river bank near this island, and elsewhere on the Gambia, are found stone circles and pillars of a "Druidical" nature, and these are venerated by the Mohammedans, but by whom or at what period they were erected is not known. There were no early towns on the river, and no trading before the advent of the white man. Mohammedan influence was first felt in the eleventh century, and in the fifteenth the Portuguese visited the river, and their embassies strove to open up trade with gold and slaves from the interior.

Both Queen Elizabeth and James I. of England gave charters and monopolies to companies of adventurers, in 1588 and 1618 respectively, to trade in Senega, Gambia and the Gold Coast; companies formed of Exeter and London merchants, who believed that at Timbuctoo a great gold mart existed. One of their ships, the *Catherine*, ascended the Gambia, but its commander was murdered by the natives. Other expeditions were sent out to the supposed land of gold, which, however, was not found, and developments of another nature followed.

The abolition, in 1807, of the slave trade, which since 1662 had been the mainstay of the colony, plunged the small community into an economic crisis, but after the Napoleonic Wars British traders established the settlement on St Mary's Isle, and some prosperity resulted. But from 1866 to 1888 the Crown colony suffered greatly by reason of the retrograde colonial policy of Britain at that period, as affecting West Africa. Later, slave-raiding potentates caused trouble and the murder of British commissioners, and primitive expeditions were sent out. The French pushed their Senegal frontier closer, and when treaty-making between Britain and France was entered upon for the settlement of limits, only a ten-kilometre strip on each side of the river was obtained by Britain.

In 1906, slavery became finally extinct throughout the protectorate; all children born of slaves were proclaimed free from birth and all slaves free on the death of their masters, and thus the humane rule of Britain was consolidated, and slave caravans were a thing of the past. The colony has been self-supporting since 1871, but a hut tax was imposed for the protectorate in 1895, to balance expenditure thereon, and this was effected without conflict with the natives, and amounts to about four shillings annually upon each family.

The native people of Gambia are generally peaceful, industrious and thrifty, the total population numbering about 146,000, made up of the negro or negroid race, with less than 200 Europeans. The chief tribes are the Jolofs, Jolas and Mandingos, with numbers of Fula. By far the greater part are Mohammedans.

The Jolofs, the leading native people, are extremely black, but are handsome and fine-featured, and noted for their powers of conversation.[1] The Jolas have a communal form of tribal government, living in stockaded villages under a patriarch. The Mandingos, according to tradition, " came from the east, riding upon horses," about A.D. 1000 or earlier. They have played the part in Western Africa that the Hausa played in the West Central region, and are a people of some historic and social interest. Their features are described as of a somewhat Mongolian type.

The Gambia river appears in the accounts of very early geographers, such as Hanno the Carthaginian, Ptolemy and the Arabians, and was once supposed to be a branch of the Nile, and later of the Niger. Mungo Park started on his famous travels therefrom, and was followed by other English and French explorers. The river is navigable for ocean vessels and sailing ships up to M'Carthy's Island, and Government steamers ply thereon, carrying passengers, whilst beyond this point

[1] *Oxford Survey of the British Empire*, " Gambia," Tremearne.

smaller craft navigate for a considerable distance. The bar at Bathurst can always be passed.

Away from the river the traveller must journey in the saddle or be carried by hammock-bearers, as there are neither railways nor roads for vehicles in the colony. Beyond the swamps the country is largely covered by brushwood, but much of the land is cleared for cultivation. In the jungle and on the hills and river banks, in their special haunts, we encounter the great fauna, famous to Africa in general, the lion, leopard, monkey, hyena, antelope, giraffe, eland, hippopotamus, crocodile and others, and a few elephants still roam in the eastern districts. Fish are plentiful, but food in the rivers being equally so they do not reward the angler's efforts by taking artificial bait. Oysters abound in Oyster Creek and are cooked by the natives for food. The usual African song and other birds are encountered, and bees are very numerous. The tree and plant life is that common to West Africa, except that the rubber-vine and the oil-palm are relatively scarce, and do not offer to the native that ready source of wealth which we have remarked elsewhere on this coast.

Nature, however, is prolific of oil-bearing material, in the ground-nut, in whose cultivation the greater part of the population is engaged. The nut—the " pea-nut," so familiar to the American public—was introduced at the beginning of last century by freed slaves from America, and its cultivation assumed large proportions. The value of the export of this article to Europe for oil extraction amounts annually to a large sum, and represents nine-tenths of the export trade. For its planting in June, after the early rains, there is a great immigration of temporary labourers, who are given land and housed and fed for their work, taking half the proceeds of the nut harvest. The Government distributes seed nuts to the native planters, with the purpose of keeping up the quality of the product, a policy which has been successful.

It cannot be said in general that much has been done by the Government to develop the agricultural resources of the colony,[1] but the people are fairly prosperous, the rate of wages is high, and the revenue of the colony since 1906 increases. There is no public debt. Land can be rented at the low rate of twopence

[1] According to the *Oxford Survey*, *op. cit.*

per acre annually for a term of twenty-one years. All vacant, unused or untilled land is the property of the Crown, and sites for " factories," as the trading stations are termed, are officially allotted at low rates.

Among other principal products of the territory are such articles as hides, bees'-wax, cotton, maize, rice, cassava, rubber, jute fibre, indigo and palm kernels. France takes by far the greater bulk of the exports, and there is a considerable trade with the adjoining French territory. The greater part of the imports are from Britain; native boat-building is carried on, and the making of " country cloths " and some leather and sandal works.

The geology of Gambia reveals rich deposits of iron ore, the ridges containing layers of hematite and limonite, but no mining in a commercial sense has yet been carried out. Gold in small quantities also exists.

The education of the Gambia people is provided for by the various missionary denominations, with grants-in-aid from the Government. The Roman Catholics maintain an agricultural school, and at Bathurst there is a Mohammedan school, managed by a Board of leading men of that faith, with the State governor as chairman, and thus religious freedom is secured. The Government pays the salaries of the school teachers. Prizes are given for the cleanest village. The Gambia natives have organised more than forty Friendly Societies at Bathurst.

From Gambia we look seaward to the Cape Verde Islands.

CHAPTER VI

THE Portuguese people have said of themselves that they "can conquer but cannot colonise," and this apophthegm has not infrequently been borne out in the regimen of their foreign possessions. The earliest knowledge of the West African coast was due to the early Portuguese navigators, who coasted and traded along this immense littoral, and Portugal has bequeathed a picturesque nomenclature in many places here. But the territory which finally remained under the Portuguese flag is not commensurate with this early activity, and, moreover, the character of Portuguese rule therein has not in general been such as to secure the economic progress of the lands so governed.

The northernmost of the Portuguese possessions upon the mainland are the Cape Verde Islands, far from the coast, and the small colony of Portuguese Guinea.

The traveller who has expected to find, upon approaching the Cape Verde Islands, that green and luxuriant appearance which the Portuguese designation of the archipelago might seem to convey, will be disillusioned, for the severe droughts which here occur, the aspect of the volcanic rocks—from one group of which smoke pours forth—the treeless and barren character of some of the islands, render uninviting at first glance a region which nevertheless contains elements of interest and worth, and supports a considerable population of over 140,000.

The *Ilhas do Cabo Verde* lie from 200 to 450 miles off Cape Verde, the ten islands being disposed in a crescent form, covering an area of about 1480 square miles. Six of the islands—the northernmost—form the Barlavento Group, a Portuguese name meaning "Windward," and the four southerly ones are the Sotavento or Leeward Group.

A hazy atmosphere hangs over the archipelago, thicker towards the African coast, but this does not betoken perennial rain, as occasionally a period of three years has passed without any rainfall. The north-east trade wind blows throughout a great part of the year, alternately with the harmattan, whose dry breath from the African coast is felt from time to time.

We first approach Praia, the principal port and capital of the archipelago, and drop anchor in its fine harbour. Praia is situated on the southerly coast of Sao Thiago Island, the largest and best populated of the group, although one of the least healthy. The picturesque old town of Velha, slightly to the west, with its ruined fort and cathedral, formed the capital until 1770, some sixty years after the French sacked it. Praia has a population of perhaps 25,000, and is the centre of a considerable trade. Here stands the governor-general's palace, that functionary being appointed from Portugal, the islands being administered as a foreign province of the Iberian republic.

Notwithstanding the somewhat uninviting aspect which the islands in many places present, Praia offers a pleasing scene to the traveller ; feathery coco-palms surrounding it, and the landscape rises in successive slopes towards the tableland, crowned by the Antonia Peak. The natural unhealthiness of the place has to some extent been improved by drainage of the stagnant marshes which lie to the west.

The island of Santo Antáo, rugged and mountainous, is the most picturesque of all, the best watered, the most fertile, and enjoys the healthiest climate. Here is produced coffee of fine quality in considerable quantities, and sugar and fruit, with some quinine and wheat. The population is over 25,000. Near at hand lies Sao Vicente Island, with 8000 people, sterile, its harbour of Mundello, an extinct crater, giving good shelter to vessels. Business is carried on mainly by Englishmen, and nearly all the inhabitants understand English.

Santa Luzia is a small island, with a good harbour. San Nicolao, with 12,000 inhabitants, is mountainous, its two highest peaks being visible far and wide. It was the earliest to be colonised, but its climate is not very healthy. The island of Sal enjoys a trade in its nature-product of salt, and dairy-farming is a prosperous industry. Whales, fish and turtles are abundant. The railway was opened as early as 1835. In much earlier times the island was once abandoned by reason of drought and famine.

Boa Vistu does not bear out its name of "fair view," as the vegetation is almost confined to a few straggling coco-nut trees, and the island in the hot season is almost an arid waste, where the impoverished islanders live principally on fish, which fortunately the sea gives plentifully. The island of Maio has as its harbour Porto Inglez, or English Port, having been occupied by the British until the end of the eighteenth century, who based their claim to this barren, treeless waste on a clause of the marriage treaty of Charles II.

We now approach Fogo, the reason for whose name, meaning fire, is apparent as we remark the active crater of its volcano, 10,000 feet high, which in the island's history has frequently wrought devastation. Two British officers were—in 1819—the first to reach the summit. Fine coffee is produced on the volcanic soil, also sugar-cane and maize, but a desert covers half the area. The island supports some 18,000 people, and Sao Felippe, or Norsa Senhora da Luz, is the capital, sheltering the self-styled aristocracy of the island in part, people descended from the old Portuguese settlers. The bulk of the population, however, are negroes and mulattoes. Drought, famine and epidemics have been frequent.

Brava Island offers some noteworthy condition in its life. Its area of 23 square miles is divided into a large number of small holdings, as the people are dominated by the desire to own their land. The population numbers some 380 to the square mile, and it is stated that the inhabitants are constantly engaged in legal proceedings arising out of the land-holding system. . There are more women than men among the population, by far, and they are renowned for their beauty. The absence of men is due to emigration to America, whence they often return rich and educated, or at least so in comparison with the native peasantry of the island. The principal industries of the Cape Verde Islands are in the production and export of its coffee, sugar, millet, spirits, physic-nuts—*Jatropha curas*, a matter of some importance — salt, live animals, hides, fish, and so forth. Tobacco, coco-nuts, bananas, the castor-oil plant, dates and other fruits and products are raised. The main article of import is coal, followed by the customary European textiles, hardware and food-stuffs.

The waters abound with fish, and across the long stretch

from the African mainland turtles come over to lay their eggs on the sandy shores of the islands. The climate and flora have perhaps been prejudicially influenced by the destruction of the forest for fuel, but afforestation has not yet been considered by the Portuguese.

The islands were first settled by negro slaves imported from Africa by the brother of Alphonso V. of Portugal, an absolute monarch who exercised a commercial monopoly thereover, and in 1461 slaves were recruited on the Guinea Coast. The discovery of the archipelago was made a few years earlier by the Venetian, Cadamorto, in the service of Prince Henry the Navigator. Towards the close of the fifteenth century considerable development and enrichment of the region took place, but slavery continued until, in 1854, the Portuguese Government abolished it, although the last slave was not liberated until 1876.

But the effects of slavery, and the enervating climate, have been stamped upon the people, among whom indolence and fatalism are marked, and due to the foregoing causes is also the vast preponderance of blacks and mulattoes over the whites. However, better conditions are following upon increased education.

We now leave the Cape Verde Islands for the Portuguese colony of Guinea, lying considerably below Cape Verde, on the West African mainland.

The island of Bulama, and others which form the archipelago lying upon the deeply indented coast of Portuguese Guinea, to the south of Gambia and the narrow intervening strip of French West Africa, was discovered by Portuguese voyagers in 1446, and in 1752 Portugal established a port here. The whole colony, covering about 14,000 square miles of a low and deltaic character, is surrounded by French territory.

The Bissagos Islands, amongst which we pass in approaching the coast, are beaten by the heavy surf and dangerous breakers of the Atlantic, Inland a vast estuary extends, into which the Geba river flows, and other streams which rise in that high, hilly region of West Africa, the Futa Jallon highlands, which we have occasion to visit upon the sources of many a great African waterway.

The larger rivers of this part of the coast give access to the interior for distances varying up to 150 miles, vessels of some size ascending them, although at times navigating with difficulty amid hidden rocks, and over shoals which the sudden lowering of the stream, fed by intermittent tropic rains, renders perilous. Terrific thunderstorms echo at certain seasons through the mangrove-lined lagoons and dense forests beyond, and the torrential rains convert into raging floods what but a short space before may have been placid waterways. The complicated network of estuaries, backwaters and lagoons which Nature has here disposed are in contrast with the bare and sandy shores of more northern Africa.

The hunter who adventures into the wild territory between the Geba and Grande rivers, main streams of this hydrographic system, will encounter the elephant and the hippopotamus, and remark how the rivers teem with the ferocious sharks and crocodiles. As he ascends to the higher lands, towards the east, the great forest closes in behind and the beautiful and valuable oil palm, which we so frequently observe throughout the Guinea Coast, the date-palm, the shea-butter tree, with the ebony and mahogany-producing timber abound. Rubber is an invariable product of these forests. The timber is densest along the river valleys, and forms the home of the baboon and chimpanzee.

These lower regions are inhabited by negro tribes of curious customs and pursuits, each tribe preserving its own dress, language and methods of life. Although neither Christian nor Mohammedan, they are described by travellers as being profoundly religious in their peculiar way, and their respect for family life and reverence of their ancestors are very remarkable. Some of them are keen traders, a characteristic we shall observe generally upon the West African coast among the primitive coloured folk, but they guard their land with jealousy from the encroachment of strangers. Others are pirates or sheer savages.

The centre of Portuguese government is situated at Bulama, on the island of that name, close against the mainland, but the white people inhabiting the town are few in number, consisting mainly of officials, traders, soldiers and convicts. The principal trading centres are at Bissao, also upon an island, part of the

D

archipelago of that name, which is inhabited to a large extent by independent native fishermen, associated with the warlike Bidiogos pirates.

The sway of the Portuguese officials does not extend much beyond these few centres, nor do the natives repose much confidence in the white men here, a condition markedly at variance with those obtaining in the West African colonies of other European Powers, and indeed it may be said that the country as a whole is little more influenced by European civilisation than it was at the time of its discovery four or five hundred years ago.

The country between the Mancoa and the Cacheco rivers, towards the north, is inhabited by the Manjak people, who also occupy some of the adjacent islands. Travellers describe them as hospitable and clever, and as good artisans and seamen, as well as eager traders, and they are said to " keep their word." Their dwellings are an advance upon mere huts, being built round a courtyard, and provided with towers and loopholes.

The country is not all forest or riverine land, for, as we pass inland, we traverse great tracts of park-like country and broad grassy areas. Here the panther and the antelope have a common home, and numerous game and other birds are found. Great ant-hills arrest the traveller's eye in these regions. The inhabitants of the interior are Mandingos and Fulas, whom we have encountered elsewhere, an advanced people superior to the mere negroes. The people in closest association with the Portuguese, however, are the Gurmettes, dwelling in the towns and forming the soldiery which Portugal employs when necessary to make war on other native tribes.

A small foreign trade is carried on by Portuguese Guinea, in such tropical products as ivory, rubber, ground-nuts, wax and other matters, but the rich and well-watered soil yields, under the backward methods of its white conquerors, but a tithe of its real possibilities, and what is really a rich field, whether for the agriculturist, the zoologist, or the trader, lies fallow to a large extent. The climate is generally unhealthy, but modern sanitation is capable of improving the conditions of life here as elsewhere.

CHAPTER VII

SIERRA LEONE

As we approach this small Crown colony and protectorate of Britain—small in comparison with the enormous areas which England controls elsewhere on the continent, for it is about the size of Ireland—whose bold coast-line faces the Atlantic waves on the western shoulder of Africa, the eye is charmed by the beauty of its mountains, which, covered with luxuriant vegetation and rising in undulating slopes from the ocean's verge to dense woods, culminates in those lofty peaks of which the " Sugar Loaf " is the monarch, rearing its summit about 2500 feet above the sea. As we remark these bold outlines from the steamer's deck we shall not readily, or perhaps only fancifully, discern the form—that of a lion—which it is conjectured may have given rise to the name of Sierra Leone, or Lion Mountain ; albeit the roar of the thunder which at times reverberates among the rocky ranges, the voice of the frequent tropical storm, might suggest to the imagination the voice of the king of Africa's beasts.

On the north and north-east of the colony stretches the French possession of Guinea, and on the south lies the negro republic of Liberia. The colony proper covers about 4000 square miles, with a total area for the whole territory of colony and protectorate of about 25,000 square miles, and a coast-line somewhat over 200 miles long.

The founding of Sierra Leone was closely associated with the slave trade, and the broad estuary of the Sierra Leone or Rokelle river, which forms a splendid. harbour—the finest, indeed, in West Africa—with its striking peninsula sheltering Freetown, the capital, was long the haunt of pirates and slavers. The colony, in fact, owes its birth to those philanthropists—among them William Wilberforce—who, in 1791, were resolved to better the lot of the unfortunate victims of the traffic in human flesh and blood.

51

Long prior to this, in 1462, the Portuguese discoverer, Pedro de Sinta, dropped anchor in the striking waterway and gave the place its fanciful name, known previously to the natives simply as Romarong, or the Mountain.

Mountainous, well watered and well forested, and healthy in the interior, the colony of Sierra Leone has nevertheless earned for itself, by reason of the unhealthy climate of its coast-land, the unenviable name of the " White Man's Grave." This notoriety more properly applies to the low-lying mangrove-covered portion of the littoral, where malaria and yellow fever prevail, proved to have been due to the bite of the mosquito which breeds in the swamps and insanitary dwelling-places of the natives. Only incessant warfare against these conditions, such as has proved a success at Panama and elsewhere, and such as has received great attention of late, due in part to the efforts of the School of Tropical Medicine of London and Liverpool, can alleviate this unhealthiness.

Freetown, standing upon its splendid estuary with the succession of wooded hills rising beyond, presents a picturesque appearance, but its fair aspect is overshadowed by the terrible reputation of its climate, which, however, the draining of adjacent marshes, good water service and campaign against malaria-producing mosquitoes may alleviate. The death-rate has decreased to 29·6 per 1000 of recent years. The houses are generally surrounded by courtyards or gardens, the town thus being spread out over a considerable area. The buildings are not of much architectural pretension, the principal being the Government residence and offices, cathedral, barracks, courts, railway station and grammar school. Some of these institutions, including the sanatorium, are situated upon the adjacent hills, served by the Government line, which runs through enchanting scenery thereto. There are about 500 Europeans in the town.

The State railway, a narrow line of 2 feet 6 inches gauge, which runs for 221 miles to the Liberian frontier, carries the traveller to the healthy interior, traversing the fertile districts of Mendiland. This line, completed in 1905, and the first built in British West Africa, cost nearly £1,000,000, and is proving of immense benefit to the country in tapping rich oil-palm belts and other resources. It is seen that the climate is dependent upon elevation. A Government railway runs from Freetown to the Government bungalows on Wilberforce Hill, which, but 900 feet above sea-level and only six miles from the city, enjoys an invigorating atmosphere and is free from mosquitoes, and here the administrator dwells, secure from the dangers of Freetown.

The considerable capabilities of the native people are note-worthy, and from the full-blooded African race have sprung such noted persons as Bishop Crowther and Sir Samuel Lewis. In their separate quarter are established the Krumen, that sea-going people who were early attracted by the maritime activity at Freetown. Here they enjoy their special reservation, which is the only place outside their own country to which they bring their wives and families. The population of Freetown is somewhat under 40,000.

Strategically and commercially, the capital of Sierra Leone is an important centre. The ample harbour might afford accommodation for fleets of any size. It contains a coaling station for the British Navy and is the headquarters of the British military forces in West Africa.

The population of Sierra Leone is estimated at from 1,000,000 to 1,500,000, and may be broadly divided under three headings : the natives, or various negro tribes ; the Sierra Leoneans, a term exclusively applied to the descendants of the first settlers and the Africans, which latter were rescued from slave ships captured by British cruisers and established at Freetown by the British ; and Europeans, who include residents, Government officials and traders, numbering somewhat less than 700.

The principal negro tribes are the Timni, the Sulima, the Susu and the Mendi. There are semi-negro races, notably the Fula, of those advanced people described elsewhere.[1] A number of Syrian traders have of late years settled in the country. The negroes in general are pagans, and each tribe has its fetishes and secret societies, which latter are very powerful, and are influential in useful purposes connected with agriculture and the great palm-oil industry.

The Sierra Leoneans largely inhabit Freetown and the penin-sula, and their language is " pidgin-English." They are nominally Christians and take much interest in religious services.

[1] See " Nigeria."

Among the negroes are many Christian converts, also Mohammedans, such as the Susu people. The Mendi people show much
skill in carving, and from their district many curious steatite
or soapstone figures have been dug up and sent to the British
Museum.

In general both natives and Sierra Leoneans are capable of
considerable intellectual progress, and many of the latter
have received a good European education, some occupying good
positions as lawyers or doctors, or in the Church. The native
progress has been remarkable in view of the savage conditions
twenty years ago, when the chiefs worked their will upon the
population of the hinterland.

The difference of physique, language and general characteristic among the various native tribes is very marked. The
Mendis are of medium stature, but of great endurance, and are
the principal carriers, bearing heavy loads over long distances.
The numerous Timnes are of somewhat similar characteristics,
and are largely engaged in agriculture, but are also skilful
canoe-men and smart traders. An exceptionally fine race are
the Mandingos. With their flowing robes and sandals, these
followers of the Prophet and vendors of Arabic nostrums and
fetishes are readily distinguished from the other peoples of the
colony.

Sierra Leone is the land of the oil-palm, and when this is said,
the wealth, resource, trade and general mainstay of the colony
have been summed up. Agriculture is backward. The main
source of wealth and occupation, and that upon which trade
principally depends, is forestal, the produce of this one tree—the
oil-palm. Trade and produce practically embody palm oil and
palm kernels, and trade is described as good or bad in accordance with the volume of these two articles brought down to the
coast for shipment.

The oil is obtained from the fruit of the beautiful oil-palm,
which grows luxuriantly and abundantly in vast belts of territory, even beyond the reach of the railway, and could these
great groves all be brought under exploitation the result would
be almost embarrassing in their yield. The oil is expressed
from the fibrous, oleaginous pericarp of the ripe fruit by the
natives by still primitive methods and sent down in large barrels
for shipment. To obtain the kernels of the fruit, which produce

a different kind of oil, and which are exported for treatment,[1] the laborious method is followed by breaking open every individual shell by hand, the shells being very hard; an experienced breaker scarcely obtains more than a quarter of a bushel of kernels, from four times this bulk of the nuts, in a day's work. The heavy demand for these vegetable fats in Europe has given rise to high prices. The natives know their value and neglect other industries to reap the richer harvest yielded by the cone of the oil-palm. With added methods of transport and scientific treatment by machinery it is considered that the wealth of the colony from this indigenous resource would rapidly increase. Indeed the whole life of the colony depends upon this beautiful and abundant tree, a gift of Nature requiring no tending by human hands.

The kola-nut production is also a valuable article of native trade, but its use is confined to the natives, who traffic with it upon the coast, enormous quantities being exported to Gambia and Senegal. As described elsewhere, this nut has hunger and thirst satisfying properties, and is chewed for this purpose. The kola-tree grows to a great size, its branches spreading wide, and its greenish pods contain as many as a dozen nuts, furnishing both food and the basis of an enormous industry for the native. Yet another palm, the cane-palm, *Raja vinifera*, yields of its abundance for native needs in the luxuriant forest belts of this rich tropical land, providing palm wine, which is largely consumed, and also giving the bass fibre known in Europe as piassava, largely used for broom-making.

The coco-nut palm flourishes everywhere near the sea, but its prolific yield has little more than local use at present. Doubtless it offers a future source of wealth and industry whose exploitations will not long be delayed.

To describe all the tropical fruits which gladden the eye and tempt the palate in this fertile land would here be impossible. Bananas and pineapples are encountered in great profusion, the orange, mango, lime, paw-paw, bread-fruit, guava

---

[1] A nutritious cattle-cake is manufactured in England and Germany from these exported palm kernels (as also from coco-nuts), a new industry in the first-named country; but the value of the cake is scarcely understood by the British farmer. Special information on this point has been published by the Imperial Institute in London.

" sweet-sop," " sour-sop " and others are common. Rice of excellent quality, cassava, ginger and maize are cultivated, but coffee and tobacco have not been remunerative. Sugar-cane is grown to some extent. Cocoa has been tried, and further experiment may bring success in its production. Ground-nuts, yielding their valuable vegetable oil, were at one time cultivated, and may again be produced, in view of the demand for their contents. Baobab, shea-butter and silk-cotton trees abound.

The rubber-vines which formerly covered the tall forest trees with their network have to a large extent been destroyed in clearing, but Para seeds have been planted. The rain forests of the protectorate have been largely destroyed by native wasteful clearing, and in consequence the climatic conditions and rainfall are said to be changing.

A feature of the prolific forests of the region are the silk-cotton trees, which yield the *kapok*, a fluffy down, largely exported and used in upholstery, and of late—in 1915—its extreme buoyancy brought it into use for making life-saving belts or waistcoats for the sailors of the British Fleet, a development of the Anglo-German conflict. The traveller, moreover, has cause to be grateful for these magnificent monarchs of the forest, which from their great height and huge crown form landmarks in the ocean of vegetation. Great buttresses or spurs radiate from the mighty trunks of these trees and spread out into the surrounding jungle, and the timber served the native in earlier times by furnishing great slabs of wood for the gates of his stockade.

Sierra Leone is administered by a governor and executive council, and the various chiefs exercise certain tribal authority and jurisdiction. A municipal council was brought to being at Freetown in 1893, its mayor being the highly educated negro barrister, Sir Samuel Lewis. Imperial military forces and a colonial frontier force and native policemen support the administration and work of the various Government departments.

There have been many changes in the administration since Sierra Leone was established, especially in earlier times, and the humorous saying of Sidney Smith in this connection is sometimes quoted, to the effect that " there were always two governors to Sierra Leone: one who had just arrived in the colony, the other who had just arrived in England therefrom."

In 1794 the governor was Zachary Macaulay, father of Lord Macaulay. In 1874, the Gold Coast and Lagos colonies, which had formed part of the great administrative unit of West Africa, were detached, and Gambia in 1888, since which period Sierra Leone has stood alone.

There are other populous centres on the coast and in the interior of Sierra Leone, such as Bonthe, the port of Sherbro, Port Lokko, reached by water at the head of the estuary, and Songo, upon the railway, thirty miles from the capital. The town of Falaba[1] lies 1600 feet above sea-level and has an interesting history of the Mohammedan Fula period. It is surrounded by a loopholed wall with gates, and is the meeting-place of various trade routes, one of which comes from the Niger.

The building of the railway and of numerous Government roads has enormously facilitated the growth of towns and the expansion of trade in the interior. It is interesting to remark the use of the bicycle over roads where but a short time ago the jungle stretched, and it cannot be long before the motor finds its way along these highways and feeder-roads. On the other hand there are vast areas of undeveloped country still lying under primeval conditions for lack of the means of transport and that stimulus which roads and railways furnish: districts through which the native staggers with his heavy loads. In these wilder regions the elephant still roams, and the panther, leopard, chimpanzee and grey monkey inhabit hill or forest, and the antelope still delights in its freedom of the plains. The rivers are the home of the hippopotamus and of crocodiles of great size. Sierra Leone, however, is not a field for big-game hunting, as the larger beasts are both scarce and shy, and the lion, notwithstanding that he has given his name to the colony, is not found. Great tarpon are found in the waters, with other abundant fish, and the mangrove oysters, which exist in great quantities, are of excellent kind.

The rivers of Sierra Leone afford some considerable means of transport, and we may ascend several of these by small craft.

[1] Which gave its name to the ill-fated steamer which was lost as a result of German methods of warfare, in 1915.

A busy scene is presented in the Sherbro and peninsular water-ways from the native traffic in their produce. Ocean steamers ascend the Sherbro river for twenty-five miles to York Town. At the mouths of certain of these streams the expert Krumen manœuvre their surf-boats in passing the bars. Horses and donkeys cannot be used for transport in the colony generally, due in part to the attacks of the tsetse fly. The limit of man-borne transport is about fifty miles, and beyond this radius from collecting and shipping points Nature reigns supreme and her lavish products return to the soil unused.

Some indigenous manufacture and arts were practised in Sierra Leone by the natives until recent times, but these have disappeared, or are fast falling into disuse. The reason of this is easily seen. So considerable a price is paid for the products of the oil-palm that the native makes money thereby with comparative ease and is disinclined for industry of another nature. Further, the easily imported foreign goods have driven out the native handicrafts. It is true that the primitive native looms still weave a little " country cloth," as the native textiles are termed. This cloth is spun by the women and dyed in the indigo pot of the village, which it is the female duty to care for, and until recently was the basis of a valuable industry, as the native cloth formed the principal currency, and much of the work thus produced was of considerable merit. Now it is difficult to obtain these cloths, which have become scarce. The cotton previously used was of native pro-duction, but it has been displaced by imported yarn of gaudy but fugitive colour. The main article of import is Manchester cotton goods. The exotic varieties of the cotton plant that have been introduced have not, so far, given good results. The economics of this system of displacing native manufacture are elsewhere discussed.

A rude kind of pottery, and some useful mat-making and excellent leather-work are the principal remaining hand industries. Among the imports, gin and rum figure largely, and these spirits bring in about half the Customs revenue. Small traders carry on the liquor traffic in every direction, and their shanties for the sale of strong drink " spring up like mushrooms near every station." [1] Another article of import

[1] *Oxford Survey of the British Empire*, " Sierra Leone," Alldridge.

is American tobacco in leaf. The people of Freetown are born traders, and the natives of the interior eagerly follow the example set.

Sierra Leone is not a land of gold or silver, and indeed the only metalliferous mineral of any value to commerce so far exported is iron. Iron ores are very abundant, and there is a peculiarly rich belt of titaniferous ore in the hills of the protectorate. Native smelting is carried on, an excellent quality of iron being produced with the aid of hide bellows and charcoal, very malleable and bright, and from this spears, swords, knives, hoes and other implements used are fashioned by the indigenous blacksmiths. It is stated that, in former times, 200 pieces of the small T-shaped pieces of iron resulting from the primitive smelting operations, whose value is one penny each, were paid customarily in the purchase of a wife.

The tribes have not always accepted British rule in such peaceful fashion as at the present time. After the abolition of the slave trade and the introduction, in 1898, of a hut tax, there was serious revolt in certain districts, and British and American missionaries, including some ladies, were massacred, and police killed. The punitive expedition which followed resulted in the death, from fighting and disease, of 600 of the British force. It was less to a dislike of the tax than the arbitrary manner in which it was collected, and to the suppression of slave raids, that the trouble was due.

But confidence in British administration was secured by preserving the authority of the chiefs, whilst safeguarding the interests of the people, and also by the modification of native habits resulting from the building of the railway. Fetishism, however, still flourishes strongly. There are a number of elementary and other schools, with special schools for the sons of chiefs, and for Mohammedans, and the scholars in general, both boys and girls, take a keen interest in their studies. The civilising influence of British rule, on the whole, may be regarded as strong, and its results as a cause of justifiable pride to the colony and the Empire.

We have but to cross the eastern frontier of this prosperous British colony to enter a republic where the black man rules.

CHAPTER VIII

LIBERIA, THE NEGRO REPUBLIC

FORMING the southernmost point of Africa's western shoulder there looms up to the traveller, on the trade route from Europe to South Africa, a mountainous land, facing upon that "Grain Coast" which took its name from the old trade in the "Grains of Paradise," as the *Amomum* peppercorns—whose flavour, we may recall in passing, Queen Elizabeth was passionately fond of—were termed : a land which forms the only republic in Africa—a negro republic, moreover, established by reason of the repatriation of negro slaves, whose forbears had been in earlier times torn from Africa's shores, borne back on the rising tide of humanitarianism which, towards the close of the eighteenth century, began to relegate slavery to extinction.

Here the West African coast, at Cape Palmas, turns abruptly to the east and north to face the Gulf of Guinea. On the west of the republic lies the British colony and protectorate of Sierra Leone, and on the east the French possession of the Ivory Coast. The coast-line is 300 miles long, and the republic covers an area of about 41,000 square miles.

The name of " Liberia " was given to the country in 1824.[1] The adjoining territory of Sierra Leone was first selected as the freed slave settlement on account of its fine harbour, as the Liberian coast does not possess a single good port. In 1821, however, Cape Mesurado was chosen by the American Colonisation Society as a suitable situation for the first party of freed negroes from the United States. At that time the Americans were still practising the abhorrent custom of slavery, and they could not admit slaves to the suffrage. During the period from that date until 1860, the great migration of negroes to Liberia took place, since when only a few thousands of these coloured American emigrants have arrived. The colony was first founded by Jehudi Ashmun, a white American citizen,

[1] The name was an " invention " of the Rev. R. R. Gurley.

between 1822 and 1828, and in 1847 the colonists declared Liberia to be an independent republic. Most of the Great Powers recognised this pretension, but the United States held aloof. Until 1857 there were two isolated black republics on the coast—Liberia and Mayland, which were not made continuous until 1892, after questions of frontier with Great Britain and France had arisen.

The coast of Liberia is of some strategic importance in that it faces the ocean highway between Europe and South Africa, but unfortunately it lacks harbours, and only at Monrovia is safe anchorage found. A kind of natural canalisation or lagoon formation connects the mouths of the rivers along the coast, and the Cavalla river is navigable for some eighty miles to its rapids, and thence beyond. Other rivers of considerable volume cross the littoral.

The interior of Liberia is largely a vast forest, except where the negroes have established their plantations, and even there the jungle disputes their control. The vast, dense Gora forest is uninhabited, and in the Nidi forest splendid rubber-trees are found. Liberia is, in fact, a typical forest country of West Africa, and these fastnesses swarm with the monkey tribes. In this and the adjoining regions under European control, due to the equatorial position, the trees and flowers of the primeval woodlands reach their most astonishing development. Beautiful aerial plants, climbers, creepers, purple and rose-coloured flowers and glorious orchids hang in festoons from the branches or grow upon the tall and mighty columns, fantastic and gorgeous in their varied hues. Birds of almost similar tinted plumage abound, and giant scorpions are found, and many deadly snakes.

There are, however, mountainous, grass-covered areas free from forest, and near the coast the trees have been cleared and burned away, as also in the north, by the Mohammedan tribes of that part. In places the land might be described as a " paradise of ferns." [1] The almost innumerable varieties of fruits, vegetables and edible plants which make up the list of tropical fauna and products are found in Liberia, including rubber, cocoa, coffee, pine-apple—which runs wild everywhere—ginger, sweet potato, indigo, cotton, the alligator-pear, banana, yam,

[1] Sir H. Johnston in *Encyclopædia Britannica*, " Liberia."

maize, sugar-cane, rice, and so forth. The coco-nut, the bombax, the piassava-fibre palm are among the trees. The Liberian coffee—the shrub grows wild—is famous for its quality.

Stories of diamond mines have come down from the mountains, but remain to be proved. Gold is found in the river sands, as is monazite, and iron exists in all parts. There are also indications of bitumen, which might lead to a petroleum formation. The mineral wealth, however, is still awaiting discovery.

The wonderful fertility of the Liberian soil has given rise to a large population of over 2,000,000 negroes, apart from those who proceeded from America. Some of the wilder tribes are of fine stature, their women still taller and handsome, but quite naked in their forest home. Mohammedan clothing is, however, being adopted, coming from the north. Some of the Mandingos are almost European in their physiognomy. The famous Krumen, known throughout the west coast of Africa as mariners, have their native home on the Liberian seaboard—the Kru coast—and these people have long furnished sailors for British warships and the merchant marine. There are cannibal races in the remote forests, and some of the negro types are coarse and ugly.

The people of American origin who have settled upon the littoral may be numbered at perhaps 15,000, and the civilised and converted natives at about 40,000. Some able men have sprung from the black and the mulatto people, prominent among them being the former president of the Liberian Republic, Arthur Barclay, a Barbadoes negro of Dahomeyan descent, but educated in Liberia, and Dr Blyden, author of books on the black race.

The republic is governed by a president, elected for a term of four years, with a legislature somewhat after the model of the United States, and the usual republican machinery of government. The official language is English, widely understood, and close relations are maintained with the adjoining English colony of Sierra Leone. The rubber trade is controlled by an English company, which is charged with the conservation of the forests. Motor roads in the interior have been made, also by an English company. In 1910, partly as a result of an

American commission [1] to report on the native life, in agreement with Great Britain, France and Germany, and in regard to the boundaries and debt, Americans were put in organising control of certain governmental departments.   American sympathy had been aroused by reports of encroachments on the Liberian borders.

The climate of Liberia is cooler in the interior and somewhat drier, being more bearable for Europeans there than on the coast.   It is noteworthy that, due to a relative scarcity of mosquitoes, fever and malaria are less prevalent than in other West African lands.

[1] Nominated by President Roosevelt.

CHAPTER IX

ASHANTI, THE GOLD COAST, THE NORTHERN TERRITORIES

UPON the southern shore of Africa's western projection, where the Atlantic waves beat ceaselessly, we approach the prosperous British colony of the Gold Coast, originally taking its name from the gold-bearing rivers and soil of the territory, which were long worked by the natives for the yellow metal, and later were appreciated by the London company-promoter and mining expert as a field for their labours.

This is a region which, although little larger as regards its western portion than England and Scotland, is perhaps the richest in the world in proportion to its size. Nature has here displayed some of her most striking forest handiwork, and has been extremely lavish in her tropical products. The history of its possession brings back the names of Ashanti and Coomassie, where, a generation ago, gallant British blood was shed in conflict with the savage potentates and their warriors of the forests. It was, moreover, the land of King Prempeh and the elusive " golden stool."

The Gold Coast Colony and its hinterland of Ashanti and the Northern Territories embody respectively areas of 24,200 square miles, 20,000 square miles, and 35,800 square miles, a rough parallelogram extending inland for over 400 miles, a few degrees north of the Equator, and cut by the Greenwich meridian. The coast is nearly 270 miles long, and on the east lies Togoland, from which it is divided by the River Volta. The French Ivory Coast lies on the west, and French West Africa on the north. The southern limit is the Atlantic Ocean—the Gulf of Guinea.

As we approach the interesting region from the sea, a strip of glistening sand unfolds, upon which we shall land direct, for as there are no natural harbours, passengers and goods both are disembarked in surf-boats, which the Atlantic rollers hurl upon the beach, the best landing-places being at Cape Coast Castle,

64

Axim or Dixcove, where rocks, jutting seawards, afford some shelter. Some artificial protection to landing has been made at Akkra—or Accra—the capital, and Sekondi, where surf-boats may unload at a jetty, and at Ada the dangerous bar of the Volta river may be passed by ship's tenders.

Bold, rocky headlands interrupt the line of the coast, and beyond the rock, sand and surf stretches a narrow belt of gently undulating country, whence arise the spurs of the hills, which, reaching to an elevation in places of 2000 feet, form the buttresses of the great plateau of West Africa. Out of the level plain of the littoral isolated conical peaks arise, giving a characteristic aspect to the landscape, and numerous rivers descend, mingling their waters with the Atlantic breakers, their mouths infested by sharks. In some districts strips of country covered with palms and shrubs intersect the coastal zone, and eastwards lie the interminable lagoons which fringe the coast as far as Lagos, in Nigeria, separated only by narrow spits of sand from the sea, and forming the haunt of crocodiles and manatees.

As the traveller proceeds inland, he remarks that the thinner vegetation of the coast gives way to a dense forest belt. At first the views are extensive, albeit the trees are close enough to afford shade from the tropic sun ; but by degrees the forest and the jungle close in upon us, with many a giant tree. Here we encounter groves of that handsomest and most graceful as well as valuable of West African trees, the oil-palm, its plume of leaves adorned by the long and drooping fringes of parasitical ferns, casting soft shadows on the uprearing column of the trunk. More magnificent perhaps in general appearance among these forest monarchs is the great bombax, or silk-cotton tree, whose soft wood furnishes the native with his dug-out or easily fashioned log canoe. Here and there, in the dim light that prevails, even at midday, in this natural crypt of the Ashanti forests the great bombax or some vast iron tree looms up, perhaps 200 feet high and 100 feet in circumference, with branches encrusted with orchids, lichens painting the trunk, and, coiling over the surface of the ground like great tortured serpents the huge roots extend, or perhaps taking the form of mighty buttresses, disappearing into the denser gloom of the surrounding jungle.

E

Notable is the effect of the gloom in the drawing up of all vegetation, from the tiniest to the greatest, towards the sunlight which beats upon the canopy far above. In this struggle upwards for life and place plant life is drawn out at times erect and slender to an absurdly disproportionate extent. Every available space above and below is filled with tangled vegetation, the earth and the sky are both hidden, and slender plant or flower stems follow the lead of mighty trunk and giant column, all being roped together in and out with the lianas or " monkey ropes," some thin as twine and " floating like fairy cradles in the air," others thick as a man's thigh, coiled about like the " aerial hammocks " of some giant forest spirit, a tangle in which no individual tree may fall, but is doomed, even in decay and death, to stand upright.

But not only is the traveller startled by the weirdness of this arboreal labyrinth. He is delighted by the flowers, the orchid and the white lily, which find nourishment here, but above all, perhaps, by the veritable fairyland of ferns, which, from dainty maidenhair to those that climb for forty feet or more up the tree-trunks, are fully at home in the dim light and moist air of the forest spaces. These ferns spread out along the ground or light up the dark recesses formed by the buttress roots of the giant bombax, or dip over with indescribable grace from the shelving banks, daintily to dip their delicate fronds in the cool surface of the forest rills, their reflections mingling with those of the giant trees above, mirrored in quiet pools and back-waters.[1]

Probably three-quarters of the area of the colony is covered with primeval forests, whose monotony at last becomes exccssive, and it is with relief we emerge from these to where, towards the north, the landscape takes on new beauties, wherein the trees are smaller, and presents a park-like appearance, with high coarse grass, dotted with baobabs, shea-butter trees and fan-palms. Still beyond, splendid open country is entered, especially in the north-east, where rich crops are raised.

In the dense forest belt there is but little animal life, and indeed in no part of the country is the fauna as plentiful as before the advent of the white man and his rifle, or before the native

[1] For a fine description of these forests *vide The Oxford Survey of the British Empire,* " Gold Coast," Tremearne.

learned to use his imported flint-lock and muzzle-loading gun. The elephant still lingers, as does the buffalo, whilst the leopard, panther, chimpanzee and other monkeys have their home in the wilds. Land snakes and water snakes of venomous kinds abound—the python, the cobra and the puff-adder. The scarcity of the elephant is due to the persistence of ivory hunters, who have almost exterminated the animal.

In the River Volta hippopotami abound, and lizards of brilliant hue and tortoises are common. Birds are not very plentiful in this region, but the general West African air fauna is represented. There are shoals of fish on the coast, including herring, mackerel, and other useful kinds, also edible turtles and innumerable oysters on the rocks and mangrove roots.

The insect life is prolific, and includes the malaria and yellow-fever propagating mosquitoes ; and in the forests those pests of West Africa, the driver ants. These insects perform their marches and deploy " much on the plan laid down for a British column in thick country."[1] These remarkable ants, it is stated, are divided into several " castes," the " workers " being a quarter of an inch long, the " soldiers " half-an-inch, and the " officers " nearly an inch, and they march in close formation, twelve abroast, with " soldiers " distributed along the line. An advance party prepares the way, and tunnels are constructed through exposed spots, perhaps thirty feet long and an inch in diameter, provided with air shafts at intervals. The ants attack whatever they encounter, even fire, and every animal gives them the right of way. By their system of communication the ants are said to be enabled to send reinforcements wherever such may be required.

The soil of this prolific region is so fertile that the native has but little contest with nature in wresting his living therefrom, but the climate is in general unhealthy, even to the native, as regards the propagation of certain diseases. The moist, hot atmosphere upon the coast is inimical to the European, who dreads the fevers and dysentery. Notwithstanding the improvements carried out at Akkra this coast town has a reputation perhaps worse than any other European settlement in West Africa. In the interior, due to the more open and ventilated

[1] *Oxford Survey*, op. cit.

country and to the absence of the excessive humidity of the coast, life is more safe and tolerable.

Violent thunderstorms and tornadoes mark the beginning and end of the rainstorms, and in the dry season the harmattan, the dry, cold wind from the Sahara, cracks the lips and fills the air with fine dust, obscuring the view and causing much inconvenience. The diurnal range of temperature in the interior, where the real harmattan blows, is considerable, amounting to 50° at times. In the coast towns the mean shade temperature is 78° or 80°, and the rainfall at Akkra averages 27 inches per annum, with a great increase at other points, as at Axini, where it rises to 79 inches.

The population of the Gold Coast and hinterland is about 1,500,000, some 2000 of which are white men, the whole of the remainder being negroes.

The first people with whom the Europeans came into contact here were the Fantis. These people, who number about 1,000,000, are of a dull brown colour, with negroid features, and the younger women are often pretty. They have many curious customs. The Ashantis are a race superior to the Fantis, but speak the same language. They are divided into a number of tribes, each with its own king, but the King of Coomassie was paramount, and he always succeeded to the " golden stool," which was the symbol of authority among these people.

The Ashantis, after being driven southwards by Moslem tribes, obtained possession, in the sixteenth or seventeenth century, of a region of almost impenetrable forest, where they founded Coomassie, and a powerful state. War took place between the Ashantis and Fantis, the ruling coast people, in which the British, at Cape Coast Castle, became involved.

Cape Coast port lies about eighty miles west of Akkra, and at the present time it has a population of some 30,000 Fantis, with 100 or more Europeans and some Krumen. The Castle is of an imposing appearance, of considerable size, and faces the sea, the town being built on a rocky promontory. The Castle is used for military purposes and as a prison. The other chief buildings are official residences, churches, school and hospital, and the brick-built houses of wealthy natives, built upon the hilly streets.

Whether the Gold Coast was discovered by the French or the

Portuguese is a debated point, but in 1364 Norman merchants may have established themselves there, whilst in 1481 a Portuguese expedition, in which probably Christopher Columbus took part (though not as leader), founded Elmina, so named from its gold washings. In 1553 English ships brought home from the Guinea Coast a large parcel of gold, and adventurers of all European nationalities, their imagination fired by the existence of the yellow metal, went out to the coast.

The Castle was built about 1661, but the English traders had previously constructed a fort at Cormantine. In the war between England and Holland, Admiral de Ruyter destroyed all these stations, except the Castle. In 1672 Charles II. gave a charter to the Royal African Company, and forts were built at various places on the coast. Thence onward until 1807, pirates, slavers and gold-seekers were among the principal elements of this fever-haunted coast, the yearly traffic in human flesh and blood reaching 10,000 annually of the unfortunate slaves who were torn from their homes. Many of these were prisoners of war of the Ashantis, who sold their captives to the white merchants. The Royal African Company, or rather its successor; the African Company of Merchants—constituted by Act of Parliament in 1750, with an annual subsidy—was ruined by the abolition of the slave trade in 1807, and the forts were taken over by the Government.

These British companies for long enjoyed a monopoly of the slave traffic. Between 1680 and 1700 the Royal African concern sent away 140,000 negroes, and by the end of 1786, more than 610,000 of the slaves had been sent to Jamaica alone, Jamaica having been a British colony since 1655. The Dutch and various private enterprises were also extremely active in this traffic, and it was a Dutch shipload of negro slaves from the Guinea Coast that first established slavery in British America, landing, in the year 1620, its human cargo at Jamestown, in Virginia. The supply of slaves was sometimes obtained by the method of setting fire to negro villages and capturing the unfortunates as they tried to escape. This was often done by their own chiefs, whose barbarity had been stimulated by the European demand and reward. Vast numbers of blacks were lost in the attacks on land, in the slave ships on the seas, in the ports, and on the plantations ; probably much less than fifty per cent. of the slaves ever coming to the point of performing labour.   It is a dark and shameful page of history, and it was a sorrowful track across the silent highway of the sea which rolls between the Guinea Coast and America, from which we may turn with relief.

British jurisdiction on the Gold Coast was defined in 1844, under which the Crown had the right of trying criminals and repressing human sacrifices.   Into the history of the conflicts between the British and the Ashantis it would not be possible here to enter at any great length, but some account will be necessary.   They resulted largely on the continued invasions by the Ashantis of the coastal districts, and despite various treaties.   The first governor of Cape Coast Castle, Sir Charles M'Carthy, was defeated, with his force of 500 men, in his expedition across the Prah, in January 1824, the enemy being 10,000 strong.   It is said that the defeat was due to an error in bringing up kegs of vermicelli instead of ammunition, but, be it as it were, the governor and a number of Europeans were killed, Sir Charles being beheaded and his skull afterwards used as a drinking-cup by the savage potentate of Coomassie.

The Ashanti War of 1873-1874 was due to alleged affronts on both sides, and the Ashantis invaded the British protectorate. Sir Garnet Wolseley was dispatched to operate against them, and advanced to Coomassie.   After severe fighting, the King was routed and fled into the bush.   With but a small force, few provisions and many sick and wounded, Wolseley decided to retire, leaving Coomassie in flames.   However, the place was again occupied and a treaty was entered into with the Ashantis to keep the peace, the natives agreeing to pay an indemnity of 50,000 ounces of gold.   But other expeditions became necessary in the ensuing years, and King Prempeh, who had been elected to the golden stool, was deposed, and the British fortified Coomassie.   The final scene in the Ashanti conflict was the siege of Coomassie by the natives, when the English governor was obliged to hack his way out, with part of the white inhabitants and the Hausa soldiers, to reach safety at Cape Coast Castle ; and this was followed by the relief of the exhausted garrison, who were on the point of destruction.   The story of the siege and relief of Coomassie, when the bugler of the beleaguered garrison sounded a salute to the approaching column, which, aided by the valiant Yorubas, had made its way through the enemy and amid the burnt houses and putrid corpses of the dead around the town, is one of the most interesting in the history of African conquest.

Ashanti was formally annexed to Britain and a Constitution granted, with a separate administration under control of the Gold Coast governor, native laws as far as possible being respected, with the maintenance of the tribal system. In 1906 it was seen that the Ashantis' native suspicion was being allayed and confidence in their British conquerors created, and thus British rule was consolidated.

The region beyond Ashanti, the Northern Territories, was constituted as a separate district in 1897, soon after the explorations of the British and others had laid bare what was an unknown territory, and under this administration its prosperity and progress have been marked.

In 1902 a railway was completed to Coomassie, cut through one of the densest forest regions known, starting from Sekondi on the coast and passing through Tarkwa, with a length of 168 miles. The cost of this important work was nearly £2,000,000, and the advent of the locomotive has had its usual civilising and stimulating effect.

Coomassie has a population of over 12,000, and is situated on a low rocky eminence in the midst of a clearing of the primeval forests of Ashanti. The fort and the house of the Resident are among the chief buildings, and there are the usual military quarters, hospitals and missionary stations. The red clay houses of the natives line the streets, some enclosed in compounds, and richly ornamented. The principal chiefs in some cases own large houses, built in European style. The town does not present now the handsome appearance it offered before the British advent, when it was characterised by some considerable culture, with a king's palace of red sandstone, which, destroyed, was rebuilt of clay. But only meagre accounts exist of the place in the height of Ashanti power, in the middle of the eighteenth century and onwards, although broad and regular streets and market-place have been described, and it may have had a population far more numerous.

The town to-day, however, has become the distributing centre for the whole of Ashanti, and a wide road, nearly 150 miles long, has been cut through the forests to Cape Coast, auxiliary to the railway. There are ancient caravan routes to the chief trading centres farther inland.

Away from these principal modes of conveyance the traveller depends largely upon the palanquin, borne by four carriers, for the tortuous paths through the forest do not admit of wheeled vehicles, although in some places such have been used, and even the motor car has been tried. But the native porter, with his load of 50 to 70 lb., is not easily displaced, nor the hammock-men superseded. Horseback journeying may be carried on in the Northern Territories and some other parts of the colony, where the tsetse fly is absent. Under the Public Works Department the roads are maintained by the chiefs of the various districts.

The remaining and important means of communication are the rivers. The Volta, curving and winding in its course from north to south—practically along the Greenwich meridian—is navigable for small vessels for sixty miles from its mouth, and is the largest stream on the Guinea Coast, its course being 900 miles long. The two main branches of this river, the Black and White Volta, rise amid grassy plateaus and break through narrow gaps in the escarpment. In its lower course through the forest the river is often over half-a-mile wide, and forty or fifty feet deep in flood-times, narrowing in one rocky pass to thirty yards, through which the great flood pours. The numerous rapids limit its value as a navigable waterway. The Ankobra lends itself to the navigation of small launches for eight miles, and it is of much value as a means of transport in the floating down of timber for the mines. The Tano, bounding the colony on the left, also has its value for navigation by light craft, for some sixty miles, and is connected with the Assini lagoon. The Prah, before mentioned, is the " sacred river " or fetish stream of the Fanti and Ashanti people.

We encounter a number of towns along the Gold Coast, and in this respect the colony differs from other West African possessions of Britain. Such towns were established by the many traders of rival nations in earlier times, who went there in search of gold. Newton lies on the great lagoon, in the west. Axim occupies the site of an old Dutch fort and was at one time the main outlet for the gold traffic. Dixcove lies east of Cape Three Points, where the heavy Atlantic rollers break in never-ending surf. Sekondi is the starting-point of the railway to Coomassie, which traverses the goldfields; Elmina lies near the mouth of the Prah; Anambi was where a handful of British successfully confronted a whole army of Ashanti warriors, who attacked the fort. Saltpard is a well-built and flourishing town. Cormantyne is the place where the English first exported negroes from the coast, so baptizing with their name the "Cormantyne" slaves taken to the West Indies. Passing thence we reach Cape Coast Castle, then Akkra, the capital, and sixty miles to the east lies Addah, on the Volta, near the Togoland frontier.

The town of Akkra has a population of some 20,000, including about 150 Europeans. The sea-frontage is three miles long, but there is no harbour, and the surf-boat is the method of debarkation from the ocean steamer, which lies about a mile out. The three forts and the various public buildings and the fine stone church on the long, straight, chief thoroughfare are among the main structures of the town. There is a race-course, Akkra being almost the only point on the coast habitable for horses. In the hills beyond are the Government sanatorium and botanical gardens. It is interesting to note that Akkra is governed by a town council, in which no racial or colour distinctions are made.

Akkra is the outlet for an important cocoa-producing district, served by a line of railway. Cocoa was first introduced by the Portuguese, but not until encouragement by the British Government was given in 1901 did planting acquire importance. Of the hundreds of plantations now existing nearly all are in the hands of native farmers, rather than foreign companies, which must be regarded as a satisfactory economic condition. The annual export of cocoa now exceeds in value £500,000 sterling, and is worth more than the gold output.

The gold-mining industry was the subject of a "boom" in 1901, upon the London Stock Exchange, when "jungles" became the rage. Numerous concessions had been taken up, and the promoter and the share-vendor in London reaped a rich, if brief, harvest, largely at the expense of the investor. The railway was built from Sekondi, partly as a result of the interest aroused, and thus some lasting benefit was conferred upon the colony. However, gold existed, and the output rose later to a value of well over £2,000,000 sterling. The Tarkwa reef was the earliest which proved to be rich, although gold—not always in commercial quantities—is found in almost every part of the colony, and the natives have worked it from time immemorial. Around Cape Coast Castle there are numerous ancient workings. The natives are dexterous artificers in gold, as they are in silver and brass. It is possible that mineral oil and coal deposits may be commercially developed in the colony.

The principal industry is agriculture, and this has been encouraged by the Government, who have established experimental farms, where natives are taught, and whence they are sent to teach others throughout the country. Displays of produce are also held. Efforts have been made by the British Cotton-Growing Association to encourage the growing of cotton for export, but the experiment has not been an unqualified success.

Ashanti produces rubber, palm oil and kernels, kola-nuts and ground-nuts, with all the other customary fruits, food products and vegetable matters, commercially valuable, of West Africa. Valuable timber is also exported, and both this and rubber should increase under greater transport facilities.

It is worthy of note that the kola-nut has had an effect almost equivalent to that of gold in stimulating commercial intercourse between Ashanti and the neighbouring native kingdoms of the West Soudan, the nut being so greatly sought after, and almost a necessity to the native, on account of its stimulating properties. Chemically this is due to the small quantities contained of theobromine and caffeine.

The rubber is of the *Landolphia* and *Funtumia* varieties, both of which exist in large quantities, notably in Ashanti, and rubber forms one of the principal exports. The Para rubber has been found to flourish exceedingly well, possibly better than the local kinds.

The oil industry derived from the beautiful *Elais guineensis*, or oil-palm, mentioned so frequently in these pages, is chiefly carried on in the Volta district. Here the oil barrels are rolled by the natives to the river, towed thence by canoes, and taken to the port of Addah for shipment. The oil forms a favourable article of diet for the native in various dishes, and is not despised by the European. In Europe the demand is for the purpose of making soap, candles, and other commodities calling for vegetable fats.

There is but little native manufacturing industry in the colony, but native cloths are produced by primitive looms, and dyed with native indigo, a process which we remark in other parts of Africa. One of the principal articles of diet is the "mealies," made from maize, of which two crops a year may be raised. There is also a considerable trade in dried fish, the coast fishing industry employing 5000 canoes, but dried fish is also imported. The sea has its own fetish and no fishing takes place on a Tuesday.

The country is governed as a Crown colony, under English law, with certain modifications, but native law is not interfered with, except where incompatible with justice and civilisation.

As regards land tenure, always an important matter in these countries, the soil is held by the tribe or the family in common, each member having a right to select a part of the common land for his own use. The land cannot be alienated without the unanimous consent of the family, and this is rarely done, although there is some individual tenure. The kings or chiefs govern the people, under the control of the district commissioners, which latter exercise judicial as well as executive functions. Slavery and peonage or debt-bondage are unlawful.

The State revenue, largely derived from custom and railway receipts, exceeds the expenditure.

The education of the people is partly in the hands of Protestant and Roman Catholic Missionary Societies, but there are a great number of Government schools. There are also Mohammedan schools, in which technical and industrial training is encouraged. A Board of Education, under the Legislature, control education. About 40,000 natives profess Christianity, but both this faith and the Moslem are making progress against fetishism, which is the prevailing belief of all the tribes. Even in this dark creed belief in a God is, it is stated, universal, as is also that in a future state. From the general intelligence of the natives progress may be expected, but the people protest against unaccustomed labour, such as modern mine and plantation methods entail.

Trade has advanced considerably of late, but there are great areas of forest and land with all their resources still awaiting development. The Government, however, wisely encourage the native agriculturist rather than the white concessionnaire.

CHAPTER X

THE West African steamer, coasting along the Guinea shore, leaves the British community of the Gold Coast and crosses the Greenwich meridian, which cuts the coast at this point, and there shortly unfolds the littoral of the small territory of Togoland, which German enterprise has well developed from its former state of a No-man's-land into the condition of a flourishing colony.[1]

The conditions we encounter here, as regards the natural character of the land, its climate, scenery and products, do not differ in marked degree from those of the British colony on the west, which we have just left. But the sea frontage is very short, being only thirty-two miles long, a straight stretch of sandy coast with neither harbour nor haven, except that reached across a dangerous bar. The shortness of the coast-line is due to the cutting off of a strip of the littoral by the adjoining British territory, and the country is much wider inland, and extends back to the north for 350 miles, the whole area covered being 33,700 square miles. On the east lies French Dahomey, part of the vast French West African possession, which also encloses Togoland to the north.

This narrow strip of coast, wherein the negro "King" of Togo ruled, during the time of the scramble for African territory, was the only portion of the Guinea Coast not claimed by a European Power, and in 1884 a German Imperial Commissioner obtained the control of the kingdom under agreement with its ruler, and Germany then pushed her claims farther into the interior, until brought up by the British and French spheres.

Behind the sandy strip we encounter a fringe of lagoons, as in the British and French territory adjoining. These waterways were, early, chosen resorts of the slavers, Porto Seguro, one of the small coast towns, having been one of the oldest centres of

[1] Togoland fell to the British in the Great War of 1914:

77

their "trade" upon the Slave Coast. This traffic was carried on from stations established by Portuguese, British, French and German traders, established at various points along the coast.

The principal lagoon is that of Togo, connected by a channel with the other lagoons and with the Mono river. The town of Togo, lying upon this sheet of water, has given its name to the colony, but the chief port and capital of Togoland is Lome, which owes its creation entirely to German enterprise. Here are situated the governor's residence, with churches, schools, hospitals and the principal business houses. A pier runs out beyond the surf, the terminus of the railway.

We encounter here the customary hot, humid and in general unhealthy climate of the Guinea Coast, although inland the harmattan induces drier and cooler conditions during the time of its prevalence. Undulating plains stretch inland for fifty miles or more, and crossing the country to the west and north a chain of mountains extends, the birthplace of the Sio and Haho rivers. These are the largest streams in Togoland, except that the Volta forms part of the western boundary. Thence to the north-west we ascend to a tableland, reaching an elevation of 1000 feet above the sea.

The river valleys are generally densely wooded, and the forests of oil-palms, rubber-vines, and various kinds of timber and dyewood are among the rich natural resources of Togoland. Along the littoral the coco-palm flourishes abundantly, having been introduced by the Portuguese more than a hundred years ago, the valuable tree extending over a region some eighty miles back from the sea. As we reach the hills we remark the baobab and other trees familiar to the African traveller, and the park-like uplands are dotted with clumps of the silk-cotton and shea-butter trees, a varied habitat in which the hunter encounters the lion and the elephant, which are among the principal inhabitants of the animal world in this tropical region.

The Germans set themselves to develop the resources of the colony on a considerable scale, and the history of Togoland has been one of peaceful progress, the only interruptions being occasioned by severe droughts, which from time to time bring loss to the agriculturist. However, Togoland was the first German colony to render itself self-supporting, and the Imperial grant earlier made towards its upkeep became unnecessary. Railways and plantations have been established, the principal railway line carrying us into the interior from Lome for seventy-six miles, whilst good roads have been built to connect the coastal towns with the principal trading centre more to the north. From the large northern towns of Yendi and Sansane caravan routes extend into Ashanti and the great trading centres of the Niger.

The plantations are largely of cocoa, coffee and cotton, which last was made the subject of modern industry in 1900, and now yields from the native, Egyptian and American varieties cotton for export to Germany. Kola-nuts are also cultivated and exported. Palm oil and kernels, rubber, maize, ground-nuts, shea-butter from the *Bassia parkii*, with ebony and fibre, are also valuable articles of the export trade. A foreign trade of about £1,000,000 is annually carried on.

The population of Togoland is about 1,000,000, of which perhaps 300 or 400 are white people, almost all Germans until recently. The coast people are of the customary African negro tribes, and those of the north, the Hausa negroid races, who, as we have seen elsewhere, are keen traders. They form small, well-organised states, largely professing Mohammedanism, and are of a peaceful character, going about the country trading with large caravans. Whilst slave-raiding among these people has ceased, domestic slavery is still carried on.

The people of the littoral—generally of Dahomey race—are also traders, as well as agriculturists, and, as customary among these tribes, fetish-worshippers.

We find in Togoland the small native industries which the African people have generally developed. There are native smiths, woodworkers, pottery makers and cotton spinners and dyers. The shawls made by these people are often of handsome appearance and good texture. They collect and spin the excellent native cotton and dye it with native pigments, indigo and others, and, as to the wood-carving, this is often of very fine execution. Notwithstanding these native aptitudes, the principal articles of import are textiles and hardware—also gin.[1] On the grassy

---

[1] Here, as elsewhere, we remark the doubtful economic condition under which native industries are ousted in the interests of foreign commercialism, as discussed in the last chapter.

savannas we shall remark the numerous herds of cattle and sheep, and horses and donkeys are bred in the highlands.

The education of the people is carried on, as customary in the European colonial possessions of Africa, both by Government and missionary schools. To the judicial system both Europeans and natives are answerable, and the administration of the colony is exercised by a nominated council of unofficial members, who advise the governors. This functionary was answerable to the Imperial authorities of Berlin.

This fairly flourishing German colony, as a result of the great European War, was invaded by an Anglo-French force in 1914.

Togoland, which is adjacent to the British Gold Coast Colony on the west, is separated from the enormous British possession of Nigeria by the narrow French protectorate of the old kingdom of Dahomey, which we have already visited and described, and beyond Nigeria we again enter a region developed by Germany, that of Cameroon.

As we approach the great right-angled bend of the West African Coast, where, in the Bight of Bifra, the beautiful Spanish island of Fernando Po reposes, we remark that Nature has disposed upon the very shore of the continent, as if marking its corner with a monument, a mighty peak, its base washed by the Atlantic waves, and its slopes and summit presenting a magnificent spectacle from the steamer's deck. This peak is Cameroon, an isolated volcano, from whose craters, 13,370 feet above the level of the sea, huge streams of lava are from time to time hurled forth upon the surrounding region.

Mongo-ma-Loba, or the "Mountain of Thunder," as the natives term this mighty uplift, is the highest peak on the West African Coast, and its fitting companion, sentinelling the approach thereto, is the high Clarence Peak of Fernando Po, thirty miles distant. The occurrence of a high mountain upon the coast is a feature unique in the orography of Africa, for nowhere else do we find a natural disposition of this character on the continent. The summit of Cameroon was reached by Burton and others, in 1861.

The possession of the great region of Kamerun (or Cameroon, as we prefer to call it) might have been British, for British commercial and missionary influence was powerful therein up to 1882. However, Britain neglected to exercise a protectorate over it, or awoke tardily to its importance, and Germany, eager, though late, to acquire territory in the scramble for African partition, obtained a signature to a treaty (at midnight) in July 1884, from King Bell, the negro potentate, five days before a British consul arrived to secure the territory for England. Six days later a French gunboat appeared, bent upon the same errand.

Cameroon has an area of 190,000 square miles, and fronts upon the sea for 200 miles, with Nigeria on the north-west, the boundary running to Lake Chad and the Sahara. The French Congo lies on the east and south, with the little Spanish Muni River Settlement, both of which we have already visited.

The Portuguese navigators, who in the sixteenth and seventeenth centuries sailed these seas, gave to the river flowing past the mighty mountain, which rises from the waves, the name of Camaroes, or the River of Prawns, after the fresh-water prawns found therein. The district at first formed part of what was known as the " Oil Coast," or " Oil River Coast," an extension of the Niger Oil River sphere, and the customary " factories " or trading stations were established. The wealthy negro merchant princes—Akwa and Bell or Mbell, exercising sway over the region, did not permit traders to enter the interior, these potentates acting as intermediaries with the native vendors of produce from the forests, who brought down rubber, ivory and other matters, eagerly desired by the white man.

Another " king," that of Bimbia, made over to Great Britain an extensive part of the bay, shore and interior in 1837, and agreed to stop the custom of human sacrifices which had prevailed. A " flourishing " slave trade grew up between Cameroon and America, about the time when, in 1845, the Baptist Missionary Society established its first station in the benighted territory. The head of this civilising movement was Alfred Saker, and, backed by British men-of-war, the work of suppressing the slave trade was stopped in 1875. A settlement of freed negroes had earlier been established, and named Victoria, at Ambas Bay, by Saker, who had been expelled by the Spaniards with the Baptists from Fernando Po. Saker was followed by Grenfell, the famous explorer of the Congo.

In 1860 a German " factory " was established on the coast,

F

two years after Saker's freed negro colony was planted, and in 1884 German sovereignty was extended over the Cameroon, and acknowledged by Britain. Certain of the tribes were disappointed that England was not to be their master, and they refused the German sway, but their villages were bombarded and the negroes reduced to submission. Considerable bloodshed also marked the extension of German power in the hinterland, extending up to 1905, and maladministration caused the recall of the German governor.

The greater part of the region is steep and broken, and as we penetrate its interior we remark the grand easterly sweep of its mountains: the great peak of Cameroon forming the north-west corner of the great plateau of Central Africa. But, in places, the picturesque ranges merge into wide-spreading plains and undulating country, with a broad and fertile valley between the main chains. Born in the high, upper region is the head-stream of the Benue, flowing south-west into the mighty Niger through British territory. The large river, Sanaga, has its source far to the south-east, and takes its way through romantic scenery, sparkling over splendid cascades and breaking through the coastal range on its way to the ocean.

The Sanaga and other streams which reach the Cameroon coast are navigable for some distance, and the Logone and the Shari, coming from the French Congo, as described in treating of that region, give access to Lake Chad, and also connect, by means of a marshy lake, with the affluents of the Niger—a peculiar hydrographic system.

Our steamer will have dropped anchor under the shadow of high-rearing Cameroon, in the bay of Ambas, a fair harbour, wherein large vessels may lie securely. Beyond stretches the Cameroon estuary, forming the mouth of the Wuri river, upon whose banks Duala, the capital of the protectorate, is situated, with other places of importance.

Duala embodies a number of trading stations and native towns, all lying together on the south bank, places earlier known as Cameroon, Bell Town, Akwa Town, and others; and forming part of the same is Hickory, whence the railway conducts us for 100 miles through a fertile region to the north. The population of Duala is nearly 25,000, of which 200 or 300 are white men, and the town is the centre of the merchant and missionary life of the colony. We shall remark the wide, tree-planted streets, and the fine park surrounding the Government Building, as also the statue to the German agent [1] to whose midnight activity Germany owed the possession of Cameroon.

This part of the African coast is indented by a number of estuaries with good harbours, and navigable waterways are afforded by the rivers emptying into them, giving access for miles into the coastal region around the base of the Cameroon mountain. Along the coast, much farther south, are the trading stations of Batanga and Campo, surrounded inland by areas of grass-land, the river mouths being lined with the mangrove, whose sombre foliage is so characteristic of the old Slave Coast. The climate of Cameroon is characteristically tropical, and malarial fever is frequent on the coast, the disease attacking even the negroes at times. But upon the slopes of the great mountain we experience conditions far superior, and here, in the higher district at Buea, the seat of government is established, 3000 feet above the sea. The flourishing town of Victoria lies in Ambas Bay.

We remark that the forest-covered slope of Cameroon ascends to 8000 feet above the sea, where the line of tree life ends, and grassy areas take the place of the timber. The mountainous region, as we cross to the interior, is also covered with primeval forest, yielding valuable woods. We pass amid groves of the beautiful oil-palm, and the wild coffee-tree spreads its foliage over the landscape. Here the elephant and other great fauna have their home, as also buffaloes and the great carnivora which prey upon the antelopes, whose graceful forms give animation to the grassy uplands. Brilliant birds and venomous snakes bear out the picture we have so continually to paint of the African wilds.

The territory is well dowered with those tropical products which furnish so much wealth to West Africa. The products of the oil-palm form the most important source of trade, but cocoa-planting has been very successful, and cocoa, with the palm oil, copra, copal, ivory, kola-nuts and Calabar beans are the principal articles of export.

The northern point of the great triangular-shaped territory of the Cameroon—nearly 800 miles long—rests upon Lake Chad,

[1] Dr Nachtigal.

where it covers 40 miles of the coast of this interesting body of water. Here we are, beyond the Mandara mountains, at an elevation of about 800 feet above sea-level, in the ancient Bornu sultanates. This district is peopled by Fula and Hausa tribes, which we shall more extensively encounter in Nigeria. The town of Dikwa, some distance from the lake, was the headquarters of the redoubtable Rabah, whose belligerent operations in this region against the British and French at length cost him his head. Dikwa is also the centre for a considerable trade with Tripoli, in silks, gold, silver, coffee, velvet, etc., conducted by caravans across the desert.

Another of the well-organised native states, that of Adamawa, lying partly in the Cameroon and partly in British territory, lies somewhat more towards the coast, and throughout this northern region the Soudanese zariba style of building characterises the houses. The more important centres are surrounded by walls, well built and strongly fortified, testifying to the martial disposition of the inhabitants. The large commercial town of Ngaundere stands upon the caravan trail from Duala to Lake Chad, traversing the colony, the inhabitants of this place being mainly Hausas. In this part of the country German occupation gave rise to severe fighting. On the Benue river lies Gama, a place of some importance.

The great Bantu races of Southern Africa people the territory south of the Cameroon mountain zone, tribes ruled by independent chiefs, and these are active traders. Whilst the Fula and Hausa of the north are Moslems, the others are pagans, but missionary societies, Protestant and Roman Catholic, are at work among them, and the schools are well attended. The European language taught has naturally been German, but it has not ousted the " pidgin-English " so common on this coast.

The population of this interesting colony is estimated at over 3,500,000, of whom less than 2000 are whites. The people cultivate the customary tropical food-stuffs, and traffic in those forestal products characteristic of the region. Cotton and indigo, bananas, pea-nuts, sugar-cane, maize, are among the cultivated plants, and the natives are known as good agriculturists, especially in certain districts. On the plateau stock-raising is an important industry, and large herds of cattle are seen among the natural pastures.

The native arts of Cameroon are of some interest. In the villages around the Cameroon estuary the people are clever wood-carvers, the highly finished workmanship displayed on the figure-heads of their canoes arousing the admiration of the traveller. The Hausas, as elsewhere mentioned, are very clever blacksmiths, smelting iron and making hoes, swords, knives, and fine arrow and lance heads with the metal. Drapery is also made from the native cotton and fibres. However, the principal articles of import are European cotton and hardware goods.

Cameroon is administered by a governor, answerable to the Imperial Authorities at Berlin, assisted by an Advisory Council formed of resident merchants, and, as we have seen, there are Mohammedan sultanates in Adamawa and Bornu. The revenue, in addition to import dues, is obtained from a native poll tax, supplemented by an Imperial grant, as the income is otherwise insufficient to meet the annual expenditure.

Operations against Cameroon by the British and French were carried on early in 1915, and the colony was lost to Germany.

CHAPTER XI

NIGERIA, NORTHERN AND SOUTHERN

THE great West African possession of Britain in Northern and Southern Nigeria takes its name from the mighty Niger river, which, flowing in a great arc through the western side of the continent, north of the Equator, its waters washing the walls of ancient towns in that part of the French Sahara which we have already visited, forms, in conjunction with its long tributary, the Benue—the " Mother of Waters " of the native—a navigable waterway into the interior of Western Africa of much importance.

Nigeria is a land of considerable interest by reason of the numerous and relatively advanced native people which it shelters : inhabitants of the famous Mohammedan emirates of the Hausa states, and the Fula people, whom we have already encountered in the adjacent territories, but whose principal home is in the region under British control. The protectorate is, further, an example of a region whose white conquerors, instead of destroying the native culture and social system and ruthlessly absorbing their heritage of the soil, have fostered these elements, atoning thus in a measure for the abhorrent practices of earlier days upon the Slave Coast, when the traffic in the negro inhabitants carried on by European and North American slave traders formed the only source of interest in these dark forests.

Nigeria covers an area of 335,600 square miles, of tropical swamp, forest and highland, extending to the French Soudan, Senegal and the Sahara on the north and west. The Hausa states, part of Bornu, and Yorubaland lie within it. Cameroon forms the south-easterly boundary, and the north-eastern corner extends to Lake Chad, in the Central Soudan. On the west lies Dahomey. The coast-line is 500 miles long, the Niger falling into the Atlantic about midway in this length.

As we approach the coast of Nigeria a region of almost

Native Canoe on the Niger

impenetrable forest and swamp unfolds, and, from the sea, the only indication of a river mouth, the enormous delta of the Niger, is a break in the line of sombre, green and monotonous mangroves, a delta which runs far out into the Gulf of Guinea, with innumerable sluggish channels. This forest belt varies in width from sixty miles in the delta to fifteen miles near Lagos, the chief port and capital. To the west a fringe of lagoons and muddy flats extends, parallel with the shore for 150 miles, providing a natural inland waterway from Lagos to the Benin river. On the delta coast we mark the constructive process of the mangrove-trees, so common on many flat tropical shores, where the arched roots hold the slime and ooze, advancing step by step throughout the ages towards the ocean.[1]

The protectorate was divided, until recently, into the districts of Southern and Northern Nigeria, and it extends back from the coast, to the north, for some 700 miles, rising from delta and forest to the plateau. We cross open, park-like plains, beyond which plateaux and mountains rise to heights of more than 4000 feet above sea-level, where the hot, moist climate of the coast, with its deadly malarial fevers, gives place to better conditions.

As regards the past of the Niger, the ancients had some vague ideas concerning the river, which seems to have been that mentioned by Herodotus and Pliny. Early Arabian geographers called it the " Nile of the negroes," and thought that both it and the Egyptian Nile had their sources in the so-called Mountains of the Moon. The river was first explored from the direction of its source, and the great network of the delta, even until the beginning of the nineteenth century, was not suspected as being the mouth of the Niger. The famous exploit of Mungo Park, in 1795, has already been mentioned. One of the most important expeditions to the Niger was that of Lieutenant Hourst, who, in 1896, navigated the river from Timbuctoo to its mouth, making a survey of the waterway and the rapids and other obstructions to navigation.

The Niger comes third in size among the African rivers. We have seen, in French West Africa, that it is navigable, with certain interruptions, for vast distances, flowing past Timbuctoo and Jenné, and that it is connected by railways with other river

[1] Very marked, for example, on the coast of Ecuador.

systems and with the coast in French territory. A navigable stretch extends from Say, in French territory, 300 miles long, crossing the Anglo-French frontier, where the river is of great width. The railway from Lagos crosses the river at Jebba, on its way to Zungeru, headquarters of the British Administration in Northern Nigeria, and runs thence to Kano.

Still descending, we reach Baro, whence runs a line to the Kano railway, and below this point is the Niger-Benue confluence, at Lokoja, a town of considerable commercial importance. Thence on its way south the Niger pours through sandstone cliffs, inundates the land below at flood-times, and, approaching its delta—the largest delta in Africa—breaks up into a network of intricate channels, forming one of the most remarkable of Africa's swampy regions, the Rio Nun being the main mouth. The Forcados river, a noble waterway with a safe bar, has, however, taken the place of the Nun as the chief means of entrance to the navigable system.

The Benue, the great eastern tributary of the Niger, rises near Lake Chad and crosses the frontier of Cameroon, traversing an enormous stretch of country and falling into the Niger at Lokoja. The river is navigable for half the year, and in 1905 a French explorer ascended it from the Niger confluence to Lake Chad, the whole journey being carried out by water, except for a porterage at Lata. Its tributaries are also navigable for long distances. The town of Yola, not far from the Cameroon border, is 850 miles from the sea, and 600 feet above it. The banks of the Benue are diversified by villages, cultivated lands and stretches of dark forest, and the names of many explorers are connected with it.

The traveller who expects to encounter throughout the whole of Nigeria a hot or oppressive climate will be reminded that even in the tropics cold and penetrating weather may be experienced. The hot, dry harmattan wind from the Sahara causes cold nights in the northern part of the colony, and even the days are cold at times, and frosts occur near Lake Chad. The tornado sweeps over plain and forest at certain seasons, with cyclones bringing rain, lightning and thunder.

However, there are evidences that a drying-up process is going on throughout the whole country lying to the north of the forest belt, and this may be the result of the excessive destruction of the forests which the natives have carried on for centuries, bringing upon themselves a castigation of nature in the advancement of the Sahara over what once were smiling plains and cultivated fields.

The interior of Nigeria is more habitable for the white man than the coast, due to a superior climate, but there are drawbacks thereto. In the north the blinding dust of the desert and the glare of the sun on the burning sands bring about affections of the eye, especially among the native labourers, who are continually exposed thereto. An unpleasing malady of the inland towns, moreover, is leprosy, which is common. The malarial fever of the coast is a scourge which science is endeavouring to overcome.

In traversing the Nigerian forests, rivers and plains, we shall encounter formidable or uncouth animals which abound elsewhere in the African continent, the lion and the leopard, the elephant, rhinoceros, hippopotamus and crocodile, together with the swift and beautiful creatures of the open—the gazelle, antelope and ostrich. The monkey forms his abundant colonies in the forests, and a host of game and other birds affords both sport and food. On the Sahara border the camel is found, but on the coast pack animals are but scarce, due to the ravage of tsetse fly. In the interior, however, there are large herds of cattle, sheep and goats. In the delta lands myriads of mosquitoes molest the traveller.

As we cross the country from the coast, the vegetation displays itself in its native zones. The sombre mangrove of the swamps gives place to the valuable and abundant oil-palm, which flourishes to about 7° north, accompanied by mahogany, rubber-vines, ebony and other varieties of timber, in the forest zone. Ascending to the plateau we remark the baobab, the shea-butter tree, the gambier, with date and other palms, and the tamarind, and numerous trees whose fruits or products form valuable articles of tropical commerce. Onwards towards the arid region the desert forms of vegetation appear, sun-resisting and water-storing. The acacia and the mimosa have their place here.

We now turn to observe the human inhabitants of this varied and extensive region of Northern and Southern Nigeria.

At the time of the Portuguese discoveries upon this coast in the fifteenth century, the Benin people were the dominating race, and their "kings" ruled over a large area. In the Niger delta dwelt a race of lower type. British traders appeared in the seventeenth century, and "factories" were established, and later on British superseded all other European influence here. The principal traffic was in slaves, but after the abolition of the slave trade palm oil was the basis of local commerce, and the "oil rivers," as the delta channels were termed, acquired great importance therefrom. Possession was taken by Britain of Lagos Island in 1861, with the purpose of terminating the slave trade, but the deadly climate led to a temporary abandonment.

In 1880 the scramble for African partition and growing British trade led to reoccupation, and the United African Company was formed in 1879, to amalgamate British interests and form a British province on the Niger.[1] The French, whose commercial interests had also been strongly established, were bought out, and they yielded the field to Britain, and in 1885 a protectorate was established. In 1886 the Royal Niger Company, under a royal charter, was substituted for the earlier British company.

In the interior, the way was barred by powerful Mohammedan chiefs, ruling over comparatively advanced peoples, but these, as well as French and German ambitions, were overcome. When further acute difficulties appeared, too serious for the powers of the Niger Company, which, nevertheless had secured such vast and valuable rights for Britain, Nigeria was placed under the direct control of the Crown, in 1900. The West African Frontier Force was established [2] and was exercised to break up slave-raiding and tribal warfare. The Ashanti War resulted in the imposing upon the emirs of the interior a humane rule and the sovereignty of Britain. Kano, a large walled and fortified town, and Sokoto, fell to British attack in 1903, and all the Nigerian chiefs, Mohammedan and pagan, took the oath of allegiance to the British Crown, and the King's Writ ran in the Hausa States.

Under the British administration to-day dwell a large population, estimated at more than 17,000,000 people, formed of a vast number of races and tribes, speaking various languages or dialects, and of different degrees of native culture.

[1] Largely through the work of Sir George Taubman Goldie.
[2] Sir Frederick Lugard's is the prominent name in this connection.

The principal type in Southern Nigeria is the negro, purest in the Niger delta and southern forests. The Yorubas of the Lagos hinterland are more cultured, as later described. The most important race of Northern Nigeria, occupying the savannas, is the Hausa, who are typical of the negroid stock, arising from the intermingling of Libyan and negro; and next, those in which Libyan blood is in the ascendant, are the Fulas of the Northern States.

The low-type negroes of the delta region early became the middlemen of the slave and gin traffic, and afterwards of the palm-oil trade, but their carrying trade is disappearing, and their degeneracy increasing on account of the liquor habit.[1] The negroes of the southern forests are generally fetish-worshippers. The Efiks of the Cross River have adopted some European customs and some learning, and among their institutions is a kind of secret society—the Egbo—or freemasonry, directed now to enforce the law and foster trade. The Ibos are described as "intelligent, cleanly, industrious and self-reliant, with a strongly imaginative and even artistic temperament."[2] The Bassas and Kokandas are, respectively, good agriculturists and watermen, and the Binis of the Benin region had developed, before the establishment of British influence, a culture uncommon to negro tribes, practising the arts of ivory-carving and brass-working. These natives had organised one of the most powerful states in West Africa, as early as the fifteenth century.

To the east of the Niger, native government had developed what is termed the House-Rule system, tribes being divided into independent Houses, and from this a clan system, with a Council of Elders, developed. West of the Niger a wonderfully organised feudal system had developed among the Igaras, and the Binis had reached the kingdom stage, with ascendancy over neighbouring tribes.

The Yorubas of the Lagos hinterland had attained to the highest culture in the south, and their state capitals are now the largest towns in the country. They are specially apt in modern handicrafts, and become skilled carpenters, engineers, masons,

[1] According to the *Oxford Survey of the British Empire*, "Nigeria," J. D. Falconer.
[2] *Oxford Survey of the British Empire*, op. cit.

and so forth.   Their past system of laws has rendered them the most tractable and useful of all the tribes in Nigeria.   The Nupes, like the Hausas, are among the principal negroid races, and they practise many useful arts, Bida, their principal town, being noted for its embossed brass and copper work.   They are of fine physique and noted as watermen and agriculturists.

The most important nation of the Central Soudan is that of the Hausas, who owe their virility to a strong crossing of Hamitic blood.   They are powerful of physique, with a black skin, but with features modified by the admixture.   Kano, a Hausa town, the great centre of trade for the Central Soudan, lying in the north of Nigeria, is a walled city with flat-roofed houses of Oriental character.   It is connected with Tripoli by a caravan route across the desert, and, by means of the Northern Nigerian railway, 400 miles long, built at a cost of £1,250,000, has direct communication with the perennially navigable waterway of the Lower Niger at Baro.   The line joins with the Lagos railway.

The Hausa language is of simple construction, but rich vocabularly, and has become the *lingua franca* of the Soudan. It is the only language in tropical Africa which the natives themselves have reduced to writing, modified Arabic characters being employed.   The Hausas are a peaceful and industrious people, and dwell in farmsteads and cities.   They are keen traders and make excellent soldiers.   As porters they carry a load of 100 lb. all day on their heads without apparent fatigue.   These people number about 5,000,000, inhabiting 500,000 square miles of territory.   About one-third of them profess Mohammedanism, their adoption of this faith dating from the fourteenth century, and large numbers start annually on a pilgrimage to Mecca.   This journey across the Sahara, or via Khartoum, sometimes occupies five or six years, and is undertaken both with religious and commercial purposes.   No important attempt to establish Christianity among these people was made until the closing decade of the nineteenth century.

A numerous and powerful people of Western Africa are the Fulas, who, in 1810, overcame the Hausa State and founded an empire at Sokoto.   These people are reddish-brown or light chestnut in colour, with oval faces, delicate lips, straight or aquiline noses and straight hair.   Possessed of intelligence and

NATIVE SCHOOLCHILDREN AT PLAY, ON THE NIGER

C.M.S. photo.

strength of character, they at first made good rulers, and established mosques and schools.   Later they degenerated, and from a just and peaceful rule descended to slave-raiding.

The Fulas are not confined to Nigeria, but we may here consider them, as they occupy a large part of the territory. They have no country of their own, but are spread over an immense region from Senegal to Darfur in the Soudan, and they generally form the dominating race in those regions they occupy. They are most numerous in the lands under French sway, such as Senegal, Segu and Massena, along the Niger, to the west of Timbuctoo, and there are large numbers within the great curve of the Niger, and as far as the Upper Benue.   They extend to the Upper Nile Valley, but their origin, which has given rise to much speculation, is not connected with the Egyptians, and they are regarded as being a mixture of Berber and negro, their features, however, differentiating greatly from the latter.   The Fulas have long been devout Mohammedans, and tradition says that in olden times " every Fula boy and girl was a scholar." They love luxury, pomp and finery, but are distinguished by intelligence and frankness, with marked strength of character, and as rulers were patient and just, if stern.   They are regarded, notwithstanding their fall, as capable of great things.

East of the Niger, Sokoto and its tributary emirates are governed by their Fula chiefs, under the British regimen, in accordance with the administrative method adopted in subject Nigeria.

The general British policy throughout Nigeria is that of governing by means of the native princes, and under this system the people respond quickly to the ideas of civilisation.   The great chiefs are appointed by the Government in accordance with native custom, after consultation with the principal men, and minor chiefs are nominated by the paramount chiefs.

The people were previously accustomed to direct taxation, and this system has been retained, although on a more equitable basis.   The innumerable taxes of an earlier period have been consolidated into a land tax, and a general, or " income " tax, and under the system the extortion and oppression practised under native rule have been abolished.   The whole system of taxation, both in the country and in the large town, has been made the subject of a careful and painstaking system by the

British, working through the native rulers. A rapidly increasing revenue is thus secured, and the general condition of life of both peasantry and the industrial element has been greatly ameliorated, and the parasitic class of extortionate publicans and middlemen eliminated.

In few places has the administrative control of Britain been more amply justified than over these savage regions of Nigeria, and this has been exemplified in the land laws governing the colony. Nationalisation of the land was declared,[1] a system in accord with native law and custom, under which the land was the property of the people, held in trust for them by their chiefs, who had no power of alienation. Thus the natives of Nigeria remain safe in the enjoyment of their land, and the European land speculator was excluded, the only restriction being the land tax. There is no freehold tenure, and land cannot be bought and sold, and the authorities recognised that their work was mainly one of administrative control.

The capitals of the Hausa States are places of considerable importance, as their population and commercial activity attest. Kano, before mentioned, has a population estimated at 100,000. It is connected by rail with Lagos. The city is built upon a plain, surrounded by a wall in places fifty feet high and forty feet thick, enclosing an area whose circumference is eleven miles. In the wall are thirteen gates, made of wood and cow-hide, set in massive towers, and around the wall runs a double moat. A great deal of the enclosed space consists of arable land, sufficient for growing food-stuffs to support the population in case of siege. Hard by the great Jakara pond within the place is the market of the Arab merchants, and here also was the slave mart. The Ameer's palace, with dome-shaped roof, supported by twenty arches and having an ornate interior, covers thirty-three acres, and the architecture of the city, although the houses are of clay with flat roofs, is not without some merit, traces of Moorish influence, with the horseshoe arch, being common.

Kano produces most of the " morocco leather " goods found in European markets, but the principal industry is the weaving of cloth from native cotton, also dyeing, and the preparations of indigo are important. The city is probably the greatest commercial and manufacturing centre in the Central Soudan.

[1] The policy of Sir F. Lugard.

Arabs from Tripoli purchase ostrich feathers and ivory, etc., for export, and bring in sugar, tea, clothing, perfumes, and other matters ; Salaga merchants exchange kola - nuts from the Guinea Coast region for cloth and live stock and leather goods, and traders from the Asben and other oases bring camels laden with potash [1] and salt, exchanging these commodities for cotton and hardware. The Hausa traders deal in live stock and horses, for trade in which Kano is an active centre. The Fulas form the aristocratic class of this interesting African town.

Another of the Hausa towns is Zaria, situated on the railway about eighty miles south of Kano. Here, dwelling in a generally bracing climate, with occasionally frosty nights, resulting from its considerable elevation of 2150 feet, we encounter a population of 60,000 people. Zaria is a healthy place, the most agreeable in Nigeria, and in respect of its climate is far superior to Zungeru, the British capital, which stands at but 430 feet above sea-level, on the railway about midway between Zaria and the Niger.

Kano also lies in the province of Zaria, which was one of the seven original Hausa States. The whole province is generally healthy, and we remark the wide extent of open, rolling plains, watered by the Niger's tributaries. Roads and telegraphs have been constructed in various directions, and a considerable trade is carried on. Zungeru, the capital, although its growth dates only from 1902, has all the appearance of a flourishing English suburban town. Its shady avenues and public gardens skirt the river which flows through it, with good houses for the Administration and settlers.

Perhaps the town second in importance of the Hausa States is Sokoto, lying 220 miles to the north-west of Kano, also on a tributary of the Niger, Northwards lies the arid Sahara, but to the south the soil is very fertile. Sokoto is the political as well as the religious centre of the Fula people, and is organised under British rule in the same way that has been successful in the other Hausa towns and provinces.

Amid these fertile plains of Hausaland, stretching to the desert, and forming the great savannas of the Central Soudan, we encounter other important native towns, such as Bauchi, with 50,000 inhabitants, and Katsena, the capital of an ancient

[1] Sodium carbonates.

state, whose manuscripts carry it back 1000 years, a town enclosed by thirteen miles of walls, and the centre of a rich and populous district. Like others of its kindred, Katsena had highly developed and elaborate systems of taxation and justice, which the British have maintained and improved. In the seventeenth century its population numbered 100,000, it is recorded, and the town boasted that it was the chief centre of learning of the Hausa States.

Hadeija, Gando, Bida, Yola and Jegga are other noteworthy towns in Northern Nigeria. Nearly all are capitals of provinces and the homes of native princes under British rule. They are surrounded by strong walls with moats, and good roads— made under British rule—connect them.

In Bornu we approach the shores of Lake Chad, lying where the rolling plains begin to sink down in the north-east, forming a contrast, in their bush-scattered, uninhabited and dreary expanses, with the park-like aspect of the more cultivated Hausa region, varied by the rounded granite domes and isolated hills, which characterise the landscape there.

The men of Bornu are famous as horsemen, and in earlier times clothed both themselves and their steeds in light iron mail. Some of their customs are singular. A huge belly is considered requisite for the dignity of a chief, and when nature or gorging does not produce this rotundity, padding is resorted to. The towns in this province, which possess populations from 10,000 to 50,000, are generally surrounded by thick, high walls, with towers and triple gates of wood and iron. The chief's palace consists of turrets connected by terraces, with separate places for each wife. The number of robes worn by the rich indicate their rank, and these are piled on despite the heat of the tropical climate. The native spearmen, however, go naked.

The province of Bornu, which is bounded on the west by the Hausa States, was divided between Great Britain, Germany and France, and so extends within the protectorates of all three provinces in this region. A well-preserved dialect of Arabic is spoken, and the population of the British area is estimated at 500,000 Bornuese. The people are largely negroes, with an infusion of Berber blood, and they profess Mohammedanism.

The foregoing is but a brief sketch of the interesting native people who occupy Northern Nigeria, and we have now to cast a glance at the more southern inhabitants of the protectorate. Lagos has a population of about 50,000 ; Calabar, known as Old Calabar, about 15,000 ; and Apobo, Bonny Town and Brass Town are other well-known places, lying on their respective rivers. Benin and Bende were famous as *ju-ju* towns, and have large populations. Owo, Abeokuta, Ibadan and Illorin are large cities and important trade centres.

The people of Yorubaland, of which some of the last-named towns are the principal centres, are of true negro stock, but have some infiltration of Arab and Berber blood. At one time they formed a powerful empire, and later they were constantly engaged in conflict with their neighbours of Dahomey. Here is a land also of large towns, the Nigerian natives being a race of town dwellers to a remarkable extent. Some of these places have populations of 40,000 or more, resulting partly upon the grouping together of the people for protection behind city walls. The Yorubans were long raided by the Dahomeyans and other coastal people, who sold them as slaves to the white traders, and it is probable that many American negroes now forming citizens of the United States are of Yoruba origin, their forbears having been torn from their native villages and carried across the Atlantic in slave ships, during that savage epoch of the slave trade. In 1738 the Yorubas turned the tables on the Dahomeyans and captured Kana, the sacred city of that barbaric kingdom.

In the native Yoruba administrative system, which displays considerable ability, a chief rules, occupying his post under a system of elective and hereditary appointment, with a council of elders, the whole controlled by the British. In earlier times, it is interesting to recall, the Yorubas used knotted strings in their mnemonic system of records,[1] and of recent times their language has been reduced to writing.

The population of Yorubaland is estimated at over 1,750,000. The Yorubas, although people of proved fighting capacity— they assisted the British at Coomassie and elsewhere—have been contrasted favourably with surrounding tribes for their courtesy and hospitality towards strangers, and for their peaceable disposition and industrious habits. They are also extremely patriotic. The administrative ability they display

[1] Reminiscent of the Incas of Peru.

G

and their capacity for industrial arts and trade are other note-worthy features of this lighter complexioned negro race.

Whilst the majority of these people are pagans, a religious feeling is strongly evident among them, and Mohammedanism made considerable progress in Yorubaland after the Fula wars. In Northern Nigeria Mohammedan schools abound, but the curriculum mainly consists of little more than learning by heart portions of the Koran. A Government school for the sons of chiefs and Moslem scholars has been opened with some success at Kano. Since 1848 Protestant and Roman Catholic missionaries have been at work in some of the large towns. It is noteworthy that a native Yoruban—Samuel Crowther—was the first negro bishop of the Anglican Church, and that he was distinguished, moreover, as a geographer, linguist and explorer, an excellent record for a man rescued from slavery. This rescue was effected by the British in 1822, and Crowther was educated at Sierra Leone, as before mentioned.

The chief town of Southern Nigeria is Lagos, lying on the island of the same name, between which and the mainland is Iddo Island, iron bridges for the road and railway connecting them, over 2500 feet long. The population of Lagos numbers perhaps 50,000. The swamps adjoining the place have been reclaimed, and Government buildings, law courts, club-house, hospital, race-course, substantial buildings and private residences have arisen, with well-kept, electrically lighted streets and good water-supply. The population of Lagos includes the descendants of some 6000 freed slaves from Brazil, who were brought over in the earlier days of British development on the coast. Of Europeans there are about 400.

Extensive harbour works have been undertaken in order to make of Lagos an open port. From Forcados, 200 miles along the coast, traffic is carried on through the lagoons for heavy cargoes. The railway from Lagos to Kano in Northern Nigeria is an important work in the development of the colony.

The town-dwelling character of the Yorubans is shown by the existence of other large centres of population. Among these, in the province of Lagos, is Abeokuta, with 60,000 inhabitants, and Ibadan, with 150,000, also Badagry, earlier notorious for its slave trade and raiding.

A YORUBA KING AND HIS RETINUE

Worthy of more passing interest is Abeokuta, which we reach by rail from Lagos, a distance of sixty-four miles, or by water, for a somewhat longer journey up the Ogun river. Abeokuta lies in a beautiful and fertile district, from which masses of grey granite lift their natural domes and turrets, the name itself, in the native language, signifying "under the rocks." The extensive town is surrounded by eighteen miles of earthen walls, behind which the village populations of the region were forced to take shelter against the incessant attacks of the slave-hunters from Dahomey and Ibadan, so making in this rocky stronghold a common front against their persistent enemy, a movement which took place at the beginning of last century. A free confederacy of the tribal groups thus grew to being, each quarter of the town preserving the name, ruler and customs of its particular settlers, and this apparently incoherent community successfully warded off the attacks of the armies sent by the Dahomey potentate and others.

The zeal of the native rulers of Abeokuta, who accepted without opposition the British protectorate—which left them their fiscal and executive autonomy whilst preventing slave-raiding and human sacrifices—largely transformed the town early in this century, and Government offices, law courts, bridges and roads were built. Sanitary inspectors were appointed, and a newspaper is published in English and Yoruba. The streets are generally narrow, and the houses of dried mud or earthen walls, and there are numerous markets for trafficking in native and European commodities. The Church Missionary Society has its Yoruba headquarters here, and both British and American missionaries have met with considerable success in their civilising work in religious and industrial schools. The people, however, are very tenacious of their independence. They are governed by a council of chiefs, who exercise legislative, executive, and to some extent judicial powers, and the president or head of this ruling body is chosen from among the members of one or two principal families.

Luxuriant forests of palm-trees, which yield the commercial oil, form the principal wealth of the Egba district, of which Abeokuta is the centre, and the Egbas, as this division of the Yoruba people are termed, are enthusiastic farmers and have striven to adopt modern methods in agriculture.

About sixty miles north-east of Abeokuta lies the town of Ibadan, whose feuds with the sister city made much of early Yoruban history. The places have some features in common, Ibadan being enclosed by eighteen miles of wall, and surrounded by cultivated lands, after the manner of these singular walled communities of Nigeria. It enjoys a considerable autonomy in its affairs, and has, in its civic governance, an interesting institution in the *iyaloda*, or "mother of the town," a woman judge to whom female disputes are referred. There are many mosques and idol-houses ; the dwellings are low and thatched, built around a court, and open spaces shaded with trees are interspersed throughout the town. Ibadan is the capital of one of the Yoruba states, and has a population estimated at 150,000.

Benin was formerly ruled by the king of Benin city, and the name was earlier included to designate the French possession of Dahomey. Human sacrifice, including crucifixion, *ju-ju* and others of the abominable "customs" of the barbaric negro potentates were carried on to an appalling extent in the old kingdom of Benin, as at Dahomey, but the arrogant independence and barbarity of the country were destroyed by the British expedition of 1897, sent to punish the king for the brutal massacre of a British consul and other Europeans. The town was found to be reeking with the vestiges of human sacrifice. It stands on a clearing of the dense forest, but to-day scarcely the remains of the ancient city, which excited the admiration of earlier travellers, exist. The houses, however, are neatly constructed with clay walls painted with ochre, with some attempt at sculptured columns. Benin was discovered by the Portuguese in 1845, and they established a brisk slave trade. But after the British occupation and the abolition of slavery and of the fetish-embargo—which existed against all articles of export excepting palm-oil—rubber, timber, ivory, kernels and so forth became important matters of commerce. The chief export is palm oil. Shorn of their barbarous customs, the chief and people of Benin have adapted themselves cleverly to the new order of life. The people were always skilled in brass-work, ivory-carving, weaving, and so forth, and both the design and execution displayed in these arts were excellent.

Trade in Nigeria has shown marked increase and activity

under British rule, and will develop still further with improved means of communication, such as are being carried out or projected. In the southern districts forest produce furnishes the principal articles of export, and palm oil and kernels are shipped to the value of much over £1,000,000 annually. Rubber, cocoa, ebony and other timber, with gum and copal, follow. Cotton goods are imported almost entirely from Britain, and spirits from Holland, the total trade in imports and exports for Southern Nigeria reaching more than £6,000,000.

The Yoruba country is the greatest agricultural district of West Africa, and model farms have been set up in order that tuition of the natives may be given, and experiment made in the culture of the numerous products which the soil is capable of yielding.

The rubber industry of Nigeria, which sprang up in 1894, was almost ruined at the close of the century, due to the ruthless destruction of the sap-yielding vines, 75 per cent. of these having been killed. The natives had discovered that a larger amount of the latex could be obtained by pounding up the roots and stems of the plant, in producing " root rubber," and to this was largely due the destruction. Ten years later, due to preservation, the industry came to life again, and now the natives are learning to plant the trees, and Government plantations of Para rubber have also been made in some districts. These, although they have not yet matured, are regarded as making satisfactory progress, although it cannot yet be said that certainty attends the introduction of this exotic.

From the various indigenous rubber-producing trees and vines several varieties of rubber are produced by the natives, the most important being that from the African silk-rubber tree, *Funtumia elastica*, which is found freely in the southern forests. Following this, in point of importance, is the *Landolphia owariensis*, in a similar situation. Adulteration is practised from the latices of other plants. " Paste rubber " and " red Kano rubber " are inferior varieties.

The product known as " shea-butter " is largely used by the natives as an illuminant and for cooking. Shea-nuts are the kernels or seeds of a tree[1] growing in Northern Nigeria and

---

[1] *Butryospermum parkii.*

Yorubaland, whose fleshy pericarp yields a white latex, and the kernel a thick oil or butter. It forms a substitute for palm oil or ground-nut oil, and a certain amount is exported. We constantly encounter this tree in West Africa.

In the Hausa States and the Soudan the kola-nuts produced here are highly valued and form an article of trade at Kano and elsewhere, among the Arab merchants and others. Two kinds of this fruit are cultivated in Southern Nigeria by the Yorubas and others, and at Bida. The use of the nut is similar to that of the betel-nut of Asia, and its chewing is said to allay hunger and thirst,[1] as elsewhere remarked.

In Southern Nigeria the planting of cocoa has been attempted, and a considerable area is devoted to plantations which are owned by native cultivators. Nearly all the exported cocoa comes from the Lagos hinterland. The climatic conditions are not specially favourable to cocoa-growing, due to the long, dry season in the Lagos district, which results in the exhaustion of the tree, but the small plantations at Old Calabar and elsewhere are situated in more favourable surroundings.

The product of the oil-palm, its oil and kernels, form the most important article of trade in the delta region, and many millions of gallons are exported annually from Nigeria, and of kernels a large tonnage. This forest industry founded on the fruit of the oil-palm almost takes the place of agriculture in Southern Nigeria, and supplies the native with an easy means of obtaining almost everything he requires. The importance of the oil as an article of export dates from the beginning of last century but it has formed an important article of diet from earliest times throughout Yorubaland and the Lower Niger Valley. In appearance and consistency the oil is reddish or yellowish, and thick, and it is extracted by the natives from the outer fibrous pericarp by boiling and chewing. The oil from the kernels is of a better quality, and is generally extracted in Europe, where the kernels are sent for treatment, as in the case of the Sierra Leone oil industry.

The palm is very abundant and the forests are exploited for it in only a comparatively limited degree. It grows most luxuriantly in a moist but well-drained soil, occupying open valleys, but soon becomes established upon cleared land. The

[1] Reminiscent of the *coca* leaves of the Peruvian Indians.

dispersion of the seeds or nuts—which are heavy—is carried on, in a natural state, by the grey parrot and the monkey. At the present time these natural fatty substances are largely sought after as food-stuffs and lubricants in Europe.

It is interesting to recollect that the palm-oil trade was that which largely superseded the slave trade.

The cultivation of cotton is indigenous to Nigeria, and Kano, the centre of the cotton-cloth manufacture, clothes half the population of the Soudan. The industry is carried out on small hand-looms, the cotton being dyed with native indigo or magenta. All the cotton grown in the northern region finds its own home or local market. The climate and soil are favourable for its production, and the population dense and industrious, although too greatly congested in the towns. There are large areas of land available for extended cultivation. Hopes have been expressed that Kano as a cotton-producing centre will become known in Europe, and that, due to easier means of transport which the railway furnishes, and with the extension of cotton cultivation outside the zone where the oil-palm and the rubber-vine flourish, cotton may be grown in Hansaland by the natives for export.

The obverse side of the matter of growing cotton for export must be briefly considered. This is the danger, discussed elsewhere,[1] of the sacrifice of local welfare to foreign trade. At present, in Northern Nigeria, especially around Kano, large cotton-fields are not planted, because the land is used for growing food-stuffs. The imports from Britain into Nigeria are largely represented by cotton goods, and in this connection it might be asked how far such import is inimical to native industry.

The cotton-manufacturing trade of Kano, it is believed, has been carried on since the ninth century.

Stations have been established and considerable experiment made for the purpose of cotton cultivation in Nigeria by the British Cotton-Growing Association, whose work in the interests of Imperial Colonial Development are well known. It was hoped to stimulate native industry and work towards making the looms of Lancashire independent of foreign sources of raw material. Whilst Northern Nigeria, as a result of this research, has been shown to be perhaps the most promising field within

---

[1] See final chapter.

the British Empire for cotton-growing, it seems also to have been established that local conditions concerning labour markets and transport would cause large plantations to be unremunerative, and the attempt to establish such, on a self-supporting or paying basis, has been abandoned. The high cost of local labour and the low market price of ginned lint are given as the principal factors against large plantations, employing native labour under white supervisors, and the Association has turned its attention rather to inducing the native farmer to grow cotton in rotation with his ordinary crops. The West African cotton is inferior in colour, lustre and length of staple, it is stated, to the best varieties of American cotton. Foreign seed has been widely distributed with the purpose of improving the indigenous kind, but these exotics do not appear always to give satisfactory results, and experiment is being made with the hybridisation of African and American kinds, at various stations throughout the country, with results which it is yet too early to decide.

After the opening of the railway from Lagos, inland, cotton cultivation by the natives greatly increased in Yorubaland, and ginneries have been set up at various places. Almost the whole of the cotton exported from Nigeria comes from the southern districts.

The general British policy throughout Nigeria, of ruling by means of the native rulers, has been eminently successful, and the state of the country generally is creditable to the empire.[1]

---

[1] The spirit animating some of the Moslem rulers here was revealed by the remarkable manifesto of the Sultan of Bornu, after the fall of the German stronghold in West Africa during the Great War (September, 1915), and the contribution to the British War Funds which accompanied it. The Sultan is a descendant of the Moslem Arabs who civilised the Soudan, after their manner, before the Dark Ages.

CHAPTER XII

THE BELGIAN CONGO

To turn from Nigeria to the Congo, which vast territory we now enter, passing Cameroon and French Equatorial Africa, which we have elsewhere described, and crossing the Equator, is to turn from a land where the native dwells under a just and reasonable rule to one wherein he has suffered the utmost measure of injustice and extortion, tardily remedied. The name of the Congo is associated in the popular mind with perhaps the most notorious system of organised exploitation of a primitive people to be found in the pages of commercial history, a royal commercialism, moreover, whose methods outraged the whole civilised world, and made of this vast heart of Equatorial Africa one of the darkest places upon the map.

Looking at a map of the territory, we remark a great curving river, which, with its innumerable tributaries, presents somewhat the appearance of a weeping-willow tree, flung across the Equator, and bending to the north, east and south. The Belgian Congo State—formerly the Free State—occupies the vast area thus enclosed, covering 900,000 square miles of tropical jungle, swamp, forest and plateau, inhabited by an enormous number of negro people of the Bantu stock, so scattered and unknown, however, in great part, that their numbers are little more than guessed at, and have been placed between the wide limits ranging from 14,000,000 to 30,000,000.

The Congo State has a very narrow frontage upon the coast, covering only the estuary of the river, with a width of twenty-five miles, with here the little Portuguese enclave of Kabinda on the north, and Portuguese West Africa, or Angola, on the south. Along the north-western and northern side of the great bend stretch respectively French Equatorial Africa, or French Congo, and the French and Anglo-Egyptian Soudan. On the east lie Uganda and German East Africa, with the great Lake Tanganyika, and on the south is Northern Rhodesia. In the north-eastern corner rise some of the smaller tribuatries of the Nile. The width of the Congo State from west to east is over 1200 miles, and from north to south it measures nearly 1300 miles.

Approaching the mouth of the Congo we enter one of the best natural harbours on the West African Coast, the port of Banana, which is capable of accommodating vessels of the largest size. The European trading stations, or "factories," as they were termed, were, in some cases, established as early as the sixteenth century. The mouth of the river was discovered by the Portuguese naval commander, Diogo Cǎo, about 1482, and he erected a marble pillar at Shark's Point in commemoration of his exploit. But for three centuries little was done to explore the great stream, a lack of enterprise hardly redounding to the credit of European nations whose traders throughout that period occupied the coast near at hand.

Ocean-going vessels ascend the Congo for sixty miles, to Boma, the administrative capital of the State, and thence to Matadi, twenty-five miles beyond, where the rapids are reached, barring the vessel's passage. This was the point approached by the British expedition, under Tucker, in 1816, and from it the Congo railway takes its way to follow the cataracts for 260 miles to Stanley Pool. Thence the river is navigable for nearly 1000 miles to Stanley Falls, twice crossing the Equator in its vast semicircular sweep, and far beyond that, to the south.

In this great territory we plunge into perhaps the most remarkable forested region upon the surface of the globe. Throughout great portions of the State primeval forests cover the face of the land, where the vegetation is so extremely dense that we can only effect a passage by forcing a way through the innumerable creeping plants and between the giant trees, whose interlacing branches shut out the sun. On every hand the land teems with life, animal and insect, displaying every form and hue.

Livingstone wrote of these forests that even the vertical sun could not penetrate the gloom except by sending down at midday thin pencils of rays into it, that for months the rain-water stands in stagnant pools made by the feet of elephants, and that climbing plants, from the size of a whipcord to that of a man-of-war's hawser, are so numerous that only along ancient paths can progress be made, and giant trees falling across the road bring with them masses of tangled ropes, a breastwork over which the traveller must climb, as cutting a path round it would take more time than he could afford.

Not all of the Congo State is thus densely wooded, the region of this character occupying the river valleys, especially in the east and north-east, and over a huge tract covering some 25,000 square miles from the mouth of the Aruwimi, into whose impenetrable forests the sunshine never enters. This last-named district is known as the Pygmy Forest, and also as the Stanley Forest, after the great explorer who first traversed it. Otherwise it is called Great Congo Forest.

At one time, it is held, an immense forest covered the whole equatorial region, of which the Pygmy Forest is only a remaining portion. What has been termed the "gallery" formations occur here, avenues of trees, "like the colonnades of an Egyptian temple, opening into aisles and corridors, musical with many a murmuring fount," as one writer has described it, appearing on the banks of the rivers amid what appears otherwise to be impenetrable forest.

From the long mountain range and steep escarpments—5000 to 10,000 feet high—along the rift valley of Lake Tanganyika and the Nyanza lakes, the land slopes gently into the Congo basin, with wooded savannas, and the colony includes, in its north-east portion, part of the Albert Edward Nyanza, and a portion of the Nile basin and part of Ruwenzori, the famous Mountains of the Moon, whose snowy slopes rise to over 16,500 feet. It also comprehends the western shore of the great Lake Tanganyika, one of the largest lakes in the world.

The native people of the Congo might of themselves alone, with their superstitions, customs, industries and modes of life, together with their melancholy story, form the subject of a treatise, but we cannot here give space to more than a brief sketch thereof.

The Bantu tribes embody various groups, some inhabiting the plains, some the forests, others the river banks. Each tribe is generally autonomous and speaks a different dialect of Bantu. The Swahili, a Bantu language mixed with Arabic, is understood by many tribes, even beyond those who come under the influence of the Zanzibar Arabs, and is a common means of communication. Some of the tribes have created considerable kingdoms or " empires." Some are skilful in elephant-hunting, some in ironwork, wood-carving, ivory-carving, copper-working, etc. Some build fine canoes and large war-boats, living much on the water. Cannibalism was largely practised among some, even those with a relatively advanced culture, although this has been partly stamped out by the State, as has the practice of slave-raiding by one tribe upon another.

Nearly all these people have tattoo markings, and, except where Arab or European influence prevails, clothing is very scanty, although absolute nakedness is almost unknown. The most savage tribes are those of the forest ; the most advanced those of the plains. The bow is the chief weapon, with spears, swords and throwing-knives in the eastern districts. The Bangala are one of the most gifted tribes, but the Lualaba region shelters savage cannibals. A curious custom among some of the coast tribes is that of placing marriageable girls, with their skins stained red, on show in small bowers built for the purpose.

In the Lower Congo, houses or huts are rectangular and surrounded by a palisade, with fetishes over the entrance. In other places, as among the Bangala, houses are built in parallel rows, and elsewhere in circular groups, with plantations of manioc surrounding the villages.

The spears, knives and shields made by the negroes often show much ingenuity of manufacture, and crude musical instruments are common. Cloth from vegetable fibre is made and often dyed into effective designs, some of it having a " velvet " pile. The Arab influence among the eastern tribes shows itself in the form of the clothing and the houses.

As to their religion, there is a vague but universal belief in a Supreme Being, of a good or neutral character, but the native pictures himself as surrounded by innumerable malignant spirits, whose propitiation requires constant magical effort. It has been remarked that the underlying concept in the native religion embodies a very close connection between this world and the next, or between the material and the supernatural.

The civilisation of the Congo State has not yet developed sufficiently to give rise to large towns, and the population of even the largest Government stations, many of which have been established, does not exceed 5000. Boma, the capital, lying up-stream, is one of the principal ports of call for the steamers on the West African coast, and a considerable trade is carried on therefrom. Bomba was a well-known slave mart in the sixteenth and seventeenth centuries, and British, French and Portuguese traders and firms established themselves at the port, although .no European Power had any control there, until, in 1884, the Boma people agreed to a protectorate of the region by the International Association of the Congo, that body from which the control of the territory sprang.

Let us descend the river from its remote upper waters, following the course of some of the great explorers. With its affluents, the Congo affords navigable waterways for more than 6000 miles through the country, and is the largest river of Africa. The great journey of Stanley down the Congo marked not only a geographical but a political and commercial epoch in the history of the continent. He started from the east coast—at Zanzibar —and reached the sea again on the west coast, having navigated the river along its enormous course from Nyangwe to the mouth, a distance of 1600 miles, overcoming vast difficulties and solving, by this greatest of African geographical exploits, the enigma of the Congo, his expedition lasting from 1874 to 1877.

Livingstone had traced the course of the Chambezi, which he thought was the headwaters of the Nile. This, the most remote of the Congo affluents, rises between Lakes Nyasa and Tanganyika, in British territory, and flows into the Luapula, which sweeps onward to the north, past Stanley's point of embarkation at Nyangwe, to Stanley Falls, situated upon the Equator. Here the steamer comes alongside the jetty. The Upper Congo, as this portion of the river is termed, is in places a mile wide, although over a thousand miles from its mouth, with flat forested banks, forming a majestic stream whose current sweeps placidly to the north-west.

But at Stanley Falls, which lies at 1520 feet above sea-level, a series of seven cataracts is reached, a fall in the aggregate of 200 feet, forming a barrier to navigation between the waters above and below, except by canoes. The enormous mass of water pouring through the narrow limits in the Falls is an imposing spectacle.

From Stanley Falls we sail down the Middle Congo for nearly 1000 miles, crossing the Equator again 630 miles from the

Falls. The river presents in places a lake-like character, and widens to four or five miles, with thickly wooded margins and scattered islands, amid which the current runs slowly. Thirty miles below the Equator we pass the mouth of the great northern affluent of the Congo, the Ubangi, a waste of waters eight miles wide spreading before us here.

The Ubangi flows for over 1500 miles to the Congo, through a generally fertile and forested region, and is navigable for steamers drawing three feet of water to the Congo rapids, 350 miles from the confluence, and for smaller boats beyond. It was first ascended by Grenfell, of the Baptist Missionary Society, in 1885. Its rise is upon the water-parting with the Nile.

The great bosom of the Congo, generally five to six miles wide below the confluence of the Ubangi river, bears us on past Lake Ntomba towards and through the narrows of Chenal, where the brick-coloured waters of the Kasai—in contrast with the black colour of other affluents—enters through a great chasm of the hills, and we reach Stanley Pool.

The Pool is a lake-like expanse whose margin is formed of peaked sierras and picturesque mountains, from 1000 to 3000 feet high. Stanley gave the name of " Dover Cliffs " to a portion of these hills, from their glistening white appearance, caused, however, not by chalk, but by masses of silver-sand. Stanley Pool is 20 miles long and 14 miles broad, and its elevation above sea-level is 800 feet, the swift current flowing thence from at a rate of 600 feet a minute. A few miles below the cataracts begin, cutting off the Middle Congo from navigation to the sea.

The mouth of the Congo widens out into an estuary, with mangrove-lined creeks and forest-covered deltaic islands. The tall cliffs on either hand have sunk, and the river at its mouth at Banana Point is seven miles wide. The brown waters of the great stream, borne down for thousands of miles through the heart of equatorial Africa, amid dense jungle and papyrus marsh and inextricable and mysterious forest mazes, flow out to sea and are distinguishable amid the blue of the ocean,. thirty miles from the shore.

Various busy shipping stations have grown to being on the Congo, scattered through the heart of Africa, and flourishing if still small towns, forming the capitals of districts. Nyangwe, among the former, was a large place when Stanley arrived, and to-day is an important centre.

The combination of rail and river systems of the Congo afford exceedingly extensive routes of travel. The line from Stanley Falls towards the Nile is to be continued through the Soudan to Lado, where the great river becomes navigable, giving a route across Africa from the mouth of the Congo to the Red Sea. The recent completion of the railway from the Upper Congo, or Lualaba, to Albertville, on the western shore of Lake Tanganyika, fills in the river-railway route from the Congo mouth, and crossing the lake by steamer to Ujiji, in German East Africa, we may take the railway, also recently completed, across the German colony to Dar-es-Salam on the east coast. A further extensive line of travel is that from Katanga, 2150 miles by railway and river to the Congo mouth.

The Congo climate varies but little throughout the year as regards temperature. The greatest heat is in March. Frost occurs on the high plateau occasionally at nights. Fever prevails on the lower rivers, and there are short but fierce storms in some seasons. It is due rather to humidity than heat that the Congo climate is unsuited to European settlement.

The Congo State contains mountains which ascend to the region of perpetual snows, in the famous Ruwenzori, the so-called "Mountains of the Moon," a constant source of interest to the traveller and scientist. Here we remark the capricious disposition by Nature, practically upon the very Equator, of high, snowy, volcanic peaks, an example of equatorial snows such as we encounter also in Ecuador.

Ruwenzori is a range of mountains some sixty miles long, lying in Central Afrca, near that point where the frontier-line between the Congo and Uganda cuts the Equator, presenting to the view a line of six snowy peaks, rising to over 15,000 feet above sea-level, and culminating in Mount Stanley at a height of 16,800 feet. Enveloped by day in dense cloud-masses, which curtain the slopes below for thousands of feet, the mountains are generally hidden from view, and were effectually concealed at first from the explorer Stanley, who camped upon the base without suspecting what lay above the mist. The clouds at times lift as evening falls, giving a momentary glimpse of the snowy summits, but with recurring day they are again enveloped. The numerous streams fed by the glaciers of the Stanley, Speke, Baker and other peaks drain into the great Nyanza lakes, so forming in part the sources of the Nile.

The "Mountains of the Moon" were thought by Ptolemy and other geographers of antiquity to feed the natural reservoirs of the Nile, but doubtless this had reference to Kilimanjaro and Kenya, lying nearer the coast, rather than to these great sentinels of the rift valley. The snow-covered area of Ruwenzori does not extend for more than ten miles, although it appears formerly to have been much more extensive.

Let us, before further examining the resources of the Congo State, cast a momentary glance at its history and the notorious conditions which came to underlie its native life. The Congo Free State, as such, came into existence as a result of the personal ambitions and force of character—however unscrupulous—of one individual, King Leopold II. of Belgium.

In 1876 King Leopold, his interest aroused in equatorial Africa, called a conference at Brussels of the leading European geographers, in order to discuss the opening up of that little-known region, and as a result the "International Association for the Exploration and Civilisation of Africa" was formed. Marked activity in the matter was displayed by the Belgian portion of the Association, other nations taking a very secondary part, and a Belgian station was established on Lake Tanganyika.

But the accounts of Stanley, after his return from the Congo, fired the imagination of Leopold, who invited the explorer to meet him at Brussels, following upon which the Association, having become almost entirely Belgian in influence, sent Stanley out as its agent, and established itself in the territory. After the Berlin Congress of 1884 the Association was recognised as a sovereign State—the United States being the first great power to give this recognition—and the Belgian Chamber authorised King Leopold to be its chief. The avowed object of the Free State was to develop the natural resources of the region by means of native help. This enterprise was at once disputed by the Arabs, and conflicts followed, resulting in the Arab influence being overthrown and the regimen of that part of Africa largely altered as a result.

The history of decrees, discussion and abuses against the

natives is a long one. The Free State Government gradually acquired a proprietary right, not only over the land, but practically over the bodies of its inhabitants. The rights of white traders were restricted and Government monopoly was established. These arbitrary acts were carried out partly under a secret decree, by which the vast wealth of rubber and ivory which was shown to exist in the enormous region was reserved to the State, and natives were obliged to sell their produce thereto, and not allowed to leave their villages. Previously, measures had been brought forward in the Belgian Chamber for annexation of the Congo to Belgium, with the purpose of securing a better administration of its affairs, to which, however, Leopold was opposed. The country had advanced large loans without interest to the Free State, and discussions in the Belgian Parliament were rendered acrimonious by reason of the serious charges brought against the Congo State administration. King Leopold had also spent enormous sums out of his private purse in the work of conquest and development.

Serious evils resulted from the system of exploitation thus established, in which the State had become a monopolist trading concern, and vast areas of the best rubber districts were the private property of Leopold. Protests by the trading companies were made against the violation of trade freedom provided for by the Berlin Act. Large areas of land were then leased to companies, with exclusive rights, in which, however, the Government participated, but this was also regarded as against the Act. Bad and cruel treatment of the natives, floggings, starving, extermination [1] and other horrors were sinister features of the regimen. At length these matters called forth vigorous protests, not only in Belgium, but abroad.

In 1903 the British House of Commons intervened in the affairs of the Congo, but other nations failed to respond to its protest, and not until 1904 was any effective action taken, when a report was made by the British Consul at Boma. [2]

---

[1] It has been calculated by one writer on this subject (Rev. J. H. Harris, in *Dawn in Darkest Africa*), that from first to last 12,000,000 coloured lives were sacrificed under forced labour and other evils in the Congo under this regimen.

[2] Mr Roger Casement.

H

Following upon this, King Leopold appointed a commission to investigate the condition of the natives, the members of which ascended the Congo. The condition of affairs, however, remained unsatisfactory, and in 1906 the British Government again expressed its dissatisfaction. The Belgian Parliament brought in a Colonial Bill, in which—after further representations by the British—the absolutism of Leopold was to be replaced by parliamentary control. Native revolts occurred, the "reforms" had been declared to be illusory, and not only Belgium but France and the United States joined their voices in support of amelioration of the condition brought about by this monopolistic and repugnant system.

It was affirmed [1] that the Congo State had "morally forfeited every right to international recognition." The "works of public utility in the Congo" included schemes for royal palaces and estates in Belgium, and for the making of a "unique bathing city of Ostend." At length, in 1908, fourteen years after its first promulgation, a treaty of annexation was passed by the Belgian Parliament. Thus the monopoly of Leopold ceased to exist, and it was proclaimed that the evils which had stained it would recur no more.

When the white man first entered the Congo the most valuable article of trade was ivory, obtained from the tusks of elephants and hippopotami, which abound in the region. The traders found that the natives had in their possession vast stores of ivory, which had accumulated from the native hunting of long periods as "dead ivory," and this was greedily seized upon, but the supply, naturally, was not inexhaustible. Then arose the demand for rubber, and to this the native owed some of his most painful history, as the annals of the Congo have recorded.

The rubber is obtained from the vines, or lianas, the *Landolphia florida*, which form the most valuable of the Congo forest flora. The impetus given to the collection of rubber by the taskmasters of the Congo was shown by the rise in the value of the rubber export, which from about £6000 in 1886 increased in 1900 to considerably over £1,000,000 sterling.

[1] By Sir Edward Grey, and in the King's Speech on the opening of the British Parliament.

The increase in the output and in trade generally was, how-ever, largely a result of the opening of the railway around Stanley Falls, which gave a necessary link of communication between the navigable portions of the river. The line was com-pleted in 1898, and the value of the exported native produce of all kinds rose later to between £2,000,000 and £3,000,000 sterling, the cost of transport, which had been previously carried on negroes' backs, being greatly reduced.

As we enter the interior of the Congo State the graceful coco-palms and the sombre mangroves of the seaboard give place to the dense forests, where, in addition to rubber, there are many varieties of valuable timber, such as mahogany, teak, ebony, African cedar and lignum-vitæ. The oil-palm, whose hand-some form we have so frequently remarked in Western Africa, yields its oil and kernels, which form increasingly valuable articles of export.

Gigantic trees, reaching a height at times of 200 feet, struggle upwards towards the sunlight through the forest canopy, and countless varieties of giant mosses, lichens and ferns, often with leaves twelve feet long, arouse the traveller's interest and admiration as he journeys through these equatorial labyrinths. The canopy which the undergrowth forms is remarkable, making a covering at times fifteen feet from the ground, and entirely excluding the sun's rays. We remark, among the other trees, the redwood and camwood, gum and resin yielding acacias, and huge euphorbias. Orchillas and orchids, aloes and the crimson-berried pepper climber are seen on every hand among them.

As the forest gives place to the more open savanna gigantic baobab-trees replace the other monarchs of the woods. Where man has established himself, the indigenous cotton and the coffee shrubs may be seen spreading their luxuriant foliage and flower, shaded by banana plants of great size.

The forests are the home of the chimpanzee, especially in the Aruwimi region, and various kinds of monkeys people the leafy fastnesses. In the jungle the python and other snakes are encountered, and crocodiles haunt the rivers, and in the Lower Congo the manatee is found. Leopards, wolves, jackals, wild hogs and tiger-cats are all inhabitants of one or another part of the Congo. A feature of the rivers is the abundant fish life, including native perch and bream, forming valuable sources of food supply.

If animal life, as represented by the larger species, is abundant, no less has Nature been prodigal of bird and insect form. The traveller remarks the abundance of parrots, also of ibises and storks, and in the Lower and Middle Congo, eagles, herons, hawks and geese, with other and stranger birds, while butterflies of varied hue and dragon-flies of gorgeous colourings bear out the tropical character of jungle, swamp and river-brink. As we ascend to the uplands of the north-east the zebra and the giraffe are met with, as also the rare and recently discovered okapi. Across the open stretches of country sweep troops of antelopes, some of which are armed with tusks, and in such districts the red buffalo is encountered. There are but few horses or mules in the Congo, although sheep and goats are plentiful.

The natives of the Congo before the advent of the white man from Europe had not developed the agricultural resources of the country, except to a small extent around some of the Arab settlements. But in the neighbourhood of their own villages they cultivated patches of bananas and sugar-cane, varied by the sweet potato, maize, rice, millet, manioc, sorghum and other vegetables and fruits, as well as tobacco. There was not much inducement, in many districts, to embark upon extensive industries, due to constant dread of attack, whether from neighbouring tribes or from Arab slave-raiders. Even of recent times agriculture on a large scale has not been established on a flourishing basis, although such important products as coffee and cocoa, with tobacco and grain, are raised for export under the supervision of the State or of private companies. Cocoa is of increasing importance for export, as are ground-nuts and timber.

The very large Katanga territory of the Congo State, which borders upon Northern Rhodesia and Lake Tanganyika, is becoming of increased importance by reason of its extensive deposits of copper ore, and to some extent of gold and iron. The best copper-bearing districts lie against the Rhodesian border, and railway-building has been carried on in that direction by a British Company, which has acquired rights over some of the immense deposits of the

GREAT BAOBAB TREE, NEAR THE VICTORIA FALLS, RHODESIA

mineral.[1] As regards iron, ironstone hills containing millions of tons of ore of superior quality have been shown to exist, and tin has also been encountered. The Ruwe gold mine lies in the south.

More than three-quarters of the native produce of the Congo is taken by Belgium. The imports are but small in proportion with the exports, and consist of clothing, textiles, machinery and metals, arms, ammunition, river steamers, food and drink. It is a satisfactory condition, in this connection, that the importation of alcohol for consumption by the natives is prohibited, although this deprives the State of a ready source of revenue.

The small proportion of the import trade is stated to be due, to a large extent, to the system of forced labour which the State maintains, and which was legalised by the law of 1903. Revenue is derived from custom dues, taxes, and the exploitation of the domain lands, and by export charges on ivory and rubber. The public expenditure of the State is higher than the revenue.

The education of the natives has been taken up to some extent by the State, largely with the purpose of obtaining recruits for the armed force, but lads up to the age of fourteen are trained in colonies, in agricultural and technical matters. The missions, whose activities have been so marked in the Congo, where there are over 500 missionaries, also have numerous schools for secular as well as religious teaching. There is no religious teaching provided for by the State, but all missionary denominations meet, under the Berlin Act, on equal ground, and missionaries are about equally divided between Protestants and Catholics.

[1] The annual report of this company gave an estimate of 40,000,000 tons of eight per cent. copper ore above water-level : net profits for first half of 1915 were £180,000.

CHAPTER XIII

THE traveller, proceeding along the Slave Coast towards the mouth of the Congo, remarks, upon consulting his map, that Portuguese, Belgian and French territory converge near this point. From a short distance north of the delta the little Kabinda enclave of Portugal forms the coast-line, whilst the southern bank of the river for about eighty miles lies in the Portuguese territory of Angola, the deltaic point of the Belgian Congo intervening.

Angola is a very extensive possession, covering 480,000 square miles of tropical territory, with a seaboard 900 miles long, the country extending inland to the east for 700 miles. The region thus covered has been dowered by Nature with rich resources, both in the vegetable and mineral world, and it is fortunate in its rights upon the Congo, large vessels reaching the town of Noki, where navigation ends. The Belgian Congo encloses Angola on the north-east, with Rhodesia on the east, and to the south lies German South-West Africa.

The small, isolated territory of Kabinda overlooks the Atlantic with a shore-line of ninety-three miles, and covers about 3000 square miles. Its possession by Portugal resulted upon the pretension of that country to obtain both banks of the Congo. Kabinda, however, lies twenty-five miles from the river.

Here, in what by reason of the beauty of its situation and the richness of its surrounding land has been termed the " paradise of the coast," lies the formerly noted slave mart of Kabinda town. The harbour is sheltered and commodious. A large trade in palm-oil and ground-nuts and other products gives occupation to the 10,000 inhabitants of the district. The Bantu negro portion of the people are known as Kabindas, who are active traders and daring mariners, enterprising and intelligent.

Loango and Kabinda retained their slave-trading later than any other part of West Africa, and until traffic in palm oil and rubber took its place. Loango was at one time included in the "Kingdom of Congo," and the early Portuguese cultivated relations with its "great monarch," Mwani Congo, to whom they sent an embassy and missionaries. This king submitted to baptism, and in 1534 a cathedral was built at his capital, which place the Portuguese termed São Salvador, an example of that pious nomenclature which we so frequently remark in various parts of the world in the wake of the Iberian discoverers.

São Salvador lies 160 miles inland, at 1840 feet above the sea. Glowing accounts of the richness and prosperity of the land were taken home by the discoverers, so brightly painted as to be incredible. Later, the fame of São Salvador declined. There was an inrush of cannibals from the interior, and from this and other causes the place decayed, the cathedral fell to ruins and the stones of the city wall were removed for other purposes.

The ancient and capital town and seaport of Loanda—dating from 1576—forms the main gateway to the interesting Portuguese colony of Angola. The bay is protected from the heavy surf which beats upon the coast by a narrow, sandy island, and is overlooked by Fort San Miguel, standing up boldly from the peaked promontory which forms the termination of its southern side.

From this town and coast innumerable cargoes of black men, slaves torn from the interior, put forth in earlier times, to cross the Atlantic to Brazil, for Loanda, like Kabinda, was for two centuries a principal point of departure for the slave trade between Portuguese West Africa and Brazil. When, however, this "trade" came to an end, the town decayed greatly, and it only rose again with the growth of a more modern and legitimate commerce. The conditions as regards black labour are, however, still strongly condemned by foreign observers, as discussed later.

The buildings of Loanda slope upwards from the foreshore, the one-storeyed houses of the European element, with red-tiled roofs, giving place on higher ground to the Government offices, hospital, governor's house and bishop's palace, and the modern growth of the place is attested by the meteorological observatory, public park, tramways and gasworks. There is a statue to Correia de Sá, who recaptured Angola from the Dutch, which people occupied Loanda for a few years from 1640.

More than half the exported products of Angola have their outlet at Loanda, and from the town the " Royal Trans-African Railway " runs for 300 miles into the interior, through Ambaca and Malanje. This railway was part of an ambitious project to cross Africa to Mozambique, the Portuguese East African possession, but the purpose has been abandoned. The line was the most expensive that has ever been constructed in tropical Africa.

To the south, upon the coast, we encounter yet another old slave centre, that of Benguella, which, established in 1617, became an important point for the slave trade with Brazil and Cuba. It has a bad harbour and has greatly declined. Some twenty miles north of Benguella we enter the excellent natural harbour of Lobito, and here steamers may lie close inshore, a condition rare upon this surf-beaten coast. Here starts the railway to the copper mines at Katanga, lying beyond the Congo-Rhodesian frontier.

An ancient route from the heart of Central Africa terminates near Benguella, at Katumbella, which further renders this district of interest to the traveller. Bihe, the capital of the plateau district which this route crosses, is a large caravan centre, and a considerable town, the residence of a native " king." Mossamedes possesses an excellent harbour in Little Fish Bay, and an important fishing industry. A railway runs hence for a hundred miles over the semi-desert region to the fertile plateau. A Boer settlement is established in the district of Mossamedes.

The broad coastal plain of Angola has but one deep inlet, that of Great Fish Bay, near the German frontier to the south. The littoral is often poorly watered, and in places sterile, but as far south as Benguella the rich West African flora is encountered, including the oil-palm and rubber-vines.

As we proceed inland, the plains give place to a mountainous country luxuriantly clothed with tropic vegetation, and, this forest being passed, we ascend to the great Central African plateau, lying from 4000 to 6000 feet above the sea. Here wide, rolling plains are displayed, generally well watered, and in this they present a marked contrast with the zones nearer the coast, where water is scarce in the dry season. This high plateau falls away towards the Congo and Zambesi basins in the east, and southwards we enter upon a barren, sandy desert. Some of the Benguella peaks—among them Loviti—rise to over 7500 feet elevation.

The climate of the littoral is not healthy for the white settler, except in the Mossamedes district, but as we pass the elevation of 3300 feet and approach the plateau, the rainfall, temperature and prevalence of malaria decrease, giving place to healthy and invigorating conditions. To the east we approach the borders of Rhodesia, and the headwaters of the Zambesi.

The plant and animal life of Angola differ little from that of Western Tropical Africa generally. Both root rubber or *Carpodinus* and the *Landolphia* are sources of rubber production, but the trees have been disastrously treated by the ruthless system of extraction carried on by the natives. The oil-palms have been mentioned previously. The fine timber includes the tacula, a tree which reaches enormous dimensions, and yields wood of a blood-red colour. Mahogany is a further valuable timber of these forests.

The export of ivory from Angola reveals the presence of the elephant in the primeval fastnesses. The lion and the leopard, the rhinoceros, hippopotamus, giraffe, buffalo and antelope are all found in their customary regions.

Some minerals are found, among them deposits of valuable copper ore, as also of iron, petroleum, asphalt, gold and salt— a wide range, and of future importance.

The inhabitants of Angola—whose population numbers over 4,000,000—are mostly Bantu negroes. Their dwellings are of the simplest construction, a small hut for sleeping and a roof of palm-leaves outside for their daytime occupations, for ease or work. Fetishism is their religion. A curious remainder of some early Christian teaching is in the crucifixes which are used as potent fetish charms among the chiefs in certain districts. We remark that the native, as in other Portuguese or Spanish tropical regions, bears a Christian name, and is generally addressed as " Dom " or " Dona "—that is, Mr or Madam.

The people cultivate manioc and mealies, also sugar-cane, cotton, coffee and tobacco. The native blacksmiths are skilled in their craft, and are esteemed throughout Angola for their ironworking capacity. Some employment is forthcoming from the sugar and rum factories, but the occupation of the people mainly depends upon the jungle or forestal products and upon the produce of the Portuguese plantations. The plantations are worked by " indentured labour," the negroes being " recruited " from the interior, generally far distant from their place of occupation. The name of Angola is unfavourably known in European circles wherein the welfare of the African natives is made the subject of study and legislation. Slavery and the slave trade flourished in the interior of Angola until a few years ago, although in general prohibited by the Government. In opposing the wild raiding tribes there was bloodshed as late as 1904.

The main cause of complaint against Angola is that the system of indentured labour—the *serviçaes*—is in reality but little different from actual slavery. The treatment of the natives has been bitterly denounced by travellers, missionaries and statesmen, and has formed the subject of constant questions in the British Parliament.[1] " The brutal exploitation of men and women to enrich the Portuguese cocoa planter and the European cocoa and chocolate vendor "[2] is a subject of unending contention. Not even the excellent relations existing between Britain and Portugal have sufficed to give full weight to British protests in this matter.

Angola is governed from the Lisbon Colonial Office, with a local governor over each district. The revenue is insufficient to meet the expenditure.

Among the Portuguese possessions of Africa are two small islands, well known to commerce by reason of their considerable output of cocoa and to philanthropists for the contentions which have constantly arisen in regard to the treatment of the imported coloured labour, the *serviçaes*, by which the cocoa plantations are worked. These islands are San Thomé or St Thomas, and Principe, or Prince's Island, lying in the Gulf of Guinea.

San Thomé is almost cut by the Equator (0°23' N.), and is distant 166 miles west of the Gaboon coast, the nearest main-

[1] At the instigation of the active and well-meaning Aboriginal Protection Society.

[2] *Adventures in Africa*, T. B. Thornhill, London. Murray, 1915.

land. Principe lies 90 miles north-east of St Thomas. The first-named is a volcanic island 32 miles long and covering about 400 square miles. Its mountains, the highest of which rises to 7000 feet above the sea, are covered with verdant vegetation and scarred by deep-cleft ravines ; and down the slopes pour numerous streams born of the heavy rainfall, leaping seaward in beautiful cascades.

The equatorial ocean current tempers the climate here, and the soil is extremely fertile, lemons, oranges, figs, mangoes, pine-apples, guavas, grapes, bananas, as well as sugar-cane, coffee, cocoa, quinine, kola-nuts, cinnamon, rubber, vanilla and camphor all yielding freely in this fruitful sea-girt garden ; and there are considerable forested areas. San Thomé is probably the most fruitful tropical island in the Atlantic Ocean.

But as in many another lovely tropical land, Nature has placed her drawbacks here to man's occupation, for malaria renders the island unhealthy, the death-rate being high. Indeed, at the time of its occupation by the Dutch, San Thomé was known as " the Dutchman's Graveyard." This unhealthiness, however, is due more to the lack of sanitary methods among its people than to any invincible defect of nature.

Again, like some other prolific tropical lands, San Thomé is extremely backward sociologically, and in fact is one of the most notorious places under the equatorial sun in respect of the conditions of native employment.

San Thomé and Principe were discovered by Portuguese navigators in 1470, as was Annobon, the small Spanish island to the south, whose name signifies " Good Year." San Thomé was then uninhabited, but colonisation was soon afterwards begun by the method of taking in a number of criminals and young Jews. By the middle of the sixteenth century the island possessed a population of 50,000, with over eighty sugar-mills at work. But French attacks, Dutch plunderings, Angola raids and slave revolts during the two following centuries devastated the capital and ruined the sugar industry, which passed to Brazil, and the state of anarchy resulting brought this fruitful island to perdition.

Towards the close of last century some revival took place, and the increased consumption of cocoa in the world's markets gave rise, in the fertile lands of San Thomé, to the exceedingly profitable industry of cocoa-planting. The population in 1900 rose to 38,000, of which 1000 were whites, mainly Portuguese. But the " natives," whether from distrust of the white man, whether from ingrained indolence, would not work, nor at the present time will they do so if work can be avoided. These people had come largely from the Cape Verde Islands, Kabinda or the Kru coast, having been brought over upon short term agreements by the planters. As a result of this inertia thousands of Angola natives were introduced, but although these people were not ill treated on the plantations the death-rate among them was appalling. Whilst they were brought in as "indentured" labour the greater part, it is stated, were carried there by force from distant regions in Central Africa, without provision being made for returning them to their homes at the expiry of their " agreements " ; the terms of which they did not generally understand. The system was in fact analogous with slavery, notwithstanding that slavery had been illegal in the Portuguese dominions since 1878 ; and both in Portugal and England it was denounced, and the authorities in Angola strove to remedy the abuse.

Public opinion abroad was aroused on the subject of these unfortunate *serviçaes* and various prominent British and German cocoa and chocolate manufacturing firms refused to purchase the San Thomé product. A demand was made that the forced labour should cease, and be replaced by a voluntary system, and, under representations from the British Government, Portugal, in 1909, forbade recruitment altogether, later improving the service ; and British consular agents were appointed in Angola and the island, charged to observe the working of the decree under which yearly contracts were permitted. But, according to those who made it their mission [1] to care for the

---

[1] It was alleged that the repatriated natives were " dumped down on the coast of Angola," that many of them could not return to their homes ; that the tracks upon which they marched could be seen by the bleached bones and shackles of those who had fallen, and that four out of every ten perished ; and that the death-rate of the workers was 120 per thousand (speech of Lord George Hamilton at the meeting of the Anti-Slavery Society of London in 1912). The Society were those who took up the matter, also *The Spectator*, and constant questions were addressed to the Foreign Secretary in Parliament by members connected with the above Society.

natives, the conditions still remained "unsatisfactory." In the similar climate and latitude of the West Indies the mortality was only 25 per 1000. It was stated [1] that 40,000 African natives taken over were largely the results of tribal slave raids and were sold into what was practically slavery under the name of *serviçaes*, or the indentured system. In the islands the labourers were said to be well housed, fed and attended to,[2] but the system was one of forced labour and should be stopped.[3]

The *locus standi* of the British Government in entering into the affair was in the Treaty of 1661, still in force between Portugal and Britain, under which Britain had bound itself to defend the Portuguese colonies. The general argument with regard to the situation is that as the plantations must be worked, it is necessary to import labourers, and if necessary to force them to work. The value of the annual cocoa export from the Portuguese islands has risen to nearly £2,000,000.

The capital of San Thomé, the town of the same name, has a population of about 8000, mainly of people descended from the varied immigrants and slaves of early times, a brown-skinned people speaking a language compounded of Portuguese and a negro dialect. There is, however, a European class, and the island is administered by a governor and a superintendent of *serviçaes*. A railway serves the cocoa plantations from the port.

The island of Principe lies 90 miles away and is of somewhat similar natural character, and its principal industry is cocoa production. Unlike San Thomé, however, it suffers from the pest of the tsetse fly, which has given rise to sleeping sickness among a part of the inhabitants.

Annobon is another very small, but, as regards nature, beautiful island, 110 miles from San Thomé, covering about seven square miles, its steep hill slopes and deep valleys covered with

---

[1] By Mr Harris.   [2] Mr Cadbury's speech.

[3] The earnest representation of their British protectors carried undoubted weight. The representative of the Portuguese Government (Dr Machado), however, denied the charges, and remarked that the system was not " slavery " and that " if the workers of the whole world, white or negro, could only have one-half the liberty and care of the San Thomé workers the whole of humanity would have attained a degree of happiness from which it is unfortunately still far away."[1]

luxuriant forest. The chief town is San Antonio, where the governor dwells, and the inhabitants of the island, who number little more than 3000, are Roman Catholic negroes. Passing vessels here take advantage of its roadstead to secure fresh water and provisions. It has also suffered from revolt and anarchy, during those times when it changed hands from Portugal to Spain.

Almost due west from Mossamedes, 1200 miles from the coast, which is the nearest mainland, lies that lonely and desolate outpost of the British Empire, the island of St Helena, which we may here consider ; and far to the north of it is the still more solitary island of Ascension. The Atlantic, unlike the Indian and Pacific oceans, is almost without tropical islands, these two being the exceptions.

St Helena has an area of about forty-seven square miles. The principal interest for the traveller who approaches the forbidding cliffs of the island will perhaps lie in the historic associations of what was Napoleon's prison.

There is comparatively little level ground, the land rising high to a great crater-like amphitheatre, with fantastic rock pinnacles in places, and the green vegetation which it is said covered the island down to the sea when it was first discovered has long since vanished, the woods having been destroyed by the goats, which were imported in the days of the Indiamen. The destruction of the forests has resulted in the denudation of the soil and the drying up of the springs. However, many exotic trees and plants have been introduced, and in the valleys there are pleasant woodland spots and cultivated areas, with some small streams.

The only landing-place is at St James's Bay, and the only town—and here more than half the island's population of about 4000 dwell—is Jamestown. Napoleon's home at Longwood, where the famous Frenchman died in 1821, is somewhat over three miles from Jamestown. The Longwood and Deadwood plains, which embody the main area of level ground, are at an elevation of over 1700 feet above sea-level. Here the date-palm and the orange grows, and the peepul-tree affords its shade in Jamestown.

St Helena was uninhabited when discovered in 1502 by the Portuguese sailor, Nova, on his voyage home from India.

The British East India Company took possession of the island in 1651, after the Dutch evacuation, and the British have held it ever since. The Company imported Chinese, and slave labour was largely employed, but this was barbarously used until in 1792 new regulations were enforced for more humane treatment of the negroes. Most of the inhabitants are of mixed British, Dutch and Portuguese nationality, with East Indians and Africans. Freed slaves from West Africa settled here after 1840, landed by men-of-war who cruised the Atlantic and the Slave Coast in suppression of the slave trade. But the population has diminished considerably since their occupation of supplying the old sailing vessels has been rendered unnecessary in the advent of steamships and in the use of the Suez Canal. The establishment of various industries, agricultural and others, has been attempted to remedy the decline caused by the abandonment of the island as a port of call, but they have not been very successful. Excellent potatoes are raised, however, and other produce for passing ships.

The climate is healthy and temperate, the death-rate being very low, and this is a result largely of constant south-east trade winds and the cold ocean current. We shall remark, in approaching these Atlantic islands, how the enormous rollers break even upon the shores when the wind does not blow, a peculiarity which is found to be due to the effect of far-distant storms.

Ascension, 700 miles to the north, has an area of about thirty-eight square miles. Upon its lee-shore the great Atlantic rollers break with terrific force, and its rocky summits are draped in mists; Green Mountain rising highest of all, to 3000 feet. Bold and lofty precipices front upon the sea to the south. The climate is very heathy and the atmosphere generally light and clear. Turtles and fish in great quantities are found upon the coast, the former being caught and kept in ponds, and great numbers of the " wide-awake " birds frequent the island, their eggs forming a useful article of food.

But neither stream nor trees are found here, and only in places is the moisture sufficient to produce grasses, fruits and vegetables for the use of sheep and man. The only spring of water is that discovered by Dampier—the British buccaneer who was wrecked here in 1701—and this is piped to Georgetown and stored for use.

Georgetown is the settlement on the north-west coast, headquarters of the Government, which consists of a naval captain, who administers the island after the manner of a ship of war. The only inhabitants are the people belonging to the Admiralty establishment and the sanatorium, institutions established about a century since.

CHAPTER XIV

"GERMAN" SOUTH-WEST AFRICA AND WALFISH BAY

WE have remarked, in the south of the Angola littoral, that bare arid lands begin to take the place of the dense forests which we had traversed on the West African Coast nearer the Equator. In the extensive region known formerly as German South-West Africa, which we now enter, we find this condition accentuated. We are leaving the equatorial belt, as the colony is crossed midway by the tropic of Capricorn.

The coast of this part of South-West Africa appeared but an inhospitable region to the Portuguese discoverer, Bartholomew Diaz, who landed upon it in 1487, in his endeavour to find his way to the Indies. But the country possesses various distinctive features, both as regards its topography and its people. It is the only possession which Germany acquired in Africa suitable for permanent settlement and colonisation by the white man. Among its native races it shelters the warlike Hereros, whose fierce, well-organised and stubborn fighting but a few years ago, in their endeavour to throw off the German yoke—for such they regarded Teutonic control—cost the German Empire so considerable an expenditure of blood and money, resulting finally in the destruction of a large part of the population, which to-day probably numbers little more than 100,000. Of these, perhaps 8000 are white people, mainly Germans. So small a population contrasts with the considerable area of 322,450 square miles of land which this colony embodies.

The Atlantic coast-line, about 800 miles long, did not belong in its entirety to Germany, the British enclave of Walfish Bay cutting a small fragment from the littoral immediately above the tropic of Capricorn. The country is bounded on the north by Angola, on the south by the Cape of Good Hope Province, and on the east by Rhodesia and Bechuanaland.

German South-West Africa possessed a curious extension of its territory to the east, reaching the Zambesi a little above

Victoria Falls, thus penetrating into Rhodesia and Bechuana-land. This tongue of land, known as the Caprivi enclave, is 300 miles long and 50 miles broad. A German residence is situated here, and in earlier times the capital of the Makololo dynasty of Barotseland was centred therein.

We observe, upon sailing along the coast, a shore generally low and with few bays or headlands of any prominence, a condition upon which follows the absence of any good natural harbour. Nor do any rivers of importance enter the Atlantic upon this long coast-line, except that the Kunene, the northern boundary of the territory, and Orange river, the southern, as regards portions of their banks, lie partly therein. These rivers are not navigable, however.

The belt of desert and sand-dunes which for a width varying up to thirty-five miles borders the sea, gives way inland to a range of mountains, whose highest elevation—Mount Omakato —reaches nearly 9000 feet. Beyond the escarpment of this range lies the great Kalahari plateau, a watered and habitable district in its western part, but merging eastwardly into the sterile Kalahari Desert, whose vast extent covers a large part of Bechuanaland.

The absence of a natural harbour in German South-West Africa has been partly remedied by the artificial one at Swakop-mund, where a breakwater protects shipping to some extent from the Atlantic rollers. The town is built on the northern bank of the Swakop river, which discharges into Walfish Bay, the southern bank forming part of the British possession. Landing here, we proceed to the Government centre of Wind-hock [1] by the narrow-gauge railway which, running eastwards for 237 miles, reaches the hilly district chosen for the building of that town. This line cost Germany £700,000.

The fine grazing lands of the uplands offer splendid pasturage for cattle, of which the Herero people, the nomadic cattle-keepers of Damaraland, raise herds of many hundreds of thousands.

Upon the coast lands plant life exists by reason of the dense fogs which afford considerable moisture, a circumstance which reminds the traveller of the conditions prevailing upon the rain-

[1] Captured by the British South African forces, under General Botha, in May 1915.

less Peruvian Coast of South America, in a similar latitude. The rivers are torrential, intermittent streams, which are often absorbed before reaching the sea. There are some curious and beautiful forms of plant life even upon this almost rainless African belt, and as we penetrate inland the handsome acacia with its useful fodder-yielding seeds abounds. The temperature is low, with a prevailing south-west wind, and throughout the greater part of the colony the climate is favourable for the settlement of Europeans.

As we approach the mountain valleys, the luxuriance of the vegetation becomes attractive, and is in pleasing contrast with the scanty coastal flora, and in the northern district we are still within the palm zone. The clear, bracing atmosphere of the uplands is delightful and invigorating, the climate being temperate rather than tropical.

We shall not now encounter in this part of Africa any abundance of the great pachyderms. The elephant and others, formerly plentiful, are gone; but there are antelopes and other typical African creatures, and many birds, whilst the reptile life of the rivers, the crocodiles and turtles, is plentiful.

The small towns scattered throughout the colony are generally Government posts or mining centres, whilst to some of the small native places the missionaries have given fanciful Scriptural names, such as Beersheba, Bethany and so forth. Of curious nomenclature we may further remark the small islands belonging to Great Britain close against this German coast, such as "Roast Beef Island" and "Plum Pudding Island."

Good roads have been made between the larger towns. About half the white settlers—consisting mainly of Germans and Boers—are engaged in agriculture; wheat, maize, tobacco, sorghum and fruits being raised. The natives in the northern region are also agriculturists, but the chief industry among them is stock-raising. The climate is not suited to horses, and camels have been imported, and thrive well. Cotton and grapes are cultivated in some districts. The immigration of German colonists into the territory has been encouraged by subsidies, and some British emigrants have entered the country from the Cape.

Some considerable mineral wealth attracts the miner and

adventurer in this part of Africa; and in 1908 a rush of prospectors was caused by the discovery of diamonds in the Lüderitz Bay district. These gems, which resemble the diamonds of Brazil, were found on the surface in a sandy soil, and various companies hastened to exploit the field, encountering considerable difficulties from the waterless character of the district. However, diamonds to the value of a million pounds have been exported in a single year. In the Otavai district, at a high elevation, there are deposits of copper ore of considerable extent and commercial value. The railway which serves the mines runs from Swakopmund thereto for 400 miles, and reaches, in passing, an elevation of over 5000 feet.

The principal native divisions of the colony are those of Ovampoland, in the north, and Damara, the land of the Hereros, in the central part, followed by the great Namaqualand regions, home of the Hottentots and Bushmen. As to native crafts, the Hottentot women are dexterous in the making of fur cloths, and some pottery is also produced, whilst the northern inhabitants are known for their good blacksmith-work, and for grass-plaiting.

The foreign trade of the country is but small, some ivory, hides, feathers and other matters being exported, in addition to the mineral products.

The Hereros, the cattle-keeping people whose name in their own tongue signifies the "merry people," who, it is recorded, had accepted the protectorate of Germany without full comprehension of the terms of the contract, took the opportunity afforded in 1904 of the conflict between their German masters and other tribes—the Bondelzwarts, north of the Orange river—to revolt, in accordance with long-laid plans. Defeated by a German column which attacked their stronghold, their main army kept up a guerrilla warfare, which rendered insecure the whole region, and which the Germans, hampered by the waterless country and lack of roads, could not subdue. The Hottentots of the south also commenced hostilities, in sympathy with the Hereros.

The causes of complaint of the natives against the Germans were in the abuses committed by the white traders, the occupation of native lands and the brutal methods of officials, but there was, among the Hereros, a general desire to throw off the white man's control, and this manifested itself also in German East Africa, on the other side of the continent, at the same time. The terrible character of German operations against the natives was shown by the proclamation of the German general,[1] who, in October 1904, incensed at his lack of swift success in overcoming the Hereros, issued an edict to the effect that every Herero, with or without arms or cattle, would be shot, that the women and children would be driven back upon their own people or fired upon.[2] Although this proclamation was repealed by order of the German Imperial Government,[3] the Hereros were hunted down, and many perished in the desert, others crossing into British territory.

It is to be recollected in this connection that German farmers in Damaraland, and their families, had been murdered and farms burnt by the savage Hereros at the beginning of the conflict, and in the south the revolted Hottentots carried on similar depredations, a number of German settlers being murdered. In both cases, however, British and Boer farmers were let alone.

A further proclamation of the German general, to the effect that all rebels would be exterminated unless they surrendered, remained unheeded, and small German detachments were cut up, their rifles serving as arms for the natives. The German general was at length recalled, after his methods had been criticised both by colonist and Imperial Government, but the serious disturbances continued until 1907.

It is recorded that the four years' war with the natives cost Germany £15,000,000 and the lives of 5000 soldiers and settlers, whilst of the Hereros from 20,000 to 30,000 perished. It was due to a conciliatory proclamation and policy that quietude was finally secured, but the struggle with the Namaqualand people was almost equally severe. The last incident in the conflict was the death of Morenga, the Hottentot chief, who had been allowed to enter British territory—the Cape—as a political refugee, but who, escaping, was pursued by a British force into the Kalahari Desert, where he and his followers were overcome on the kopje upon which, in the midst of the waterless plains, they were endeavouring to hold out.

[1] Von Trotha.
[2] An early example of German " frightfulness."
[3] Bulow.

The colony had at its head an Imperial Governor, answerable to Berlin, assisted by a Council, and a grant-in-aid was necessary towards the annual expenditure.   There are schools established by the Government, in the chief towns, but education is largely in the hands of the missionaries.

The whole region under German control might have been British had the treaty made in 1876 with the natives, at their own request, been ratified by Britain.   Only Walfish Bay, however, was at that time retained.[1]   German South-West Africa was lost to the German Empire during the Great War, falling before the British South African forces under General Botha.   This conquest was one of the most stirring and complete of the war.   The colony came into German possession in a time of weakness and indecision earlier displayed by Great Britain.   It was the first of Germany's colonies, the German flag having been planted at Angra Pequena at the behest of Bismarck, and was long a centre of intrigue against Britain's influence in South Africa and of machinations made in the hope of disaffecting the Boers.   By the irony of fate, German rule was overthrown by the most famous of the Boer generals, the Boer element fully recognising the difference between German and British colonial methods and rule.   The whole German force and government in the colony, which had retired to the difficult mountainous country in the north, surrendered unconditionally to Botha's army in July 1915.   The total foreign trade of the possession at that time amounted to about £4,000,000.   " Süd West " was regarded by the Germans as the principal nucleus of a possible African empire, but, apart from all national questions, it may be doubted if German rule was such as could have held or developed a permanent oversea empire in Africa.

Walfish Bay, the small British enclave upon this coast, lies a few miles north of the tropic of Capricorn, and covers a piece of territory about thirty miles long and ten or fifteen broad.   Its value is almost entirely due to the fact that it is the only good natural harbour upon a thousand miles of the coast of South-West Africa.   The name was derived from the whaling industry, of vessels which formerly frequented the harbour.

[1] Due to the work of Sir Bartle Frere.

The sand-buried, dune-covered coast of Africa, here a desolate littoral, is for vast distances practically desert. The great dunes rise 400 feet high in places, scattered inland for several miles. Water is very scarce, and at the Walfish Bay Settlement drinking water is obtained by condensation. The Swakop river, which forms the Anglo-German boundary on the north, is but an occasional water-course, floods occurring perhaps once in ten years.

The days in this region are hot; the nights may be cold. Except in some small valleys there is little vegetation, the most noteworthy plant perhaps being the wild melon, which furnishes food for the Hottentots for part of the year, such as we shall remark in the Kalahari Desert, to the east. The elephant and the ostrich, which formerly inhabited the region, have been killed off or driven away.

The climate is influenced by the cool Benguela current, and, due in part to the action of this ocean stream, shoals of fish abound, also whales, seals, especially on the islets that fringe the coast, and which belong to Britain; and guano deposits result from the presence of swarms of birds, pelicans, penguins and gulls, which frequent the dreary coast-line and islands.

The district is governed as part of Cape Colony, and there are missionary stations of long standing therein.

We have now to leave the coastal lands, as the southern extremity of Africa, which includes the Cape of Good Hope, lies beyond the tropics, and so is out of our sphere in this survey; and in penetrating inland we reach Rhodesia, a name which falls pleasingly on British ears.

CHAPTER XV

RHODESIA

THAT vast territory of Rhodesia, lying under the sun of the southern tropics in the heart of South Africa, a territory larger than France, Germany and Holland combined, stands in a unique position among the colonies of Britain, in that it is almost the only tropical land under the Empire wherein a white people have formed a permanent community[1] and the basis of a nation: a condition arising from causes which we shall presently remark.

The name of Rhodesia carries with it a sense of open vastness, the vision sweeping across veld and kopje and fertile plain, generally bathed in the sunlight from a turquoise sky, its horizons of granite hills painted a pearly grey. It is a land where the hut of the pioneer and the primitive ox-trail have rapidly given place to the municipality and the railway, with a civilisation British in its inception and development. No other part of the tropics, whether in Asia, Africa or America, contains a white community of so modern a character, wherein exists comparative social equality and common security in the development of the natural resources.

Whilst this description may justifiably be applied to a great part of Rhodesia, it is, however, to be recollected that a region so vast is of greatly diversified character, and such conditions do not obtain throughout the whole territory.

Rhodesia is made up of Southern Rhodesia, North-Eastern Rhodesia and North-Western Rhodesia. The total area—which on the map takes somewhat the form of a three-leaved clover—is about 450,000 square miles, the northernmost point being about 8° south of the Equator, and its southernmost edge near the tropic of Capricorn, a range of nearly 1000 miles, with a greatest width of over 600 miles.

The country is cut off from the sea, Mozambique or Portu-

[1] British East Africa, of course, offers somewhat analogous conditions.

136

RHODESIA, THE VICTORIA FALLS

guese Africa intervening in the south-east between it and the Indian Ocean, the great Zambesi river crossing both countries on its way thereto, as does also the railway, which terminates at Delagoa Bay. On the south lies the Transvaal, across the Limpopo river, which waterway also takes its course eastwards to the Indian Ocean. On the south-west is the British protectorate of Bechuanaland, itself an enormous territory. Westward lies Angola, and a long, thin tongue of German South-West Africa presses in to the Rhodesian border on that side. To the north-west is the Belgian Congo, and to the north-east Lake Tanganyika and German East Africa. The eastern extension is cut off from the great Lake Nyasa, and in part from Portuguese East Africa by the narrow territory of the British Nyasaland protectorate.

Rhodesia, it may be remarked in passing, covers about the same belt of latitude as northern or tropical Australia on the east, and Peru and Central Brazil on the west.

The outstanding topographical feature of Rhodesia, and that to which must be referred the conditions of settlement and permanent white civilisation which distinguishes the colony from other lands within the tropical British Empire, is that the principal part of the country lies high, consisting of an extensive, habitable tableland, whose elevation is from 3000 to 5500 feet above sea-level. This tableland is part of the great South African plateau, which forms the interior of the continent south of the Congo river and extends eastwards beyond the great lakes and along the boundary of Portuguese East Africa, where we shall remark its edge in ascending by rail from Delagoa Bay.

It is owing to the elevation and consequent temperate climate of this great upland that the white man can live and work in comfort in Rhodesia, and that his wife and children may flourish by his side: conditions such as are rarely possible in other regions which come under our survey in these pages.

Whilst this is true of Southern Rhodesia, Northern Rhodesia is a region of a different stamp, and its suitability for permanent European colonisation and settlement has not been demonstrated, and possibly cannot be so demonstrated. The natural line between these two divisions is the Zambesi river, with its low-lying, broad, hot and unhealthy valley, and Northern

Rhodesia belongs rather to Equatorial Africa than to temperate South Africa, both as regards its geography and its climate.

The Zambesi flows through Rhodesia with a great semicircular sweep, and it is crossed by the railway at the Victoria Falls, but is of no commercial use for navigation here, the point to which river boats may ascend being in Portuguese territory, 250 miles from the mouth of the river.

Southern Rhodesia is a land of modern towns, of a large railway mileage and many farms, with a white population of 30,000. Northern Rhodesia, with its vast, undeveloped area, has only 1500 whites, equal to one European to every 200 square miles, and there is no town that may be called such in the whole of the trans-Zambesi region, although it embodies two-thirds of the colony's total area. Southern Rhodesia is rapidly approaching the time when it will develop into a self-governing colony, but Northern Rhodesia is more in the nature of a large native dependency.

Rhodesia is not a mountainous country, but there are numerous ranges of hills and a good deal of broken land, the greatest heights being along the eastern side. The grass-clad heights on the edge of the tableland, which overlook Mozambique, may remind the traveller of the English Cheviot Hills, or possibly of the Lower Alps of Switzerland. Broad stretches of granite country unfold as we traverse the plateau. Immense rounded hills, sometimes in smooth masses with a gentle slope, at others steep and shattered, characterise the landscape, and in places gigantic boulders, weathered in that peculiar form which the elements impart to granite, are piled high upon each other, often in seemingly perilous equilibrium, frowning down upon the trail like fortresses and battlements. Of this character is the scenery around Matopo, where the grave of Rhodesia's founder lies.

Elsewhere geology works marked changes upon the landscape, and saw-like ridges and innumerable peaked summits, divided by profound gorges, mark the occurrence of the schistose rock. Or, again, a red-white sandstone formation offers its characteristic horizons, or great basaltic sheets, such as those carved through by the Zambesi in the magnificent Victoria Falls.

But the rocky structure of Rhodesia is by no means generally unclothed. Luxuriant plant life and wooded areas distinguish certain districts, with, in places, tropical forests, contrasting with the rugged, inanimate grandeur of the granite lands. This is the case notably along the eastern border, and between Bulawayo and the Victoria Falls one of the chief features of Rhodesian scenery is the wide expanse of grass-covered veld, the numerous grasses ranging from a few inches to fifteen feet in height.

The conditions of natural pasturage here displayed are highly advantageous for cattle. Colonising agencies, anxious to secure further settlers, depict Rhodesia as a land in which fat kine stand knee-deep in luxuriant natural fodder, the picture backed by a gorgeous sunset, with an athletic Englishman in white or sporting garments in the foreground, the owner of the convenient residence which stands near at hand, and the master of the attendant negro. This pleasing picture is not drawn entirely from imagination, and the information generally adds the useful commentary that only by hard work can such desirable conditions be attained to.

It is also urged as necessary that the intending settler in Rhodesia should look to his resources. The conditions underlying settlement on the land and farming, or ranching, differ considerably from those prevailing in such countries as Canada and Australia. Rhodesia is not a "poor man's land"; it does not offer much prospect to the white labourer, nor invite him, as do Canada and Australia, to carve out an independence for himself, without more capital than his muscles. The prospective settler in Rhodesia is advised that he should possess from £500 to £700 before he treads its soil, and those without capital are warned that the colony is no place for them.[1]

Stock-raising may become the chief agricultural industry of Rhodesia, it is considered. The diseases, such as the rinderpest and fevers, which carried off the cattle in 1896, are now largely understood and controlled ; the cattle have been increasing with rapidity, due to natural propagation, and almost every part of the colony is suited thereto. The conditions of rainfall, soil and climate are more favourable than farther south in Africa, and cattle do not require to wander so far for water as in the South African Union.

The earlier occupants of the land, the Mashonas and

[1] South African Company's pamphlets.

Matabeles, possessed vast herds of cattle, and at the present time the natives own hundreds of thousands of head. Adverse criticism has pointed to an alleged slow natural increase and a heavy death-rate among cattle, which those who are in favour of the expansion of the industry disprove. It has been remarked that a danger lies in unsuitable class of stock, of heavy British type developed by the large land companies whose directors live in England, as contrasted with the hardier native type, or a suitable crossed variety.

Be this as it may, it is shown that over 90,000,000 acres of land suitable for cattle ranching exist in Rhodesia. The present average is less than one beast per hundred acres. Land is cheap and may be purchased from the company at about seven shillings an acre. The market is still a home one and is likely to remain so for some years, due to the demand in the colony. Meat is imported in large quantities to cover the demand. The many successful ranches attest the condition of the industry.

As regards sheep, the native kinds are woolless. Horses suffer from sickness in the rainy season, but mules thrive well.

The richly wooded hills of the Selukwe and Umtali districts afford a strong contrast with the more southerly parts of Africa, and Rhodesia as a whole may be regarded as fairly well wooded. Along the rivers palms and tree ferns abound, with other luxuriant plant forms, reminding us that we are within the tropics. In general, however, the trees in Rhodesia do not attain to any great height, and this is ascribed in part to the prairie fires, which from time immemorial have swept across the land and still periodically traverse it. Also, the rock is comparatively near the surface of the ground, and added to these causes are the long droughts of winter.

A curious condition in the growth of the veld flowers takes place during the droughts, which last from five to seven months, the blooms beginning to appear all over the plain before any rain has fallen, as if they were heralding the coming change. Even the grass shoots through the hard-baked veld before the rain appears, and the gladiolus, solanum, convolvulus, asparagus, orchid and other flowers burst forth. Many of these plants are of bulbous varieties, or, having large roots, have stored up nourishment during the preceding wet season.

This varied growth of brilliant-hued bush and wild flower is

NATIVE CATTLE IN RHODESIA

often exceedingly beautiful, and to its effect is added the splendid colouring of the September foliage which, the winter past and the sap running anew, springs into glowing shades of red, yellow, copper and bronze, the colour of an English autumn replacing, in this tropical spring, the green tints of the temperate lands. The long, undulating hollows filled with the rich green and blue of marsh grasses, swaying in the breeze, and the purple, scarlet, golden and azure flowers, backed by the evergreen trees of the uplands, presents a singular and charming picture, and " neither the pen nor the camera can do justice to it." [1]

It must be recorded, however, that most of these flowers are without perfume. In this they remind the traveller of the wild flowers of some Spanish-American lands within the tropical zone.

Many curious and useful trees, some of which have no equivalent English names, are encountered in the Rhodesian forests. Among them the mapane, useful for its poles ; the umkamba, a sort of mahogany ; the umasasa, gathered as fuel ; the umvagaz, or blood-wood, so named from the blood-red sap which springs therefrom when cut, and the umkana, which " hisses like a serpent " at the axe-stroke.

The baobab uprears an enormous soft-wooded trunk, its fruit yielding a cooling drink and its bark a fibre employed in net-making. The nectar-stored flowers of the " sugar-bush " are also among the most abundant. The wood from the forests is generally either very hard or very soft, and building timber is imported, although the native varieties serve for mining props.

In this sunny land of Southern Rhodesia a wide variety of food-stuffs are now raised, and in almost every district agriculture in one form or another may be profitably undertaken. The staple crop is maize, which is reliable, easily cultivated and giving the greatest returns of any cereal ; and it is to this product that the settler gives his first attention. The consumption of maize is very large, the grain being the principal food of the great native population. More than 2000 farms are engaged in its production, and the surplus, it has been shown, may command a profit after export to European markets.

Tobacco is the product next in favour with the planter,

<hr>

[1] *Oxford Survey*, " Rhodesia," L. M. Foggin. .

although only of very recent years has a large yield been obtained, the amount disposed of at the official auction sale reaching over a million pounds. Under this system of marketing the company advances the greater part of the value to the planters before the crop is ready. Tobacco is not exported, the home demand at present exceeding the supply, but the excellent Virginian variety which the soil produces will doubtless compete in the world's markets later on.[1]

In Southern Rhodesia the orange groves are a pleasing feature in certain districts, with their clusters of golden fruit. The orange grows well, and its cultivation is being extended, the climate and soil proving more suited to citrus fruits than other varieties. Rhodesian oranges are considered to be equal to the finest imported into the London market, according to test shipments which have been made.

Coffee and rubber both give some success, and it may be proved that cotton is capable of cultivation. Under irrigation in winter, wheat, barley and oats are raised. Vegetables of all kinds are grown and other useful tropical products are under experiment by scientific agriculturists.

The climate of Southern Rhodesia on the whole, it has been shown, is suitable to the bringing up of British families. The most trying circumstances are the occasional heat and drought. At such times all eyes turn from the parched country-side to the clouds, and the topic of conversation is as to whether the rainfall will ensure a good season, or whether produce will be hampered by the drought.

The constant bright sunshine is perhaps the most noteworthy feature of the Rhodesian climate. The rainy season coincides with the summer months, mainly from September to March, and winter is a dry time, extending throughout June, July and August, a condition which makes for a certain equableness in the climate. The spring months of September, October and November, just before the rains begin to fall, are the hottest parts of the year, but the nights are almost always cool. The shade temperature rarely reaches as much as 100° (except in the hot Zambesi Valley and at Tuli), and the winter frosts are sufficient to benefit farming operations. The mean temperature may be taken at 65° F.

[1] Reports of British South African Company.

The rainfall is sufficient for the summer crops, but in winter irrigation must be practised. The mean annual fall varies from 19 to 27 inches in the western districts, to from 24 to 44 inches in the eastern, but is considerably less in the south-west corner of the country.

The inevitable African malaria occurs in Rhodesia in certain districts during the wet season, but malaria is now being successfully combated in all advanced tropical lands. The relative healthiness of Rhodesia is shown by the fact that the death-rate among its white population is only about 15 per 1000. Sleeping sickness, which has appeared in Southern Rhodesia, is not, it is held, likely to become serious, and the whereabouts of its germ-host, a fly akin to the tsetse, is circumscribed. The most prevalent diseases, of malaria, pneumonia, dysentery and typhoid, are largely due to the lack of sanitary precautions among the native population, and might be eliminated were these people cleaner in their methods.

Let us turn now to the people of the country. There are 1,500,000 Bantu natives, almost equally divided between Northern and Southern Rhodesia. The white population is in vast disproportion, numbering perhaps 32,000, all told. Of course this is a large white population for an African state, but it is to be recollected that Rhodesia is a " white man's land " as regards its climate and soil, as before remarked.

The chief defects of the natives—to deal first with their adverse side—are indolence, sensuality, and moral, although not physical, cowardice. When removed from European influence the native tends to relapse into his primitive condition, and when aroused from his generally good-natured and sociable, if improvident, frame of mind and somewhat philosophical temperament, the Rhodesian negro is capable of the wildest acts of cruelty and unreason. Witchcraft and sexual affairs are largely responsible for his changed temper at such times.

On the other hand, the native is remarkably submissive to the control of the white man, and he recognises the security under which he is now enabled to live, as against the times when rival chiefs and tribes upheld their prestige by periodical visitations of " frightfulness " upon each other. The work of implanting European ideas upon him does not yet penetrate very far below the surface, and it is easier to destroy what was useful in his native mind than to replace it with anything new. The influence of Government and missionary is at work all through the colony in the endeavour to combat the gross and debased practices which dominate the negro mind.

The dwelling-places of the native Rhodesians are huts, circular in form, somewhat different from the beehive huts of the Zulus, and the whole family, with its dogs, fowls, goats and sometimes cattle, inhabit it together, although the latter generally have their separate kraals. It is usually kept clean, however, and a part of the earthen floor is raised for the sleeping-place, and there is a special place for the fire.

The tobacco plantation is close to the dwelling, generally, and some distance away are the " mealie " or maize fields, which supply the principal food, the grain being stored in upraised granaries, against the depredation of vermin—especially the ubiquitous white ants. Like many natives of the tropics, the negroes in some districts are adepts at basket-making, the close weaving of which is such that the baskets are water-tight.

One of the principal objects in the life of the native would seem to be that of growing sufficient maize for his food and for making native beer. Yet this great native population, whilst presenting the most insistent problem of government, is the greatest of the country's assets. Regarded as " labour," the nearly 750,000 natives of Southern Rhodesia form the only source of manual work, as white men do not perform such, a condition which distinguishes Rhodesia from Canada and Australia. It would seem, or at least it is so stated,[1] that the native is growing more and more to realise the advantages of working for the white man, but the insufficiency of labour in the expanding industries of the colony is a condition to be reckoned with.

Beyond the Zambesi, the natives live their own primitive life, and the tribal system has scarcely been disturbed. To meet the shortage of labour in Southern Rhodesia, a native labour bureau has been established by the farmers and others, with the purpose of recruiting negro workers from the adjacent territories. The members of this body are supplied with labour on payment of a capitation fee of two pounds, with a monthly fee of a shilling in respect of each native employed for one year. An adult native farm labourer is paid from fifteen shillings to

[1] British South African Company's pamphlets.

The Tropical British Empire: a young orange grove in Rhodesia

one pound per month, but youths are given much less. In addition to this wage, rations of about two pounds of meal per day are given, with, sometimes, an allowance of beans, ground-nuts and sweet potatoes. Occasionally a little meat is added. The native drivers from the south earn perhaps three pounds per month, with rations.

The white settler's household work in Rhodesia is carried on entirely by native "boys," as the men-servants are termed, who are paid from fifteen shillings to two pounds per month, and given their food. They make, in general, fair cooks, and are quick to learn, and they acquire a little English. But in everyday intercourse with their employers the "Kitchen-Kafir" is talked, easily learned and understood by native and white. However, some knowledge of the native tongue of Matabele or Zulu and Mashona—which are the native divisions of Southern Rhodesia—is necessary for the planter in control-ling his labour. This is not regarded as difficult, and the intelligence of the native enables him rapidly to grasp an order and meaning conveyed with a very limited vocabulary.

The employment of male negroes about the house and in domestic operation does not seem to cause any particular embarrassment to the white mistress of the household. How-ever, some criticism of the system, or rather of the carelessness which familiarity has bred in this respect, has been made of late, arising out of the occurrence of cases of negro assault on white women, with the reprisals following thereon. Yet on the whole the balance between the two peoples is remarkably well held, and must be regarded as a tribute to both, contrasting strongly with the terrible affairs which from time to time arise in those parts of the United States where negro labour is employed.

Whilst all the necessaries and most of the luxuries of civilisa-tion are forthcoming to the traveller or the resident in Rhodesia, the cost of living is extremely high, the rent of houses being, to European eyes, out of all proportion, and food and clothing are very expensive. White workers require an extremely high wage in order to cope with the exorbitant cost of living. Rhodesia, in fact, is an instance of that not uncommon and curious condition of a "new" land, which, although abound-ing with natural resources of food supply and material necessary for life, is an extremely expensive one to inhabit. We

K

encounter this phenomenon in Australia, Canada, South America and elsewhere. It is due to various causes, and in general can only be ascribed to the lack of a true economic science in the conduct of human affairs, in whatever country of the world be it.

The chief towns of Rhodesia reflect the modern and energetic life of the white community. The largest town, although not the capital, is Bulawayo, Salisbury being the seat of Government of Southern Rhodesia.

Bulawayo is of historic interest. It has received its baptism of earlier conflict and bloodshed, and its site was formerly the kraal of the redoubtable Lobengula, the chief of the fierce Matabeles, whose conquest involved Britain in a large expenditure of lives and treasure, and the exercise of considerable strategy against rival European ambition for its possession or domination. The name in the Zulu tongue is interpreted as the " Place of Slaughter," a place founded by the father of Lobengula in 1838, some distance to the south.

The town stands near the south of Southern Rhodesia, in a central position on the Matabele portion of the tableland, between the Limpopo and Zambesi rivers, at an elevation of 4470 feet above sea-level. From Cape Town the distance by rail is nearly 1400 miles, and Beira, the nearest seaport, lies nearly 700 miles away, the line winding greatly in its descent to the coast across Mozambique. North-west from Bulawayo runs the " Cape to Cairo " railway link, which crosses the Zambesi near the Victoria Falls. The population of Bulawayo is about 10,000, equally divided between white and black people.

It was towards the close of 1893 that the predatory Matabele, a fearless, proud race of warriors whose generation had never come into conflict with the European, nor suffered defeat in inter-tribal wars, collided with the British settlers. Cecil Rhodes and Dr Jameson—the latter appointed administrator in the service of the then recently established Chartered Company—were the moving spirits in confronting the natives, and it became necessary to cease from offering peace to the warriors who, by their continual raids upon the inoffensive Mashonas, showed that there was no peace in their minds, and to impress them with the effect of superior force. But Jameson's force was a slender one. He sent to consult

Rhodes, who replied by telegram, saying : " Read Luke fourteen thirty-one."

It is recorded that Dr Jameson thereupon opened his Bible and read : " Or what king, going to make war against another king, sitteth not down first, and consulteth, whether he be able with ten thousand to meet him that cometh against him with twenty thousand." Jameson then telegraphed : " All right," and, the situation not admitting of delay, he set forth with 700 volunteers, 700 natives and 225 British Bechuanaland police. The two columns into which the force, in conjunction with reinforcements, was divided, under Major Forbes and Major Wilson respectively, came into contact with nearly 5000 Matabele, and the natives were mowed down by the machine-guns. In a second fight 7000 Matabele attacked the white men, the picked troops of Lobengula dashing up to the very muzzles of the guns, which again swept them away.

As a result of the campaign Lobengula was put to flight, and the British entered Bulawayo. Major Wilson was later the leader of that band of thirty-four heroes who, having crossed the Shangani river, which rose behind them and cut off reinforcements, were surrounded by the Matabele. They made a magnificent stand, forming a square and fighting until they fell. Their exploit sent a thrill through the whole British Empire. But it must be recorded in the interests of historic truth that a regrettable incident marred the glory of this campaign. Lobengula had earlier sent to the column under Forbes an intimation that he would surrender, with a packet of gold dust worth about £1000, but the gold was kept and the message suppressed by the two troopers to whom they were entrusted, a crime for which, when discovered too late, they were punished with fourteen years' penal servitude.

In modern Bulawayo no vestige of its former savage masters remains save the tree under which Lobengula sat when giving judgment, which has been preserved. From Government House, the site of the potentates' kraal, runs a wide avenue a mile and a half long, planted with shady trees, and from the market square in the centre of the town roads and avenues are oriented to the four points of the compass, with Rhodes Street in the centre. There are public and private buildings which may be described as handsome, although the

architecture in general is of that somewhat nondescript character common to colonial towns.   There is a large park and a
botanical garden, and a residential suburb.

Fronting the Stock Exchange stands a memento of a later
sanguinary incident in the history of the place, a monument to
the 250 settlers killed in the rebellion of 1896, in which the Matabele murdered defenceless men, women and children.   The
Matabele were suffering from certain grievances, notably that
they were treated as conquered people, and from an irksome
reputation requiring them to work for the white farmers and
miners for a certain part of the year at a fixed wage, which was
enforced in a tyrannical way by the native police, and also the
Company had seized the Matabele cattle, claiming that they
had belonged to Lobengula and therefore were lawfully theirs.

This last matter, however, had been arranged before the
rebellion broke out, and was less the cause of revolt than the
action of the chiefs, who were restless at their loss of power
and believed they could throw off the white man's yoke.   After
fierce fighting peace was arranged by Rhodes, who, with his
companions, went, unarmed, to a parley in the Matopo Hills.
Later, the Mashona revolt took place, also the terrible rinderpest of 1897, which carried off nearly all the cattle in the colony,
and thus, with the destruction of grain in the wars, nearly
reduced the inhabitants to starvation and the whites to penury.

The rebellion, which cost the Company £2,500,000, and the
Jameson Raid, caused a change in the form of government of
the colony to be brought about by the Imperial authorities, and
under more settled conditions gold-mining and railway-building
developed rapidly, largely under the guidance and enterprise
of Rhodes.   The name of Rhodes was perpetuated in that of
the colony, and the grave of the great Imperialist organiser lies
in the spot chosen by himself, at Matopo, thirty-three miles
from Bulawayo.   A colossal bronze statue of Rhodes stands in
the main street of the town.

We reach Salisbury by rail 300 miles to the north-east, the
town lying in Mashonaland at nearly 4900 feet above sea-level.
The population embodies about 3600 white inhabitants, and is
the commercial centre for a wide mining and farming country.
Here is the seat of the administrator and resident commissioner,
and the town itself, like Bulawayo, is controlled by a town

THE CLUB, SALISBURY, RHODESIA

council. The tobacco factory is the principal industrial institution, but there are engineering works and other evidences of the activity of Salisbury's British spirit, which has established this pleasing town where formerly the wilderness stretched uncompromisingly, with the lion and the Kaffir as its principal denizens. Smiling orchards and mealie-fields press up to the foot of rocky ridges, and beyond the hills arise, a horizon tinted with the pearly grey familiar to the traveller in tropical uplands.

From Salisbury the railway runs south-east to Umtali, on the border of Portuguese East Africa, and thence descends to Beira. The view of the Umtali from its neighbouring hills is soft and picturesque.

We shall remark the condition in Rhodesia, that the urban, in proportion to the rural, population is small, an advantageous condition which indicates, in a tropical colony especially, the permanency of occupation of the whites upon the land.

The progress of Southern Rhodesia must be regarded, on the whole, as encouraging. Yet it is to be recollected that in a period of about thirty years the white population has only come to number about 30,000. It is true that the Rhodesian ideal is to conserve the colony for people of independent resources, or such as may readily render themselves independent, and not to throw it open to the entrance of an indiscriminate white immigration. But in view of the millions of fertile acres awaiting settlement and cultivation, and the need for a more numerous population, it may be that this policy is a somewhat difficult or even selfish one. Whether any part of the British Empire can permanently shut itself off from the great number of poorer would-be emigrants from Britain, is doubtful.[1]

A somewhat materialistic element has been remarked in the character of the Rhodesian community, but doubtless this is inevitable in every land which depends for its progress to a large extent upon the industry of gold-mining. It is also to be recollected that Rhodesia is a land where the white man is master, finding a servile race ready to do his bidding. The white Rhodesians are almost entirely British in type, and strongly "Imperialistic" in sentiment. About half their

: The author enlarged on this subject in his lecture in 1910 before the Royal Society of Arts in London, and before the British Association ; also in his book, *Our Imperial Commonwealth.*

numbers are people born in South Africa, with thirty-five per cent. from Britain, the remainder coming from other British colonies or European countries. Somewhat less than a quarter of the children are of Rhodesian birth, and there are over 3000 white children at school throughout the territory.

Southern Rhodesia is administered under a system of partially representative government, with a Legislature consisting of elected and nominated members. The government is in the hands of the British South African Company, based upon the Royal Charter of 1889, with various modifications. The Company possesses a right of veto over the Legislature in financial matters, and the Secretary of State for the Colonies has a similar power in matters of general legislation. The Charter of the Company expired in 1914, but was renewed.

The administration of Northern Rhodesia is centred at Livingstone, near the Victoria Falls, which is the gateway to that part of the colony.

The Rhodesian railway system embodies over 2500 miles of line, a large mileage for the size of the white population, and since 1910 has yielded a good revenue. But tariffs are high, and Northern Rhodesia has but a small service. Large sums of money have been spent upon roads, but " drifts " or fords are greatly used in place of bridges, which are too expensive generally. The roads, as we shall remark in passing through the great stretches of granite country, are deep in sand, heavy and slow for traffic in the dry season, whilst the deep mud or the washing away of the surface in the wet months make travelling in wheeled vehicles very difficult. It is upon the narrow paths, tramped hard and flat by generations of barefooted natives, mere trails a foot wide, that the lonely settler must depend in districts away from other means of communication. Along these tracks a waving line of dusky bearers is often to be seen, pursuing its steady march hour by hour beneath the beating sun, each carrying his load, balanced on his head or shoulder, of fifty to sixty pounds.

The slow-moving ox has largely given place to the mule and the donkey, as cattle transport frequently spreads disease among the animals, and in Northern Rhodesia the tsetse fly prevents the use of any beasts of burden, whilst motor traffic demands roads which are too costly to build at present.

The traveller in this generally pleasing land of Rhodesia, if his errand be that of big-game hunting, will suffer disappointment except in the region where the energetic farmer or the shriek of the locomotive has not driven away the characteristic animal life of Africa. In the wilds of Northern Rhodesia he will encounter the elephant, hippopotamus, rhinoceros, buffalo, zebra and antelope, but all these have become extinct in the southern part of the country, except where preserved.

Remote from the range of civilisation the lion and the leopard prowl nightly around sheepfold and cattle-pen, and indeed Northern Rhodesia, whose few settlers have not yet availed to disturb natural conditions, is a vast natural game-preserve. On the plains of Southern Rhodesia the wild ostrich, which is protected by law, freely traverses the high ranges in large numbers, but does not thrive in captivity, although the country is regarded as suitable for ostrich-farming. The people of Southern Rhodesia hunt the jackal with packs of hounds, after the manner of fox-hunting in England.

Rhodesia has its full share of insect pests, which, however, science is attacking. There are also deadly snakes and great pythons in their special but remote haunts. The ravages of the locust are being brought under control, but the white ant is a relentless enemy. Small birds of brilliant colouring, as well as certain kinds of game birds—bustards and so-called grouse and pheasants—abound, and the naturalist finds a wide field for his work. The chain of muddy pools which the rivers leave in the dry season are infested by the dreaded crocodile, but there are also numerous fish in Rhodesian streams, yielding good sport for the angler.

Rhodesia, like the whole of the continent, excepting Northern Africa, is a land without ancient history or the profusion of buried temples of an earlier race such as we encounter in other tropical regions. Yet, nevertheless, there are ruins whose origin is lost in doubt, and which are not without elements of awe and mystery. A large population may have had its being in prehistoric times in what is now Southern Rhodesia, between the Limpopo and the Zambesi, for here we encounter extensive ruins of walls and buildings, the extent of the ancient villages being very remarkable, whilst the long, well-planned irrigation channels among the hills attest to a considerable practice in agricultural hydraulic art by the ancients. There are also ancient and extensive gold and copper mines, and widely distributed rock paintings and carvings.

The Zimbabwe ruins are among the most noteworthy of the early Rhodesian remains. Some writers have ascribed a remote antiquity to these, even associating them with King Solomon and the Queen of Sheba, but others regard them as not more ancient than the eleventh to the fourteenth century of the Christian era. Interesting as are these ruins, it cannot be said that their builders displayed an advanced structural knowledge therein, as neither arch, beam nor column, nor yet hewn stone, bond, plumbed wall nor the use of mortar have entered into them. The material used is the natural, unshaped slabs and blocks of the neighbouring diorite or granite, roughly placed. However, Zimbabwe embodies a great so-called " Elliptical Temple," and an " Acropolis," rows of monoliths, conical towers—possibly of phallic significance—and the large group of " Valley Ruins." The culture of their builders may have been of a " great negro empire," it is considered, which at one time traded with Egyptians and Mohammedans.

The ancient gold mines had evidently been worked with considerable vigour. Some of them are 120 feet deep and of great length, and old gold and copper ornaments are found in the ruins. These old workings have proved a valuable guide to the modern miner, indicating the richest deposits, and upon their sites some of the most flourishing mining concerns of the Rhodesian shareholder and London company promoter now stand.

The early hopes in regard to Rhodesian mining were found to be over-sanguine, and due to great difficulties of transport, and other matters, the London " market " lost faith in the country's mines. But after 1904 a great revival took place, and in the last few years the output of gold has reached a value of more than £3,000,000 sterling, the total amount of gold taken from Rhodesian soil being calculated at over £25,000,000 for the modern period.

It is, however, to be noted that a large part of the output is from the smaller mines, such as offer returns to the individual miner and his associates, rather than to the great joint stock concern, and this provision of Nature, which offers a reward to

Rhodesia : a view of Umtali

the humbler miner as well as to the greedier magnate, may be regarded with satisfaction.

The coal formation in Rhodesia is best shown by the Wankie Colliery, an important enterprise producing coal to the extent of hundreds of thousands of tons annually. The coal-field lies some 200 miles from Bulawayo, and covers a large area, with a modern plant giving employment to a considerable number of people, forming a valuable industry for the colony.

The copper-bearing district, lying towards the Congo frontier of Katanga, is of considerable importance. The Broken Hill deposits of lead-zinc ores are of a remarkable character. Other metalliferous minerals of commercial importance are found in Rhodesia, and there is also diamond-bearing ground. The development of mining and agriculture simultaneously is a wise process, and its accomplishment in Rhodesia has obviated the growth of the savage if picturesque conditions of frontier life such as grew to being in other lands, among the mining communities.

Taken as a whole, the life of Rhodesia must be regarded as full of energy and promise, and the misfortunes and depressions through which the colony has passed are beheld in a due spirit of proportion. It is a land towards which the thoughts of all British Imperialists go out affectionately : whose ordered conditions and settled white man's homes and sentiments are veritably a refreshing oasis amid those savage areas of tropical Africa which we have been called upon here to tread.

The fall of German South-West Africa, if that vast territory be retained by the British Empire, may give Rhodesia an outlet to the Atlantic Ocean.

CHAPTER XVI

THE BECHUANALAND PROTECTORATE, AND NYASALAND

To the west of Southern Rhodesia lies the great British protectorate of Bechuanaland, occupying the heart of South Africa and the central part of the great plateau which stretches north to the Zambesi. It is cut midway by the tropic of Capricorn, which traverses the great Kalahari Desert, and covers about 225,000 square miles of territory, with German South-West Africa upon its western side.

A curious region unfolds before us in this part of Africa, a land of considerable elevation, averaging 4000 feet above sea-level, covered in places with grass or acacia shrubs, alternating with vast marshy depressions into which rivers fall, with no outlet to the sea, forming inland lakes whose waters stretch to the horizon in times of flood, but which, receding in the dry season, leave a baked plain where the mirage deceives the traveller who adventures over these inhospitable wastes.

Here and there are the remains of ancient forests, and deeply furrowed river-beds in which no water now runs, natural indications which, together with the evidences of man's earlier work in the ruins of ancient settlements and stone dykes, seem to point to a period when Bechuanaland was a well-watered country. It may be that the demolition of the timber and the custom of burning up the grass, practices which have been carried on both by the white and the black occupant of the land, are conducive to the change of climate, a desiccation which is still going on.

The climate of this high plateau region is such as we shall have expected to find in this latitude and elevation—that is, healthy and bracing, the heat offset by the dry air, the nights cool and refreshing, due to the diurnal range of temperature, and indeed at night intense cold prevails. Thunder and hail storms occur in summer, and in the dry winter season the storms are of dust, which in choking clouds " mingles the ravaged landscape with the skies." The western portion is the driest, with a rainfall of only ten inches, increasing to two or three times that amount in the eastern districts.

But the soil of the great plateau, like that of many other semi-desert lands, is extremely fertile wherever artificial irrigation can be brought to bear upon it. Even the unirrigated land of Bechuanaland yields, it is said, under native cultivation, twelve bushels of wheat to the acre. In the river valleys, however, and upon the great marshes, malaria prevails at times. On the " sweet veld," as the best natural cattle pasturages are termed, in distinction from the " sour veld," the Bechuanas amass wealth from their herds of cattle, oxen, sheep and goats, in whose care they are expert, and water for these can generally be obtained by sinking comparatively shallow wells.

The great marshy sinks of which we have spoken are principally those of Ngami, in the west, and the Makari-Kari salt pan, both draining vast extents of territory, and united by the Botlette river. As before remarked, into these depressions great streams of water pour, there to be lost in the desert. The Kalahari Desert lies to the south of these depressions. The term " desert " scarcely does justice to this region, for, although it has been termed the " Southern Sahara," these are portions of it which are covered with grass, scrub and even forest. The thirsty traveller, moreover, may suddenly be delighted by coming upon patches of wild water-melon, with whose luscious fruit he may slake his thirst. Indeed this remarkable plant, a singular gift of Nature in such a spot, acts as a " water supply " to both man and beast. The melon is of two kinds—sweet and bitter, the former being that sought after by the wayfarer.

Among other curious plants and flowers in the Bechuanaland deserts are the scarlet cucumber, the white mimosa, and the " wait-a-bit " thorn, and, if rarely, the great baobab-tree and the palmyra are found.

The Kalahari is a country in which big game still abounds, and the roar of the lion startles the traveller as he traverses it or camps at night. The elephant, with the other typical or ferocious animals of Africa still have their home in the desert, as do the buffalo, zebra, quagga, ostrich and baboon. Elephant, eland and giraffe hunting is now prohibited, and there is a close season for all game.

But a drear and desolate waste unfolds to the view over great portions of the desert, where the sandy surface, undulating to the horizon, looks like the suddenly arrested surface of a billowy sea, alternating with patches of salt.[1] The Kalahari covers some 120,000 square miles of territory, and its principal human inhabitants are wandering Bushmen, who live by hunting, killing game with poisoned arrows. The cattle-keeping Hottentots dwell upon the western border. These people have a passion for cattle-breeding, and they also carefully cultivate the melon, the only plant Nature permits their small gardening operations to produce.

Worthy of note is the method some of these people have of obtaining water for their desert home. A reed about two feet long is sunk in the ground where water is known to exist below the surface, a bunch of grass having first been tied to the buried end. The woman—it is part, perhaps, of her household duty—draws up water by sucking through the reed, and discharges it from her mouth into the empty ostrich-shell which serves her as a receptacle. The precious commodity is then often buried, not to be produced even to the thirsty traveller or under the threats of an enemy, save possibly out of native hospitality.

Timid, peaceful and grave are these people of the desert, loving the scanty gardens of their villages. It is recorded by Livingstone that even their children are never seen at play: perhaps the subduing influence of the desert. In earlier times large parties of travellers have perished in the Kalahari, of thirst, and the main routes through the desert are known locally as "the thirsts," although the Government has sunk some wells.

The population of the protectorate of Bechuanaland is about 125,000. This is distinct from British Bechuanaland, part of the Cape Colony, which lies outside the torrid zone. Many of the Bechuanas have become Christianised under missionary influence. It may be remarked that their language, which the missionaries reduced to writing, was rich enough for the translation of the whole Bible. These people have many curious customs, both in regard to their life—often good customs—and

[1] Reminiscent of the deserts of Tarapacá, Chile, in the same latitude. Both lie near the tropic of Capricorn, on the western side of their respective continents, and at about the same elevations above sea-level.

RHODESIAN LANDSCAPE

as regards death. They are excellent craftsmen, smelting and working iron and copper, and carving wood with dexterity.

The missionaries, active and influential, here [1] came into opposition with the Boers, who had endeavoured to control or tax the natives of Bechuanaland. To Pretorius and Kruger, who made the natives an invitation to combine their territories against the British, the reply of Maitoria, a native chief, was: " Does one inspan an ox and an ass in the same yoke? " Later the people complained to Britain of what was in effect the system of slavery carried on by the Boers. Bechuanaland, which owes much to such men as Livingstone, Moffat, Mackenzie and Rhodes, was formally taken under British protection in 1885, and the natives have shown great loyalty to the Empire.

To the east of Northern Rhodesia lies the Crown Colony of Nyasaland,[2] a narrow strip of country bordering the western shore of Lake Nyasa, and covering about 39,800 square miles. It is bounded on the south-east by Portuguese East Africa, through which it has a natural outlet to the sea by means of the Shiré affluent of the Zambesi. To the north lies German East Africa.

This essentially agricultural land lies mainly within the savanna region of Central Africa, and the landscape is often of a pleasing and park-like character, covered in places with very long, coarse grass, and other vegetation, and generally well watered. We do not encounter those forests of lofty trees here which are so striking a feature of West Africa, but the country may be regarded as fairly well wooded. The cypress, or so-called cedar, abounds on the upper plateau, with good building timber of smaller kinds here and there, including mahogany and other tropical varieties. There are no desert areas in this part of the British dominion.

The somewhat isolated territory of Nyasaland may be entered through the port of Chinde—which has a British " concession " area for transport purposes—in Portuguese East Africa, ascending the Zambesi and Shiré rivers, which form the

[1] Especially the London Missionary Society.

[2] Originally part of British Central Africa, which owed so much to Sir H. Johnston's work.

main artery of transport to and from the coast. From Port Herald, on the Lower Shiré, near the Portuguese border, light-draught, flat-bottomed steamers, or barges—depending on the state of the river, which is low in the dry season—make regular journeys. From the same point a line of railway runs to Blantyre, the commercial centre of the colony, situated upon the Shiré highlands at an elevation of 3000 feet above sea-level, at a total distance by rail and steamer from the sea of 300 miles. Blantyre was so named after the birthplace of Livingstone, and the town was founded by the Church of Scotland Mission, in 1876. The fine, red-brick building of the Church of Scotland, with its lofty turrets and white dome, approached by a cypress and eucalyptus avenue, was erected by the hands of natives throughout. The town has a population of about 6000 natives, with little over 100 white people.

The surrounding Shiré highlands have attracted the largest number of white settlers in this part of Africa, and several hundred Scottish planters are here engaged in coffee-growing, which branch of agriculture formed the first attraction to the white settler in Nyasaland. The industry, however, suffered from severe difficulties in its early stage, but nearly 1,000,000 pounds of coffee are annually exported.

Although Nyasaland cannot be termed a "white man's country"—that is, as regards the carrying on of manual labour by the European—its resources and possibilities are promising and attractive, especially in the cultivation of such products as cotton, tobacco, rubber, maize, rice, coffee, and even tea. Considerable success has been obtained by the white settlers in their plantations, and the free and prosperous life attracts many who have a small capital at their command.

The Government fosters cotton-growing among the natives, as also tobacco-planting, notwithstanding that it has been argued by the settlers that the encouraging of the native to work his own plantation militates against production for export and decreases the available labour supply. It is asserted, moreover, that planters who feed and house, and generally attend to the wants of their labourers on a sufficiently generous scale, can always count on a sufficient supply.[1] The work of

[1] *Oxford Survey*, "Nyasaland," J. B. Keeble.

women and children on the holdings frees the men for outside labour at times.

The natives own large herds of cattle—perhaps 50,000 head —and sheep. The Angoniland plateau is the best cattle district, but cattle thrive elsewhere. The stock-raising industry, both of the European and the negro, will doubtless become even more important when the railways are extended to reach Lake Nyasa and the sea-coast, at Beira, and agricultural export may then be more readily entered upon. This line will be the solution of the low-water problems of the Shiré and Zambesi.

The great, deep Nyasa, occupying the south end of the rift-valley, is the third in size of Africa's lakes. It is 360 miles long, and lies 1645 feet above sea-level. Small steamers and gun-boats navigate this lake and Tanganyika, forming an important link in the water-route from the Zambesi's mouth to the heart of the continent. A German gunboat was earlier launched thereon.

Roads suitable for ox-wagon traffic are plentiful, and roads for motor cars and motor bicycles have been made in several places. On account of the tsetse fly and its effect on cattle transport in certain belts, light carts are drawn by natives, and form the only means of transport in some cases.

A number of cotton ginneries of modern type have been established by cotton-planting companies and the British Cotton-Growing Association, for preparing the cotton crop for shipment, and considerable quantities are exported. The production of tobacco and ground-nuts has also largely increased. A large plant for dealing with the tobacco crop has been erected at Limbe.[1] The tobacco cultivated is both native and exotic. Coal and iron are both found in Nyasaland, as also some gold, silver, lead and copper, and traces of petroleum. The iron ore appears to be plentiful.

The climate of Nyasaland, as a whole, cannot be described as unhealthy, and the children of Europeans thrive in the higher districts. But in the low valleys malarial and black-water fevers are dreaded, although the occurrence of the latter is diminishing considerably. There are frosts at times on the plateau.

The fruits of Europe do not flourish generally, but oranges

[1] By the Imperial Tobacco Company.

and lemons, planted by the Arabs earlier and the British later, do well, as does the mango from India, and the pine-apple. Splendid wild blackberries grow in profusion. The flowers and the bees of the colony are productive of honey and wax, which latter is exported.

In the wild life of Nyasaland we encounter the elephant, roaming abundantly in his peculiar haunts, whilst lions and leopards are relatively common, as are zebras, antelopes, gazelles and other of the well-known upland fauna. In the waters and upon the river banks numerous crocodiles and hippopotami are found.

The native population numbers 1,000,000, the people being the Bantu-negro, with some admixture of Hottentot and Bushman, and there are somewhat less than 1000 Europeans and several hundred Indian traders in the country. The negro Mohammedan Yaos are a reserved and intelligent people, well built, making good soldiers, craftsmen and workers generally. Other tribes have less useful attributes, and are dirty and lazy, and again others equally merit praise for their character and disposition. Progress has been remarkable since, under the British regimen, tribal peace and security has been attained to.

The Crown Colony Government of Nyasaland has given exceptionally good results, with a governor-in-council and other executive officials, a legislative council and district residents and magistrates, which latter have become endeared to the natives from their accessibility and close touch with them, as contrasted with the former venial native governance. Native councils have been instituted, as a result of the spirit of independence and individuality that has sprung up among the natives.

The people are multiplying and seeking the development of their country. The young native is quick to learn, not merely as a labourer, but for clerical and even engineering work. " It is true that his avidity and passion for education, evinced by other races all over Africa, are due to an inherent commercial instinct to better his lot and satisfy the created desire for creature comforts "[1] and he is fast leaving behind the old savage conditions.

The missionary societies carry on the educational system,

[1] " Nyasaland," *op. cit.*

and to their excellent methods the good sociological results are largely due. Hundreds of village schools, in charge of native teachers under white superintendence, exist in the colony, and in these over 120,000 natives are instructed, native artisans also being trained by the missions. The native has a strong leaning towards religion and hymn-singing, and rivalry between missionary societies and Mohammedan teachers causes practically the whole population to be reached. It is to be noted that a great attraction of the missions is the industrial and educational training, apart from the religious teaching, a matter of some value.

The scene of our survey again changes from British to Portuguese territory, as, to the east, we enter Mozambique.

CHAPTER. XVII

PORTUGUESE EAST AFRICA OR MOZAMBIQUE

A COUNTRY watered by many rivers, chief among them the Zambesi, queen of East African waterways, of fertile lands and quaint towns, a community upon which the Portuguese have stamped the qualities and defects which they ever carried overseas from the Iberian Peninsula; such is the region that unfolds before us as the steamer drops anchor in Delagoa Bay, or other seaport of the long Mozambique coast.

Mozambique, as Portuguese East Africa is officially termed, does not lie wholly within the tropics, its southern portion being cut by the tropic of Capricorn. Across the Mozambique Channel lies the great island of Madagascar, and its shores are bathed by the Indian Ocean, which extends its vast expanse towards Australasia. To the north lies German East Africa; west, the British possessions of Nyasaland and Rhodesia; and south, the Transvaal and Natal. A vast land, 1500 miles long and 500 wide in its greatest dimensions, Mozambique has an area of 293,000 square miles.

This broad and diversified territory shelters a relatively sparse population of somewhat over 3,000,000, consisting, with the exception of some ten per cent., of negroes of the Bantu tribes.

The first occupants of the coast were the Arabs, who dealt in gold and ivory with the negro potentates of the interior, and they built some fine towns. Vasco da Gama, the Portuguese, appeared in 1498, on his voyage to India, and in 1510 the Portuguese built forts of stone brought from Portugal (on Mozambique Island, from which the country takes its name) and mastered the Arabs. Lured on by tales of gold, the Portuguese penetrated the Zambesi, but the expeditions suffered disaster.

From 1560 onwards the Jesuits and Dominicans strove earnestly to bring the native races of the region to Christianity, and gained thousands of converts—at least nominally. But their efforts waned in the eighteenth century, soon after the middle of which they were expelled. The Portuguese were unable to bring about effective development of the regions under their control. Africa was neglected for the more valuable possessions in India and the Far East; the enervating and unhealthy climate was unfavourable to European settlers, and the officials sent out, and the planters and others, led a life marked to a large extent by idleness and debauchery. Quarrels between native tribes were purposely instigated by the Portuguese, who, however, failed themselves to profit by the disunion, and the decadence of Portugal at home resulted in Africa in the capture by the Arabs of the Portuguese territory north of Cape Delgado.

The Dutch attacked Mozambique at various periods in the seventeenth century, and trade was menaced by the British and the French. The export of slaves to Brazil, from the middle of that century, had furnished a new source of wealth to the merchants of Mozambique and Sofala, and this nefarious traffic continued until towards the middle of the nineteenth century, slavery having been finally abolished in 1878.

The character of Portuguese rule in East Africa is thus seen not to have been conducive to efficient economic development or civilisation, although good work was done by one of its governors, Francesco Laceda, a man of high talents, who strove, at the close of the eighteenth century, to bring about reforms in the administration. But after his efforts apathy and decay spread over the colony, and, during the nineteenth century, the territory was devastated by the attacks of Zulu savages. It was not until the discoveries of Livingstone on the Zambesi, from 1850 to 1865, leading to British settlement on Lake Nyasa and influence in Mashonaland and Matabeleland, that Portugal was aroused to the necessity of taking her part in the scramble for the partition of Africa.

In 1875, following upon arbitration with Great Britain, Portugal secured entire possession of the valuable harbour of Delagoa Bay, a portion of which Britain had laid claim to. The rivalry of other Europeans caused Portugal to take action towards the development of the territories, and charters were given to the Mozambique and Nyasa companies, and these, as well as the Zambesi Company, undertook to carry out such development, British capital being largely interested therein.

The Government was removed from Mozambique to Lourenço Marques, and in 1909 an agreement was made with the Transvaal concerning the import trade through the port, and the recruiting of natives for the Rand mines, from the Bantu tribes, which furnish the main source of labour for the Rand.

Portuguese East Africa has been well dowered in its river system. Six rivers converge upon Delagoa Bay. The Limpopo and the Sabi and other streams cross the country to the Mozambique Channel, and the great Zambesi flows through it for nearly 600 miles, and is navigable throughout, except for the rapids 400 miles from the mouth. By the Shiré affluent direct steamer and rail connection is given with British Central Africa. The Limpopo is navigable for 100 miles from its mouth, for shallow-draught steamers, and the Sabi for 150 miles, and there are other navigable streams.

These rivers generally flow placidly through alluvial valleys, watering land which lends itself to irrigation, and towards these tropical coast-lands enterprise is being directed. From the Rovuma to the Zambesi, a vast stretch of littoral, the soil is of great fertility. The conditions of available river transport upon the seaboard, and the situation of the most fertile districts thereon, are in contrast with those prevailing in British and German East Africa, whose fertile districts depend largely upon the long line of communication of the Uganda and Ujiji railways.

The natural resources are very considerable, as might be expected of a country possessing both a tropical and subtropical climate, and stretching over sixteen degrees of latitude. The climate cannot be described as healthy, as, along the coast and on the banks of the Zambesi, malaria is endemic. But Europeans, with proper care, enjoy good health here, whilst on the plateaux and uplands the climate is healthy and temperate. In the afternoon the prevailing east wind often blows from the sea, refreshing the littoral, and also constantly builds up the sand dunes which parallel the coast. During the monsoon a fairly even mean temperature of about 76° F. is experienced in the districts on the Mozambique Channel.

Lourenço Marques, the capital of Mozambique, is not a

tropical seaport in the strict meaning of the term, as Delagoa Bay, upon which it is situated, lies south of the tropic of Capricorn. Situated thus, near the southern end of the province, its activities are somewhat centralised, and, being the nearest seaport to the Rand mines in the Transvaal, and slightly under 400 miles from Johannesburg, its business is closely connected with that British possession, half of whose imports find entry there, whilst the coal from the Transvaal collieries at Middlesburg is practically the only article of export. Cheap Portuguese wine for the Kafirs—a deleterious trade worth £500,000 annually—is the main article of import for local consumption.

There are elements of progress in the capital, however. The harbour is well equipped, steamers being able to discharge alongside the railway terminus, and Delagoa Bay is accessible to large vessels at all seasons. Lourenço Marques has a population of over 10,000, half of whom are Europeans. The surrounding country is low and unhealthy.

The town of Inhambane lies almost upon the tropic of Capricorn, and enjoys a good reputation for its healthful climate. It is beautifully situated on the bank of the river, which falls into the Bay of Inhambane, and is the natural port and outlet for the extensive and fertile region between the Limpopo and Sabi rivers.

Here we encounter a community of considerable architectural merit, superior in this respect to any town of Portuguese East Africa, with the exception of Mozambique. The fort, mosque, churches—the latter built with stone and marble from the Motherland—are the principal structures. The town dates from the sixteenth century, but possesses a population, little more than 5000, of a mixed character, including Europeans, Arabs, Indians, half-castes and negroes. Its ancient mainstay was in trade with ivory and slaves, and it is now the centre of the best recruiting ground for Rand mining labourers, whilst one of its chief imports is cheap wine for Kafir consumption.

Beira, lying nearly 450 miles to the north of Lourenço Marques, is a comparatively healthy town, and forms the terminus of the railway from Salisbury and Bulawayo. The population is little over 4000, a third of which are Europeans, the business community being largely British.   The town is built upon a sandy promontory of the river shore, and has few architectural pretensions, but its public gardens are fine.   It occupies the site of one of the ancient settlements of the Arabs upon this coast, long since forgotten.

A little to the south lies Sofala, also a place of bygone importance, for here at one time stood the chief town of a wealthy Mohammedan state, dating from the twelfth century.   The harbour, which formerly might have given shelter to 100 great vessels, has silted up, and the ancient ruins of its forts speak eloquently of vanished authority.   The early Portuguese traveller Lopes, who accompanied Vasco da Gama to India, in 1502, identified Sofala with the home of the Queen of Sheba and with the Ophir of Solomon, and Milton alludes to Sofala in this connection.

Quilimane, which we next approach, is a town of some 2500 inhabitants, lying in the midst of perhaps the most fertile districts of the East African coast, with a river system affording great possibilities, both for transport and irrigation.

The Quilimane river, during the wet season, becomes a deltaic branch of the Zambesi, which great river enters the sea to the south, and there is ample and deep anchorage at the mouth, although the entrance is obstructed by a bar.   A long acacia-shaded street skirts the river, and here are built the homes of the European merchants, the Indian or Banyan inhabitants occupying another street, with the native town behind.   The trade of the place has somewhat declined since the rise of the town of Chinde, not far to the south.   The railway, about eighteen miles long, serves a richly planted tropical district.

The seaport of Chinde lies at the mouth of the navigable Zambesi, and has a population of 3000, with a few white men. Large steamers cannot cross the bar.   The British concession or " British Chinde " covers a small acreage for purposes of the transport trade from Nyasaland and North-East Rhodesia, with which there is water communication, and upon this trade the town largely lives.

From the Kebrabasa rapids to the ocean, a distance of 400 miles, there is no obstacle to navigation on the Zambesi, except that the stream in places becomes shallow in the dry season, the current spreading out in a broad valley. The banks are low and fringed with reeds. The soil is fertilised by the inundations of the river, and sugar and rice plantations abound, and wheat and other cereals are produced. Although as a navigable waterway this great river suffers by reason of occasional shallowness, and by reason of its bar, like all African rivers, nevertheless it forms a fluvial highway into the interior, with a vast aggregate navigable length.

Much farther to the north we reach the seaport and town of Mozambique, the latter on its coral island. Its forts contain some modern guns. The mixed population of 7000 people includes Indians, Arabs and Europeans, but is mainly of Mohammedan negroes.

Along the coastal districts of Quilimane there is much land suitable for coco-nut cultivation, and some small planting has been done. Land favourable for sugar-cane cultivation extends behind, followed, as the interior is gained, by land where sisal hemp of good quality is being produced. Some fifty miles from the coast an undulating, forest-clad country is reached, traversed by numerous streams, and great masses of granite rising into mountain peaks give a weird appearance to the landscape. Native clearings encountered here yield manioc, sweet potatoes, tobacco, bananas and other common tropical products, supporting, in some places, a considerable population. A large extent of land is suitable for Ceara rubber, and about 1,000,000 trees have been planted in various estates, which it is considered may lead to the development of profitable rubber culture. Ostrich-farming and cattle-rearing will depend for success upon local conditions—the last named upon the presence or otherwise of the tsetse fly.

Inland of Lourenço Marques the dense native population raise good crops on the very fertile soil, but the magnificent plains of the Limpopo are often flooded. The sugar industry is very important, especially in the Zambesi and Inhambane districts. Coco-nut planting, rubber, hemp, tobacco and cotton are capable of much development. Maize is a staple crop with the natives, but its planting has not been taken up by Europeans on a large scale. Excellent grazing lands for cattle-breeding and dairying exist in the district of Lourenço Marques. There are some possibilities for dry farming in the drier regions. A large part of the colony is timbered, but due to cost of transport the timber will doubtless long remain of value principally for local consumption.

Portuguese East Africa possesses some valuable mineral resources, notably in its vast coal deposits, as also in ironstone, whilst copper ores are also found, as well as petroleum. Ancient gold workings abound in some districts, and, as elsewhere remarked, some writers have endeavoured to identify the auriferous region north of Delagoa Bay with the land of Ophir.

One of the principal obstacles to the economic development of Portuguese East Africa is the relative scarcity and condition of labour, due to the sparse population and the natural indolence of the people, and, it is stated, to the inexperience of the white employees.[1] Moreover, a considerable cause of complaint among planters and others is in the extensive recruitment of native labour for the Transvaal mines. Under an agreement between Mozambique and the Transvaal the mining authorities, in return for certain considerations, are permitted to recruit labour here. It is argued that this recruitment is a drain upon the labour market, and that the mortality at the mines is a serious loss. About 100,000 men from the province are continually employed on the Rand, and it has been shown that this labour, if employed, for example, on the sugar plantations, could produce 300,000 tons of sugar per annum, of a value of £5,000,000 sterling. On the other hand it has been argued by those who uphold the system that the loss is more apparent than real, as only a small proportion of the population is employed on the sugar estates, and that the Rand labourers can well be spared. The Kafirs return from their work at the gold mines, it is stated, with so much money that the low rate of wage offered by the planters has no attraction for them.

Furthermore it has been shown that plantation labour is equally or more difficult to obtain in those districts—such as the Mozambique Company's territory—where recruiting for the mines is prohibited, the native there preferring to work in his own small affairs. It is also remarked that the market and streets of Inhambane, where formerly hundreds of natives congregated daily, bringing their products for sale, have almost

[1] *Mozambique*, R. N. Lyne.

been emptied by recruiting for the mines, and it is argued that if this were stopped the small industries and plantations of the natives would no longer suffer destruction. But in the Mozambique district and the Nyasaland Company's territory there is less native industry, and it is stated that the heavy recruitment of many thousands of natives does not injure the landed interests.

Under the *prazo* system, which obtains in Portuguese East Africa, the land is divided into large estates, where the natives work for the benefit of the European planter, who is also empowered to collect taxes in labour or produce, partly by "farming-out" the tax collection; and serious evils are said to result therefrom. In the Quilimane district vast areas are subject to the system, which results in the land being largely locked up, whilst the *prazo* companies are enabled to exploit the natives, who are obliged to bring in loads of produce over long distances without payment, as tax work, suffering hunger and privation, which the appearance of their emaciated bodies bears out.[1] On the other hand it is contended that good arises out of the system, as the *prazo* establishments teach the natives trades and crafts, in a way said to be "superior to the missionary management." The condition of native labour in this part of Africa would seem to call for impartial investigation.

The present condition of Mozambique shows that the greater part of the rich and valuable territory lies fallow. This is partly ascribed to the complicated taxes and tariffs, the labour and land systems, and, it is stated, to some prejudice against the foreigner, whether settler or capitalist. In foreign capital danger is scented, a condition possibly brought about in part by the "capitalistic methods" of the Rand, which, to some extent, were looked upon in Mozambique as leading to the suppression of Boer independence in the former neighbouring republics. Further, the growth of the surrounding progressive British colonies leads to a certain amount of apprehension, as if absorption thereinto might be a threatening element in the future. However, the Portuguese are, in general, well disposed, and the Government, knowing the danger of the backward economic condition of the colony, puts no

[1] *Mozambique*, Lyne, *op. cit.*

obstacle in the way of the real colonist, of whatever nationality.

Here we leave the African mainland for a space, to visit those tropic islands, curious, beautiful and fertile, which lie in this part of the Indian Ocean, and which may conveniently be considered at this point.

CHAPTER XVIII

MADAGASCAR, MAURITIUS AND SEYCHELLES

In the great island of Madagascar, the third largest in the world, lying 260 miles off the East African coast, we approach a land of many peculiar attributes, both as regards its natural history and the life of its human inhabitants. Here dwell a people not originally indigenous to the island, extremely clever in various native industries, dominated by ideas which at times have been extremely barbarous, at others marked by a desire for progress, and—a curious circumstance for such a nation—long governed by female sovereigns.

Of the area of Madagascar six-sevenths lie in the torrid zone, the tropic of Capricorn cutting its southern end. The island is nearly 1000 miles long and reaches a width of 360 miles, covering an area of 228,000 square miles. We remark its regular outline, with few bays except upon the western side. But a great length of narrow lagoon has been built up on the eastern coast, a result of the contest between the waters of the rivers and those of the sea, and these lagoons form a natural canal nearly 300 miles long, navigable for considerable distances.

Broadly, the island embodies a high mountainous region, reaching 3000 to 5000 feet above the sea, around which are extensive lower plains, broken by hills. Here lie many fertile areas and smiling valleys, in some cases occupying the sites of lakes long since dried up. Bare, rolling, moor-like country is characteristic of the interior uplands, covered with a reddish clay soil, but in the valleys there is a rich alluvium. Except in the south and south-west, there are no arid or desert regions.

As we ascend to the high interior we remark the weird forms into which the crystalline rocks are broken and weathered. Here arises what fancy might present as some Titanic castle, and there, beyond, stand pyramids and cathedral towers, sculptured domes and symmetrical spires of Nature's building. Higher than all these arises " That which mists cannot climb," as the lofty Tsiafajavona mountain's native name implies. This is an extinct volcano, rising 5000 feet above the adjacent land, and 8635 feet above the level of the sea.

Few of the rivers on the eastern side of Madagascar are of service for navigation, except for the shallow native canoes, for, as the water-parting of the island follows the straight east coast at only some eighty miles therefrom, the eastern streams are generally short and torrential. On the west, one at least of the rivers, the Betsiboka, can be ascended by small steamers for 100 miles, and there are other considerable streams here and on the south coast.

Giant cascades falling into magnificent gorges clothed in dense forest mark the course of the eastern rivers, the water descending in successive rapids and cataracts, having cut their way through the rocky ramparts of the uplands. The Matitanana rapids are described as descending at one plunge for 400 feet, and there are fine falls near the line of railway which follows the Vohitra river.

Madagascar, since 1896, has been a French colony, and its population numbers about 2,750,000, including 11,000 Europeans, with some Indians, Arabs and Chinese, which latter are found in almost every town.

The native inhabitants are collectively known as Malagasy, and embody a large number of tribes. The lighter complexioned natives are derived from the Malayo-Polynesian people, and the darker from the Melanesian, and these two divisions account for the majority of the population. This derivation is inferred from similarity in appearance, custom, and language with the people of the Pacific and Indian archipelagos, and traditions also go back to this far-off source. An Arab element is found on the coast in places. The vestiges of an autochthonous or original people are but scanty.

One language alone is spoken throughout the island, and it has been shown that this is allied with the Malayo-Polynesian languages. Its character, numerous proverbs, folk-lore, songs and legends are indicative of strong imagination. The dominant people of the island are the Hova, who appear to have been the last comers, and are the lightest complexioned and most civilised of the island peoples.

The principal seaport of Madagascar is Tamatave, lying on

the east coast, facing the Indian Ocean towards Mauritius. The coral reefs enclose a fairly commodious harbour, the town standing upon a sandy spit. Here are some of the foreign consulates, and many of the very numerous French officials which assist in governing Madagascar dwell in the town, whilst good shops, offices of merchants and houses line the main street, the population being about 5000. Due to the unhealthy situation, however, the greater part of the native population has been established in a new place some distance off. Perhaps half the trade of the island is carried on through this seaport.

From Tamatave we embark upon one of the small steamers which run for sixty miles through the navigable lagoons along the coast, to reach the railway that ascends to the capital, Antananarivo. This, the "Town of a thousand," as the native name interpreted means, stands some ninety miles inland, in a very commanding position on the summit of a lofty ridge rising 700 feet from the rice-cultivated plains around it, themselves lying at over 4000 feet above the level of the sea.

From every point of view as we approach this Hova capital we remark the lofty roof and towers of the great palace, and surrounding this are numerous stone and brick buildings, many of considerable architectural pretension, including royal palaces and houses of former ministers and nobles, the Protestant and Roman Catholic cathedrals and churches, the French Residency, schools, hospitals, law courts, and numerous good dwelling-houses.

The French, since their arrival in 1895, have made excellent roads, connected in the steeper places by steps, with a handsome tree and flower adorned central square. The planting of flowers and trees in various parts of the city, and the pretty gardens, have rendered the general aspect attractive and bright. There are over fifty churches in the district, and a Mohammedan mosque. The population of this interesting capital is over 70,000. Antananarivo has developed thus under the Hova chiefs, and later under the French, from what was in early times a wooden and rush-built village.

Royalty was abolished in Madagascar when the island became a French colony, in 1896. The earlier Hova sovereigns lived in much state, surrounded by ceremonies and the taboo. About 1877 the great mass of slaves—mostly from

Mozambique, brought over by Arab dhows—were freed, under agreements with the British, and they are now merged into the population. Slavery was of a patriarchal nature and not generally cruel.

The first European arrival on the island shores was the Portuguese Diogo Diaz, on a voyage to India in 1500. The Dutch, and later the British, at the time of Charles I., endeavoured to form settlements, and tyrannical French commandants in the seventeenth and eighteenth centuries also strove for dominance. The British from Mauritius endeavoured to obtain some control, but the Home Government preferred to recognise the Hova supremacy. An enlightened Hova king, Radma I.—" a kind of Peter the Great "[1] of Madagascar—saw the need for a modern civilisation, in the early part of the nineteenth century, and he treated with Mauritius for aid in establishing such, and at this period the London Missionary Society took an active part in the spiritual affairs of the kingdom, imparting knowledge of arts, introducing printing and the Bible.

But reaction set in on the advent of the female sovereign Ranavaloria, and hundreds of the Christian Malagasy suffered cruelly for their faith. Rebellions, wars, and the exclusion of Europeans followed, a reign of terror, which came to an end when Radma II. appeared. Almost equally foolish, this monarch was killed in his palace, and Queen Ranavaloria II. came to power. Under this woman great progress was made. Christianity was publicly recognised, foreign consuls were established, and agreements regarding their respective influences made between France and Britain. But as time went on wars with France took place, followed by a French protectorate, and, in 1895, this not being recognised to French liking, a French invasion and conquest resulted.

Under General Gallieni, the French governor, who was animated by a just, kindly and vigorous spirit, the Hova rule was overcome, and economic development entered upon. The queen died, and her successor of the same name carried on her rule, but she was at length exiled by the French. Errors of government, however, under the French regimen, have made themselves severely felt, such as excessive taxation and inter-

[1] Sibree, " Madagascar," *Encyclopædia Britannica*.

ference with religious matters, whilst the great number of French officials have been described as much in excess of requirements.

The natural history of Madagascar is of extreme interest, and palæontological research has revealed the remains of curious creatures, and of animals which have become extinct in the human period. Much of the present animal life has Asiatic rather than African affinities. None of the large quadrupeds are found, however, except that the introduced African humped cattle has multiplied well. There are deadly serpents on the island. Madagascar may have formed part of an Afric-Asian land connection in former geological times.[1]

The abundant rainfall gives rise to varied and sometimes beautiful forest trees and flowering shrubs, about one-eighth of the island being forest-covered. The " traveller's tree " found here, *Urania speciosa*, has a graceful plume of leaves surmounting its tall trunk, and " yields a supply of fine cool water." The climate of the interior is temperate, with frosts at night occasionally. Malaria is often fatal on the hot coast. There are terrific thunderstorms in their seasons, water-spouts and cyclones.

From this very cursory reference to the interesting human and natural history of Madagascar we must turn to consider the industries of the people. Agriculture is the principal occupation, and in some branches of this, notably in the terracing and irrigation of the rice-fields, the people are very ingenious, and indeed it has been said that in agricultural engineering and hydraulic work connected therewith no people surpass some of the tribes of the island.

The work of preparing the grain is mainly carried on by the women, and their great dexterity is further shown in their weaving, with the simplest appliances, of silk, cotton and hemp cloth, elegant both in pattern and colour.- They also make mats and baskets, and the grass hats are almost the equal of those of Panama. This dexterity in the use of fibre seems to indicate a link with Polynesia.

The men are clever metal-workers, and with a few primitive tools they fashion beautiful charms and filigree ornaments of gold and silver. Further, the native ironwork and copper

[1] Wallace.

and brass work are excellent, and much inventive ingenuity is displayed in these crafts. Iron is found abundantly, especially magnetite, and the natives have always smelted and worked it, with but the simple appliances derived from their original Far-Eastern home. Gold is freely found.

Some of the tribes build their dwellings on the tops of lofty hills, for security, or in villages enclosed in thick hedges of prickly pear. The high wooden houses of the chiefs, in the towns, have been largely replaced by brick or stone since the European advent, but the forest and coast natives still make their dwellings of wood, covered in with the stalks of the "traveller's tree." The men wear a loin-cloth, the women a kind of long apron, and both sexes cover themselves with a garment "not unlike a Roman toga."

The people of Madagascar are described as often immoral, and cruelly disregardful of human life in war. But it is to be noted that women have a high place in the native life and regimen. The people are courteous to foreigners, courageous, and loyal to their tribes, affectionate in family life and respectful to the aged.

As to their "native" religion, there has been no native priesthood or temples, but there is a recognition of a Supreme Being, known as "The Fragrant One," or "The Creator," but witchcraft and fetish enter into their beliefs. The famous "poison-ordeal" of these people, under which thousands perished each year, was abolished under the Anglo-Malagasy Treaty of 1865.

Cattle-raising and rubber-collecting are among the industries, and fibre, bees'-wax and hats are exported. Gold-mining has been carried on since 1897, some of the mines being worked by the British, and gold to the value of some hundreds of thousands of pounds annually is exported. Various European factories have been established. Carriage roads connect the towns and seaports, used largely by bullock wagons.

Midway between the northern part of Madagascar and the mainland we encounter other French possessions : those of the Comoro Islands, the four largest of which cover an area of about 760 square miles, with a population of 82,000 ; and there are numerous coral islets in the group. In these fertile lands flourish forests of coco-palms ; and they produce sugar, cotton, coffee, vanilla, rice, maize, sweet potatoes, paw-paw and other fruits. Turtles, which abound on the coast, are exported ; there are numerous sheep and cattle ; sugar and rum-making, also saw-mills, give employment to the people, and sugar and vanilla are the chief articles of export.

Great Comoro, the largest of these islands, is about 38 miles long, and rears its dome-shaped volcano upwards for 8500 feet above the sea, seen for 100 miles away, as we approach from the south. The activity of this mountain from time to time has resulted in forming a lava-strewn desert plateau around it ; but it is clothed, save for its last 1500 feet, with dense vegetation. Maroni, the chief town, with about 2000 people, is the seat of French administration.

Anjuan, the next in size, also presents a series of richly wooded uplifts, whose culminating peak is 5000 feet high. The former capital, Mossamondu, is built of solid stone, with a surrounding wall and half-ruined citadel. Here dwells the Sultan and the governor. Mayotte, the third in size, is similarly broken in configuration, and Moheli, the fourth, is of less abrupt formation. The islands were discovered in the sixteenth century, and Arabs, driven thither by contrary winds, exercised, as petty slave-trading sultans, control for a long period. In 1841 a treaty was made ceding the domination of Mayotte to France, and later, fearing German aggression, the other islands were taken over.

The people are of Malagasy, Negro and Arab blood, generally Mohammedans, with perhaps 600 French and 200 British Indian traders. The climate is warm but by no means inimical to Europeans. The French navy has a coaling station here.

Chief of the Mascarene Archipelago, and lying 550 miles east of Madagascar, arises the beautiful island of Mauritius, the old *Ile de France*, taken, with its companion island, Rodriguez, by England, in 1810, when, under Napoleon, it was a centre from which privateers preyed on British vessels.

The importance of Mauritius formerly depended upon its transit trade, which the building of the Suez Canal superseded. At the present time its prosperity depends upon the sugar crop, and when the profits of this are annulled by hurricanes—for

M

Mauritius lies in one of the most terrific hurricane zones in the world—or by bounties or other fiscal matters, disaster befalls its dense population.

The island has a length of thirty-five miles and a width of some two-thirds this, and covers an area of 720 square miles. Three ranges of mountains traverse its surface, and this hilly character endows it with a varied and picturesque scenery. From the sea Mauritius spreads to the traveller, as it were, a bright green carpet, a peculiarly beautiful and striking aspect, due to the sugar-cane fields with which the plains are covered, for the island is practically a vast sugar-cane plantation.

Overlooking Port Louis, as we approach from the Indian Ocean, we remark the singular and lofty peak of the Pouce, so named from its fancied resemblance to a human thumb, a tall, bare, natural obelisk pointing upwards to a height of over 2600 feet above sea-level. St Louis is the capital and seat of Government, lying in an excellent harbour—one of the finest in the Indian Ocean—formed by an inlet on the north-western shore, and capable of accommodating the deepest vessels.

The pleasing appearance and natural advantages of the approach to Mauritius are not attended by similarly good conditions of health. Due to the low rise and fall of the tide the drainage of Port Louis is difficult, and the situation is not healthy, notwithstanding all that has been done in its amelioration. The population has declined from 70,000, twenty-five years ago, to considerably less, and the death-rate is given as over 44 per 1000. However, at Curepipe, the favourite residential town, twenty miles away in the high hills, reached by railway, the conditions are as pleasing as in the south of France. But elsewhere malaria is often deadly among all classes, and epidemics of other kinds occur. The climate is pleasant in the cool season but oppressively hot in summer, except in the hills. The severe cyclones at times destroy the houses.

The architecture of St Louis is marked by no particular style; the Protestant cathedral was formerly a powder magazine, and the Roman Catholic is described as of a pretentious but tawdry character. The town hall, royal college, theatre, public offices and other structures are among the principal buildings.

Mauritius is one of the most densely inhabited of countries, notwithstanding the absence of manufacturing industries; the population numbering about 400,000, with over 520 persons to the square mile. The people embody two main classes, the descendants of the French and English, with many half-castes, and that of their African slaves, and the 250,000 Indians, the greater part of which are native born. A French patois is generally spoken. The Creole or European people are distinguished for their culture, and although more numerous than in any other tropical colony, are diminishing, the Indians appearing instead.

The condition as regards over-population is regarded as dangerous,[1] especially in view of the dependence upon the sugar crop. The rapid increase of the Indians—37 per 1000—will necessitate some outlet elsewhere. The Indian coolies were brought in to work the sugar plantations after 1834, when slavery was abolished, but many abuses were committed. The tendency to an Indian peasant proprietorship in the land is marked. The Indo-Mauritius people are dominant in agriculture and commerce. All available land has been mapped out for sugar production, but there are fewer estates than formerly, due to the acquisition of small areas by Indians.

Sugar forms ninety-five per cent. of the exports from Mauritius. It has been greatly affected by beet-sugar competitions and bounties elsewhere. In 1878 the value of the export was nearly £3,500,000, but in 1900 it decreased to about £2,000,000, although a rise has taken place since. South Africa and India take the bulk of the exported sugar, followed by Britain and Canada.

The forests of Mauritius have been almost destroyed in clearing the land for sugar, but there are some useful timber trees, and the coco-palm has been extensively planted. The "traveller's tree" also flourishes. Many varieties of fruits are produced, including the mabolo, whose combined qualities of exquisite flavour and extremely disagreeable smell are remarkable. Tea, rice, fibre and other tropical products are raised. The Pamplemourres Royal Botanical Gardens have always been famous for their collection of plants.

Mauritius was the home of the dodo (*Didus ineptus*), but the unfortunate bird, from its ineptitude, being absolutely defenceless and wingless, soon became extinct from the attacks of the

[1] *Oxford Survey*, "Mauritius," Gardiner.

first colonists. The enormous tortoise, such as we shall remark in the Galapagos islands off the coast of Ecuador, existed formerly, and its fossils are found. Sea-slugs are exported by the Chinese of Mauritius to Singapore. Mauritius, like Madagascar, is a land " where silent bats in drowsy clusters cling," the fruit-eating bat being plentiful, and the only " mammal " originally found on this oceanic island.

The government of Mauritius is that of a Crown colony, with a governor and executive council. Good roads, railways and tramways exist throughout the island. There are about 7000 electors on the registry. The Roman Catholic faith predominates, Protestants numbering somewhat more than a twentieth part of the Catholics. The Indians are mostly Hindus, but there are over 30,000 Mohammedans. Government and church schools supply primary education. Mauritius occupies an important position strategically in the Empire, lying as it does between Africa and India, and its splendid harbour is of great importance.

Notwithstanding its economic difficulties, the cyclones, fevers and other visitations, Mauritius has succeeded in keeping up its importance as a centre of sugar production. The colony made a handsome gift to Britain early in the great European War, in 1915. A strong garrison is maintained in its forts, at a considerable cost to the Empire.

The Mascarene Islands were first discovered by the Portuguese sailor Mascarenhas, in 1505. Rodriguez Island, one of the largest of the group, lies over 300 miles east of Mauritius, and is eleven miles long, a mass of volcanic rock, like its sister island. A coral reef surrounds it, with a boat passage intervening, the common natural structure of such islands. It was discovered by the Portuguese Rodriguez in 1510, and the Dutch—with a colony of Huguenots—and the French both occupied it at various periods, the British dominion beginning in 1809.

This now parched and naked pile of rocks was, when the Europeans first landed upon it, beautified with forests of lofty trees, and was the home of myriads of dodos and other flightless birds, and of the giant tortoises. These are now extinct, only their bones remaining in the limestone caves. The cutting of the timber, fires and the introduced goats and pigs have all been factors in the destruction of the vegetation, whilst the defenceless dodos were killed off and the great tortoises killed or exported. The native deer were also hunted out of existence.

The physical conditions of the island are in some respects similar to those of Mauritius, but the climate is much healthier. Some parts are still well wooded, and there is good pasturage. Sugar-cane, cotton, coffee, rice, tobacco, the sweet potato, maize, ground-nuts and many kinds of fruits flourish. Fishing and cattle-rearing are the chief industries, and salt, fish, live stock and tobacco are exported.

The population numbers about 3000, mainly grouped in the districts of Port Mathurin and Gabriel, the principal settlements; the French Creoles of mixed or negro descent predominating. Rodriguez forms a station on the " All British " cable route between South Africa and Australia, the telegraph company maintaining a considerable staff of assistants. A magistrate appointed from Mauritius governs the community, and there are schools, hospitals and a Roman Catholic church.

The smaller dependent islands of these groups are often rich in coco-nut groves, and thus may be of more commercial value in the future.

The French island of Réunion, in this group, lies between Mauritius and Madagascar, 130 miles west of Port Louis. It is 45 miles long and covers an area of 965 square miles. Its earliest inhabitants were a dozen mutineers sent there from Madagascar in 1646.

The island has no natural harbour, and from the narrow coastal belt the hills extend upwards to where they culminate in striking basaltic massifs, with huge craters, flanked in places by vast masses of rocky debris, such as that which, towards the close of last century, slipped down and overwhelmed a neighbouring village. The naked crater of Piton des Neiges rises to over 10,000 feet above the sea, and its companion, Le Volcan, displays thirty miles of a mountain wall so remarkable in its regularity as to be unique in mountain structure.

Upon these great uplifts the mists hang all day, but clear off soon after sunset. Snow falls every year upon them, and ice forms upon their summits, and upon the windward side as much as 120 inches of rain annually are recorded. Terrific cyclones sweep fiercely over the island from time to time, and have devastated the sugar plantations during its history. The climate is generally healthy, but there is malaria on the coast.

Upon the maritime belt various towns and villages are situated, St Denis, the capital, lying on the north side, with a population of about 30,000. It contains some fine public buildings, and a pure water supply, but the harbour is an open roadstead. The botanical gardens are a noteworthy feature, and the town is the seat of a bishopric. Another town is St Pierre, with a population of about the same number, and St Louis and St Paul are smaller towns. These places are united by a railway, eighty-three miles long, and there are some good roads.

The Creole population of Réunion—or Bourbon, as the island was earlier termed—is descended from Norman and Breton immigrants, and mulattoes, negroes, Indians and Asiatics make up the total of perhaps 175,000 inhabitants. The French arrived in 1638 and annexed the island some years later. Slavery was abolished in 1849, but previously the colonists— after the French Revolution—resisted the desired abolition. The British attacked the island in 1810 and held it for five years. In 1860 coolies were imported to replace the slave labour, but owing to their ill treatment by the colonists the Government of India prohibited this recruitment. The population tends to decrease.

The white Creole people of the towns, who form the most typical upper-class inhabitants, are described as quick-witted and charming in their manners, brave, and very proud of their sea-girt home. We shall remark the astonishing wealth of flowers and shrubs, native or exotic, in the gardens of the coast towns ; plants brought from all parts of the world.

The cultivated area of the island reaches about 230 square miles, of which nearly half is under sugar-cane, with maize, coffee, cocoa, cloves, vanilla and other food products, and there is a little quinine. The beet-sugar competition and bounties of other lands very adversely affected the sugar industry, as did the scarcity of labour. Rum is largely made and exported, as is also the geranium essence, a product peculiar to this island.

The country is governed as a province of France, sending representatives to the French Chamber, and all inhabitants, of whatever shade of colour, have the franchise. A local governor and council conducts its affairs.

In the Seychelles Archipelago, scattered to the north and north-east of Madagascar—the largest nearly 1000 miles east of Zanzibar and about a similar distance from Mauritius—we approach veritable island-gems of the Indian Ocean. These lands of great beauty and fertility formed, in the opinion of General Gordon, the ancient Garden of Eden.

This outlying part of the British Empire includes the Amirante, Cosmoledo, Aldabra and other islands, in all some ninety islands under the Seychelles Government, with a total area of 156 square miles, apart from some 50,000 square miles of coral banks within the area. The islands are widely separated : the Aldabra group lying perhaps 600 miles from the Seychelles.

The largest island is Mahé, covering fifty-five square miles. Its hills rise in general abruptly from the sea to nearly 3000 feet, and Frigate, Curieure, Félicité, East Sister and North are the next in importance. The islands were discovered by the Portuguese in 1502. In 1742 the French explored them, and some cultivation was entered upon, but they were captured by Britain—by H.M.S. *Orpheus*—in 1794.

Mahé and Silhouette still display in places dense jungles and tropic rain-forests, with vast trees hung with trailers, orchids and ferns, the ground beneath being gay in their season with begonias and other brilliant flowers. The shores are covered with coco-nut palms, interspersed with vanilla plantations and coffee and clove gardens, and the coral reef of purest green with a white fringe of everlasting surf broken only by the channels which penetrate the barrier, surrounds the whole.

The principal harbour of Mahé, Port Victoria, lies amid such a scene, being approached by a deep coral-bound channel, giving access to an excellent harbour. The blocks of coral of which the walls of many of the houses are built glisten in the sun like white marble, affording a beautiful contrast with the many hues of feathery palms, and the blue sea and sky. The port is a naval and coaling station, and native sailing craft and Government steamers keep up the means of transport.

The population of the whole group is estimated at somewhat over 22,000, of which about 600 are white people, the remainder being largely descendants of slaves—introduced by planters from Mauritius—and freed slaves from East Africa. Large numbers of Indians and some Chinese also form part of the population, and the greater part of the valuable town property has been acquired by the former, and the retail trade by both. A rude Creole patois, of French with Indian, Bantu and English words is spoken.

These islands, in most cases, are of granite formation, and are regarded as the remains of the continental land supposed once to connect South Africa and India. Their flora and fauna include species peculiar to them, among the former being the singular Maldine double coco-nut or *coco de mer*, which, at first believed, by reason of the sea-borne specimen cast up, to be of submarine origin, was esteemed as an antidote to poison, fetching high prices in the East. The nut grows on a very tall palm, each nut weighing over forty pounds and taking seventy-two years for a generation. It is like an enormous double coco-nut in form.

The ancient fauna has disappeared, but the gigantic land tortoise—*Testudo elephantina*—is found on the Aldabra Islands, and nowhere else in the world except in the Galapagos Islands. The occurrence of this great tortoise has been advanced as an argument for the former existence of an inter-continental land.

The granite soil affords a luxuriant vegetation. Rubber, cloves, cinnamon, nutmegs, cotton, coffee, tobacco, sugar, maize, rice, bananas, yams, cocoa, vanilla and other tropical products are cultivated. The vanilla was earlier the principal article of export, and was famous, but copra and other coco-nut products, and rubber, have somewhat displaced it. Careless and slothful agricultural methods have been responsible at times for serious losses in trade. Banana flour is produced, and the *coco de mer* nuts are exported, as is also tortoise-shell. Cotton goods and hardware from Britain are among the principal imports. Guano is a product and export of considerable value.

Much of the land is divided into small holdings, and small islands often belong to single proprietors. French customs prevail, and most of the people are Roman Catholics. The climate is generally healthy, tempered by the sea breezes, and indeed for a tropic land the temperature is very moderate, and there is no fever. Good roads and riding tracks have been made in many directions.

The Government of the Seychelles group is that of a Crown colony, and its economic and strategic value may be of growing importance in the future. The chief port is a valuable naval centre, and the inhabitants are famed as good seamen. The affinities of the colony are largely with East Africa, and the two regions may prove of much value to each other.

Far to the east, in the centre of the Indian Ocean, lies the Chagos Archipelago, a group of coral atolls belonging to Britain and governed from Mauritius. These lonely islands, which cover an aggregate area of about 76 square miles, are disposed in circular form, and the coral barriers enclose what is one of the finest natural harbours in the world. Some of the islands have disappeared beneath the sea. On Diego Garcia, the largest, there are about 700 people. The only commercial product is coco-nut oil, and under the earlier French possession the Archipelago was known as the Oil Islands.

Leaving these remote sea-girt tropical isles, our survey takes us again to the African mainland.

CHAPTER XIX

In the great protectorates and colonies of German and British East Africa we approach perhaps the most interesting region of tropical Africa. Nature has here developed some of her most striking topographical features, and the great explorers, their imagination fired by the wonders of mighty equatorial lake and snowy mountain, concealed from the white man's knowledge until the middle of last century, made this coast their starting-point on adventurous journeys, which in their period excited the utmost interest.

Following upon these explorations the German and the Briton, in their respective spheres, set their hands to an industrial development which has resulted in the establishment, in a remarkably short time, of ordered towns and seaports, and lines of railways penetrating forest, plateau and scarped valley, rendering accessible the picturesque and savage wilds to the march of civilisation. Steamers plough the waters of the great equatorial lakes, whereon the first explorers looked relatively so short a time since.

The story of the acquisition by Germany of its great East African colony, a territory twice the size of Germany itself, has been described as one of the romances of the continent, and embodies an enterprising trick, which, however, appeals to the spirit of daring and adventure.

The area of the colony is about 364,000 square miles, and on its longest axis it measures nearly 1000 miles, reaching to within a degree of the Equator. It shelters a vast native population of over 8,000,000. The sea-coast, more than 400 miles long, is washed by the Indian Ocean, with the islands of Zanzibar and Pemba lying off shore. In the north the colony stretches to the Victoria Nyanza lake, with British East Africa forming the northern frontier. To the west lies Lake Tanganyika, one of the largest fresh-water lakes in the world,

and the Congo State, Rhodesia, Lake Nyasa and Nyasaland.
The southern frontier is that of Portuguese East Africa.
Almost upon the British frontier arises the giant peak of
Kilimanjaro, the highest mountain in Africa, sending down
a myriad streams from its glaciers and perpetual snows, to
swell the Pangani river.

The narrow coastal plain, extending inland in places for
thirty miles, is built up largely of coral, and is low and sandy
generally, but covered in places with fertile soil clothed with
the dense and sombre mangrove bush. The various harbours
give access to interesting towns, whether of Arab origin or
modern developments of commerce and administrative life.

Crossing the coastal zone we encounter the precipitous
eastern slopes of the interior plateau of Central Africa, from
which several high peaks arise, the general elevation of the
tableland being from 3000 to 4000 feet above sea-level, traversed
by the deep gorge of that depression known to geographers as
the "eastern rift-valley," and terminating on its western side
with the steep walls of the still vaster depression of the rift-
valley of the great Central African lakes, Tanganyika and
others.

Wherever the early Arabs established themselves upon the
coast in this part of Africa they made the desert fruitful with
the coco-palm and the mango, and with other vegetation,
giving beauty and variety to the landscape. The entrance to
the harbour of Dar-es-Salaam—whose name in Arabic signifies
the "Harbour of Peace," from its perfectly sheltered passage
—is clothed on either hand with palm-groves, and the scanty
ruins of the palace of a former sultan of Zanzibar, forming part
of an unfinished town which that potentate had planned,
stand on the shores of the bay, which opens from within the
narrow sea entrance.

Dar-es-Salaam is the capital of German East Africa, and was
built by the Germans in European style upon the site of an
insignificant village. The town has some large and handsome
buildings, including the governor's house, post office, hospital
and barracks, also various churches, and the schools of the
Government and the Missions. The public buildings face the
quay, overlooking the anchorage of the mail steamers from
Europe, and adjacent to the governor's residence stretch the

botanical gardens, where scientific experiment in the acclima-
tisation of European plants has been carried out.

This important centre of European control was laid out
by the Germans in the anticipation that an extensive trade
would follow their flag, and the head offices of the Deutsch-
Ostanfrikamsche Gesellschaft, the largest German trading
concern in the colony, owning enormous areas of land, are here
established. But commerce was of slow development. A railway
runs into the interior for about 180 miles to Mrogoro, near the
route over which the British began, in 1876, the construction
of a road designed to run to Victoria Nyanza, from what, at
Dar-es-Salaam, was meant to become an important seaport
—a project which, however, failed. A year afterwards the
Germans occupied the bay, but fighting with the Arabs resulted.
The railway recently completed from Dar-es-Salaam to Ujiji, on
the eastern shore of Lake Tanganyika, is an extremely important
route, likely to compete with the Congo systems as an outlet from
the heart of Africa. This line is 700 miles long and replaces
the caravan route to Lake Tanganyika, which took sixty days
to traverse.

A short distance to the north of the capital lies Bagamoyo,
a port but twenty-six miles distant from Zanzibar, whose coast
may be seen on a fine day therefrom. It possesses no natural
harbour. A large number of British Indians are included in
the population of 18,000 of this town and district, which is
Oriental in character. A mosque and Hindu temple are among
the principal buildings, which include the usual administrative
offices, barracks, schools and missions, the mission of the Sacred
Heart possessing a large coco-nut plantation of its own.
Bagamoyo was the starting-point in African explorations for
such famous travellers as Burton, Speke, Grant and Stanley.

The port of Tanga lies opposite the island of Pemba, near
the northern extremity of German East Africa, and possesses
the customary buildings and institutions of administrative
character. It is the centre of a thriving plantation district,
and a railway runs from it for over 100 miles inland. A
broad, deep channel leads to the harbour. Fighting was
carried on here during the British invasion, early in 1915.

On the southern part of the coast we enter the seaport of
Kilwa, with its ancient and modern towns, places of consider-

able historic and commercial interest. The modern Kilwa is on the mainland, the old on an island, separated by a narrow channel, with a deep harbour sheltered from every wind by coral reefs. Here we remark, amid the ruins of this old Arab stronghold, which may have been founded in A.D. 975, massive walls, bastions and forts, and the remains of a palace and of two large mosques, whose domed roofs still retain their form. To the south, upon a small adjacent island, are the ruins, buried in dense vegetation, of an ancient city, whose origin is unknown. Chinese coins and fragments of porcelain have been found on the beach.

It is recorded that over forty Arab sovereigns reigned at Kilwa through a period of five hundred years, and at the time when the first European explorer, the Portuguese Alvares Cabral, visited the place with his fleet, on the way to India, Kilwa was a large and wealthy city, with three hundred mosques. Vasco da Gama took the town in 1502, but its destruction resulted upon the subsequent fighting between Arabs and Portuguese, numbers of the latter dying of fever. Kilwa was later an important slave-trading centre.

The new town, eighteen miles to the north, is laid out in European style, with large and substantial houses, fortified Government and military buildings and fine public gardens, forming the outlet for a fertile and closely populated country, with a considerable trade, largely in the hands of Indians. From Kilwa runs the caravan route to Lake Nyasa. The population is mainly of Swahili people, and numbers about 5000.

Let us cast a brief glance at the story of German domination of this old Arab land. In the partition of Africa Germany came late, and it was largely due to British apathy that she obtained her foothold on the continent. But, by the close of 1884, she had established the foundation of a colonial empire in South-West Africa. Colonising activity was the result of private enterprise rather than of governmental initiative, for the general opinion and governmental disposition of Germany were opposed thereto. The " Society for German Colonisation," established early in 1884, was inspired by young and active men, and preparations were secretly made for the acquisition of rights on the East Coast of Africa, in the region under the influence of the Sultan of Zanzibar, and were largely designed to circumvent British influence. At this time Zanzibar was in semi-dependence upon India, and the influence of Britain predominated. Earlier in 1877, moreover, the Sultan of Zanzibar had offered a lease to a British representative,[1] of all his mainland territory, but unfortunately British enterprise had not awakened to its value, and this was refused.

The method followed by the German adventurers was daring. A few days before the Berlin Conference of November 1884, three members of the Colonising Society, disguised as mechanics and equipped with German flags and " a supply of blank treaty forms," landed at Zanzibar, and although their enterprise had been discountenanced by the German Government, the adventurers crossed to the mainland, held a palaver with the native chief at Mbuzini, signed a " treaty " with him, and raised the German flag for the first time upon the east coast of Africa. This act was carried out practically at the moment when the diplomatists at Berlin were " solemnly discussing the rules which were to govern the game of African partition."

In 1890, the important treaty between Great Britain and Germany, defining the respective spheres in Africa, was signed, providing, among other things, for a British protectorate at Zanzibar and the cession by Britain to Germany of Heligoland. The Sultan of Zanzibar, a little later, ceded to Germany the mainland territories that had formerly been leased to a German company, receiving £200,000 as consideration.

It was in 1888 that the German East African Company had taken up this lease, but the coast Arabs immediately revolted against the Teutonic administration, great hostility being aroused by the rude and clumsy treatment of the usurped rulers, and it required a naval and military expedition to crush them. The colony was then taken over by the Imperial Government, company control being thus superseded. Conciliatory measures were adopted, and the development of the territory was carried on.

Within a few years, however, further trouble arose with some of the native tribes, and in 1891 a German force was annihilated thereby. Very harsh treatment was then employed

[1] Mackinnon.

by the Germans, including the execution of women.[1] The administrator[2] responsible for these barbarous methods was censured by the Berlin authorities, and displaced, and again for some years the development of the colony followed peaceably, and government was carried on as far as possible through those native chiefs able to maintain any organisation. But Berlin grudged the expenditure of money for railway-building.

In 1905 the Bantu negroes revolted, largely as a result of resentment against the methods of German " kultur," which included compulsory work on the plantations, and it is stated that the Hereros of South-West Africa had been partly instrumental in instigating the revolt. A number of priests, missionaries and nuns were murdered by the natives near Kilwa. The Germans brought in an army of Masai and Soudanese, and natives from New Guinea, with naval and military contingents from home, to crush this uprising. The official returns of deaths, before this conflict was over, was given as 120,000, men, women and children. It was not until the Imperial colonial secretary[3] visited the colony, in 1907, that a more humane treatment of the natives was adopted, and a more vigorous economic development of its resources entered upon.

The territory thus controlled by Germany now shows many instances of order and growth, and its seaports, in some cases, compare favourably with the somewhat neglected, if picturesque, appearance of Mombassa in British East Africa. The region is in general well watered, but except for the Rujiji, the large river which intersects the colony midway, across its eastern slope, flowing from the plateau to the Indian Ocean, the waterways with small exceptions are not navigable. This river we may ascend in small steamers as far as the Pangani Falls, and the Ulanga, its tributary from the south-west, is also navigable through a great part of its course. This fluvial system traverses, in general, broad, level plains, which are inundated in the rainy season.

It was upon this river that the German cruiser *Königsberg* strove to conceal itself from British cruisers in the Great War,

---

[1] A local example of German " frightfulness."
[2] Carl Peters.
[3] Herr Dernberg.

but was blocked in and at length destroyed by British monitors.[1]
The vessel had been covered with the branches of trees, and so
lay almost unseen amid the tropic jungle, but its position was
located by aircraft.

The Pangani river, whose source is amid the snows of Kilimanjaro, enters the Indian Ocean upon the coast-line between
the isles of Pemba and Zanzibar, and at its mouth is the port
of the same name, the outlet for a rich tropical district, carrying on a thriving trade with the islands. The Rovuma river
forms, in part, the southern boundary with Mozambique, the
frontier having been deviated so as to allow its mouth to lie
within the German sphere.

Somewhat to the north of this we shall have remarked the
port of Lindi, picturesquely situated upon its bay. Formerly
an Arab slave centre, the trade is now mainly in ivory. A fine
sculptured Arab gateway has been preserved here by the
Germans.

The climate of the littoral is greatly influenced by the warm
currents from the Indian Ocean, which bring both heat and
moisture, and malaria is often dangerous. As we ascend to the
plateau, however, the region becomes colder. The precipitous
slopes intercept the moisture-laden winds, which deposit their
burden thereon, and on the higher parts of the plateau the
nights are cold. Grassland covers the plateau, and the steppe
is often congested with dense and sometimes impenetrable
mimosa bush, and on Mounts Kilimanjaro and Meru we remark,
well exhibited, the vertical zones of climate and vegetation.

Dense belts of forest line the river banks, yielding useful
timber and varieties of gum-producing trees, and the broken
face of the precipitous slopes, which lead upwards to the tableland and look towards the rain-drenched monsoon, are clothed
with primeval forests, also containing very valuable timber. It
is in this belt that the best rubber-yielding bush, the *Landolphia
Florida*, is encountered, as also the silk-cotton tree, *Bombax
Ceiba*, and other useful or beautiful tree forms.

In their various districts herds of antelopes are features of
the tropic landscape, and the hippopotamus, rhinoceros,
chimpanzee, giraffe, ostrich, buffalo and zebra are encountered
in their respective haunts, whilst the lion and the leopard range

[1] In July, 1915.

throughout the whole country. Crocodiles abound in the rivers, venomous snakes are plentiful, as are land and water turtles, the tsetse fly infests certain districts, and locusts are a further source of trouble to the agriculturist.

Various large native towns are found in the plateau region, notable among them being Tabora, with a population of perhaps 40,000, and here the trade routes from Lake Tanganyika, Lake Victoria Nyanza and the coast meet. The railway now serves this place, which will become an important railway centre, as other lines crossing the country in opposite directions have been projected.

The Usambara highlands have become an important agricultural region, private land companies and individuals having set out extensive plantations therein. The railway from Tanga reaches Korogwe, the chief town of the district, lying upon the Pangani river. Large numbers of cattle, sheep and goats are raised in these plateau districts.

Upon the lakes German activity has established various settlements, or contemplated the development of native centres existing on their coasts. The southern half of the Victoria Nyanza belongs to the colony, and wholly or in part the eastern shores of Lakes Kivu, Tanganyika and Nyasa, deep down in the great rift-valley.

Mwanza, on the southern shore of Nyanza, Bukoba, on the western shore, and Schirati, on the eastern, are places destined for further development. The Victoria Nyanza is an important feature, topographical and economic, of German East Africa, and Government steamers are maintained on all the lakes, making connection with the lake terminus of the Uganda Railway. The trade of the Victoria lake district was largely taken by this line, whose building gave rise to an economic crisis in the German colony. The projected line from Kilwa to Lake Nyasa, in the south, would, however, give the most rapid means of communication between the coast and that part of British Central Africa and the Congo State.

Upon the long German coast of Lake Tanganyika, Ujiji, Usumbura and Bismarckburg are the principal stations. The first-named has a population of some 15,000, and is a meeting point for traders over a wide area, and the terminus of the Dar-es-Salaam railway route. Formerly it was a great centre

N

for the trade in slaves and ivory, and the headquarters of the Zanzibar Arabs, who had penetrated there. Burton, Speke, Livingstone and Stanley were the first Europeans to behold the bosom of Lake Tanganyika, viewing it from Ujiji, which thus was strongly associated with British enterprise, although it came within German control in 1890. There are, upon the lake shores, four stations of the Algerian "White Fathers," with stone-built churches, schools and other edifices.

The people of the East African seaboard are mainly the Swahili. They come of long crossing between the negro and the Arab and involve every hue and stature from the pure negro to the Semite. The Swahili are generally a powerfully built people, often handsome and broad, with Semitic features, and their native intelligence and activity have caused them to take a prominent part in the development of commerce and industry and in the exploration of nineteenth-century Africa. Even south of the Equator their widespread language became the principal medium of speech, and the great explorers profited largely by this circumstance on their expeditions from the coast into the interior. The Swahili language is a mixture of Bantu and Arabic, with Indian, Persian, English, Portuguese and German words or phrases interspersed. The Swahili people, whom we shall encounter in British East Africa and Zanzibar, number over 1,000,000, and are Mohammedans, although negro in disposition. Portions of the Bible have been translated into their language, but missionary work has made little progress among them.

The indigenous Bantu tribes of the interior of German East Africa are generally peaceful agriculturists, but some of them have the martial character of the Zulus, of whom there has been some infusion. They worship spirits, but have a vague belief in a Supreme Being—held by this people elsewhere, a Being whom they regard as being "indifferent" to the lot of their race. The Masai, possibly the handsomest and finest of negro races in Africa, whom we also encounter in British East Africa, are wandering cattle-breeders, warlike, living in square, mud-covered houses, which they fortify at times. A powerful negro state was formed at one time by another native people—the Bahima—on the shores of the Victoria Nyanza.

Both the Government and the missions of German East

SWAHILI WOMEN HAIRDRESSING, BRITISH EAST AFRICA

Africa—Protestant and Roman Catholic—are engaged in the civilising of the natives, but the number of Christians among these is small, and the Mohammedans are vigorous in their proselytising. The slave trade has been abolished, but domestic slavery is still permitted, although the children of slaves are now automatically freed. A system of compulsory labour is maintained by the Germans in connection with public works. The authorities have endeavoured to teach the natives the German language, partly with the purpose of fitting those with adaptability for small administrative posts, and native chiefs take some part in the dispensing of justice.

Among the native industries we remark the numerous small smithies, where iron is worked, a useful indigenous craft so commonly found throughout tropical Africa. A number of minerals are found in the colony, such as the iron deposits of the Victoria Nyanza, where mining began some years ago, and gold, coal and copper are also found, with precious stones of various kinds.

But the principal source of wealth is agricultural and forestal. Coffee, tobacco, sugar-cane, cotton, vanilla, ground-nuts, coco-nuts and great quantities of bananas, with most of the fruits, cereals and vegetables common to tropical Africa, are raised. Sisal hemp, fibre, rubber, skins, ivory, tortoise-shell, pearls, copra, coffee, ground-nuts and cotton are all exported. There are flourishing trading stations in nearly all the seaports, and a number of companies are engaged in trading, planting and mining. The modern growing of cotton was begun in 1901. The Government has reserved to itself large forest preserves, and the plantations are worked entirely by native labour.

The various administrative districts into which the colony is divided are controlled by chiefs, and upon the district councils white merchants and planters are represented. The revenue, derived from custom and other dues, and from the hut tax, has not, so far, sufficed to meet the expenditure of the colony. The foreign trade is valued at about £2,000,000 per annum.

CHAPTER XX

THE British possessions, which, broadly, are grouped together under the general name of British East Africa, include the British East Africa colony or protectorate, Uganda, and the islands of Zanzibar and Pemba.

We here enter a region of striking interest and variety. The scenery, natural resources, topographical structure, teeming wild animal life, and possibilities for settlement and economic development by the white man which are here displayed are such as render this portion of the continent of peculiar value and attractiveness. The picturesque Oriental life of the old Arab towns is, moreover, an added source of interest to the traveller.

The protectorate of British East Africa covers an area of 240,000 square miles, with a seaboard on the Indian Ocean 400 miles long; and Uganda, on its western side, widely separated from the sea, but with a coast-line on the Victoria Nyanza and other lakes, covers about 110,000 square miles. These two regions are cut by the Equator. Zanzibar and Pemba, as we have seen, lie off the coast of German East Africa.

The country is bounded on the north by the old kingdom of Abyssinia, and Uganda by the Anglo-Egyptian Soudan. Both have as their southern neighbour German East Africa, and to the west of Uganda lies part of the Anglo-Egyptian Soudan—the Lado Enclave—and the Belgian Congo.

The topographical formation and scenery of British East Africa are to a large extent well revealed to the traveller upon the Uganda railway, running from Mombassa to Lake Victoria Nyanza, traversing in succession the coastal plain, the plateau, the great "rift valley" and its escarpments, and descending to the lake, whence the steamer is taken for Uganda. The way through the East African uplands is very picturesque. Standing against the blue of the tropical sky the snowy summit of Kilimanjaro arises, and herds of antelopes and zebras graze peacefully as the train traverses the great game preserve, passing which the park-like country, as of England, is reached, between Nairobi and Kikuyu. Suddenly the line drops, in almost amazing fashion, into the great "rift valley," ascends thence through the cedar forests of the main plateau, and once more descends sharply to the lake, a total distance of 584 miles from Mombassa.

The old seaport of Mombassa, the gateway to British East Africa, is built upon a coralline island, divided by a channel from the mainland, and as we approach presents a picturesque appearance. The town has an Oriental aspect, and its maze of narrow, irregular streets shelters a cosmopolitan population of some 30,000—Arabs, Indians, negroes and Europeans. The Anglican, Roman Catholic, Mohammedan, Parsee, Hindu and other temples bear witness to the varied faiths and religions of the inhabitants. The Dutch fort, built in 1593, stands up boldly from the coral hill, and forms a landmark from the sea.

The history of Mombassa is an interesting one. Its name is taken from Oman in Arabia, and a Perso-Arabic settlement was made on the island in the eleventh century. After the advent of Vasco da Gama, in 1498, Mombassa was subjected on various occasions throughout the centuries to battles, burnings and massacres. In 1823 the city was placed by its ruler under British protection, but Britain repudiated this, and left it to Arab bombardment. Later General Gordon and Ismail Pasha wished to annex it and the vast region now forming the East African protectorates, to Egypt, but Britain prohibited the transfer. However, in 1887, the Sultan of Zanzibar handed Mombassa over to British administration. Mombassa appears in *Paradise Lost*, as do certain other places on this coast.

The port is but a small one, not accessible to large vessels, and it is to Kilindini, a magnificent harbour on the south-west side of the island, connected therewith by rail and tramways, that we look for the principal commercial movement. The name of Kilindini, in the native tongue, means "deep waters," and here the largest steamers find accommodation.

The main street and Government square of Mombassa are not of the general Oriental character of the place, but contain the Government and other modern official buildings. There is a bronze statue to Sir William Mackinnon, to whom Mombassa owes its renaissance, and whose activities on this coast were so noteworthy. The European suburb overlooks the sea, and the climate is fairly healthy for Europeans. Mombassa is no longer the capital of British East Africa, the seat of Government having been removed to Nairobi, in 1907.

The Uganda Railway, which conducts us from Mombassa westwards, is a work which may be included among the romances of engineering. Its scenic value is very marked, and its economic effect has been of the utmost importance. By doing away with the carriage of goods on men's backs, into and from the interior to the coast, the railway gave the death-blow to the slave trade in this portion of East Africa, and rendered possible the development of land suitable to the white settler, firmly establishing at the same time the British position in Uganda, matters which before its advent were impossible. A prosperous colony has been won from what, in effect, was formerly a desert, by the building of this railway.

A bridge half-a-mile long connects the terminus at Mombassa with the mainland, and this past, the line ascends to the plateau, reaching an elevation of 7600 feet above sea-level, where the line crosses the Kikuyu escarpment, on the edge of the rift valley. Thence it descends by curve and viaduct to the valley floor, and again rises to the main escarpment at over 8300 feet elevation, its highest point. The cost of the line was £5,330,000, equal to about £9500 per mile, the work having been carried out by the British Government. The first locomotive reached the lake at the close of 1901.

The railway yields a fair return on the capital expended, and being therefore a financial as well as an administrative success, has negatived the prophecies of those who insistently regarded its construction as an ill-judged enterprise.[1] The building of the Uganda railway soon revealed the valuable fact that a "white man's country" was being passed through, for, as the high plateaux were reached, it was discovered that there were large areas of land, very thinly populated by natives, where an excellent climate was enjoyed, and where the conditions were favourable for white colonisation.

Applications for land were soon made by syndicates—among

[1] Especially in the Parliamentary debates at the time the project was brought forward, where the customary lack of imagination was displayed.

BRITISH EAST AFRICA. A PARAMOUNT CHIEF AND HIS SIXTEEN WIVES

them the East Africa Syndicate—some of whose financial magnates were members of the Chartered Company of South Africa, and this company negotiated for 500 miles. There was also a Zionist settlement scheme. In 1903, however, the decision was arrived at not to grant further large areas of lands to syndicates, partly in view of the arrival of numbers of prospective settlers, and of the existence of certain native pasturage rights of the Masai people. Some acute questions arose in the matter of granting land to syndicates, but by the close of 1905 more than 1,000,000 acres had thus been disposed of. However, a Colonists' Association was formed, in the interests of the white settlers. A plan was also drawn up for the settlement of British Indians adjacent to the coast and the Victoria Nyanza.

The distance from Mombassa to Nairobi, the capital, is 327 miles, and thence to the Port Florence terminus on Victoria Nyanza, 257 miles. The town commands magnificent views of Mount Kilimanjaro in one direction and Mount Kenya in the other, and lies at the foot of the Kikuyu hills, at an elevation of 5450 feet above sea-level. There are European, Indian and native quarters, and although the site was selected for a town and the first buildings erected only in 1899, a population of 5000 has come to being, forming the head of a progressive highland region and an important administrative and social centre, where, but a few years ago, the lion was monarch of the hills.

Thus a considerable part of the country consists of high plateaux between 5000 and 8000 feet above sea-level, with a healthy climate suited to the settlement of Europeans, although lying upon the Equator. On these splendid uplands fever is almost unknown. The average temperature is about 66° F. in the cool season, and 73° in the hot season. Frosts are experienced as we ascend to 7000 feet. The average rainfall in the highlands is between forty and fifty inches.

The coast region naturally is hot, but in general is more healthy than the seaboard of other tropical countries, due to the prevailing breeze blowing from the Indian Ocean, and to the dryness of the soil. The rainfall on the coast is about thirty-five inches.

The wide level upland plain, known as the Nyika, extending

for a considerable distance, is formed mainly of quartz, and contains large waterless areas, of which the Taru desert is typical. Thence the grassy uplands extend to the eastern edge of the rift valley, varied with cultivated ground and forest. Here we encounter the highest elevations in the territory, culminating in the mighty Kenya, almost on the Equator, and the Saltima range. Continuing westwards, we reach the edge of the rift valley, with its line of cliffs, forming the Kikuyu and Laikipia escarpments. On the west of the valley the main escarpment reaches 8000 to 9000 feet in elevation, cutting the Equator, its crest covered with dense forest.

On the northern boundary of the protectorate, the frontier of Abyssinia, arises the steep Goro escarpment. Westward of this lies the large lake, Rudolf, and south is an arid, desolate region strewn with great lava stones, which in its turn gives place to uplands affording pasturage for thousands of camels. The country bordering upon the Victoria Nyanza is typically tropical, with a rainfall of over sixty inches, and a climate generally unsuited to Europeans. In the plateau region at times severe droughts occur.

The two summits of Mount Kilimanjaro's great volcanic domes, the highest of which reaches 19,200 feet, are, as elsewhere remarked, in German territory, the boundary crossing the base of the mountain. The British flag was the first to be hoisted at the summit of Kilimanjaro. The second highest peak in Africa, Kenya, reaching 17,007 feet, lies almost in the centre of the protectorate, and some of the great escarpments and extinct volcanoes reach very high elevations.

The population of British East Africa is estimated at between 2,000,000 and 4,000,000, of which some 25,000 are Indians and about 3000 Europeans, among whom are emigrants from South Africa, with hundreds of Boer families. The white folk are generally agriculturists, and, in addition to the Kikuyu uplands, have settled mainly around Kenya and in the rift valley. The African people are of various stocks—a borderland between the negro and the Hamites. In the south the Bantus are largely represented, and a naked race dwells by the northeast shores of the Victoria Nyanza. The Masai of the uplands are a race of cattle-rearers. They long inspired the semi-Arabs and negroes of the coast with fear, and thus prevented the exploration of the hinterland, although they later became the allies of the white man. Between the Tana and Juba rivers are Somalis. There are also scattered hunting tribes of small stature. The clothing of the Swahili is notable for its finely patterned robes, of native design. These now, however, are largely made in Manchester—a usurpation of native industry. On the coast we find the Swahili and the Arabs.

The settlers in the uplands have largely developed the growing of vegetables, and considerable crops of potatoes are raised, also wheat, oats, barley and coffee, in their respective zones. Large flocks and herds are possessed by the white settlers, the uplands being excellently adapted for breeding. Merino sheep have been brought in from Australia, and ostrich farms have been established. The soil is in general easily cultivated, and the native tribes are not, in most cases, averse from work thereon, although the number available for employment by the whites is limited, and in some of the most favourable districts there is no native population at all.

In the " tropical " districts—the coast and the Victoria Nyanza —exist numbers of plantations owned by British people, but cultivation is mostly in the hands of natives or Indian immigrants. Maize, rice and other cereals are largely produced, also cotton and tobacco, and from the coco-nut plantation a fine quality of copra is exported. Sugar-cane is cultivated by the natives, and the collection of rubber from the valuable *Landolphia* rubber-vines, found near the coast and in the forests, gives employment to numbers of people. A profitable trade is done in the beans of the castor-oil plant, which flourishes in abundance inland. Timber and fibre-producing plants are a further source of profit. The weaving of cloth and the making of mats and baskets are native industries.

On the Mau escarpment we enter vast forests of olive and juniper, and on that of Kikuyu the cotton, fig and bamboo are characteristic of the vegetation, amid dense forests of enormous trees whose lowest branches are at times fifty feet from the ground.

The vertical zones of vegetation are very marked, the short grass upon the treeless areas, up to 3000 feet, and the palm-fringed river banks, give place to the varied tree and shrub life which flourishes up to 8000 feet, with bamboos and yews above, whilst at 10,000 feet we enter upon an Alpine flora, with giant groundsels, and tree-heaths fifty feet high. Tree-ferns clothe the sides of the mountains somewhat below this level, and the vegetation is often festooned with the graceful orchilla lichen. Much lower down, towards the swamps, the papyrus rush is found, covering extensive areas, and white lilies and the beautiful blue lotus water-lilies adorn the quiet waters of the lagoons. In almost every month the landscape is gay with bright flowering plants, save on the arid uplands of the north, and flowering shrubs are ablaze with their bloom, above the carpets of ground orchids and gladioli.[1] In the more arid districts below 3000 feet, are huge baobab-trees and dense acacias, wild figs, with wild dates in the water-courses, and valuable fibre-bearing plants and curious fruits.

Here in brief is all the variety and charm of tropic plant life, rendered profuse in flower partly from the absence of a marked wet and dry season, due to the situation within the equatorial zone, and to the varying elevations of the land.

Not only do we encounter a richness in the vegetation, but the animal life which has its home here is of the most plentiful and varied, or at least as regards the large mammals. Nearly every African kind of antelope is found, and herds of buffaloes, elands and giraffes still linger, with the elephant, which is being driven more and more to unfrequented regions. The rhinoceros is very abundant and dangerous, and hippopotami are common. Elephants with enormous tusks are encountered, and the finest ivory in the world comes from this great region of British East Africa.

In the rivers, and in the Victoria Nyanza, crocodiles are common, and the deadly puff-adder is prominent among the relatively few snakes. It would appear that mosquitoes, as also the other small noxious vermin, are less abundant than in other tropical lands. Bees are very numerous in some districts. Cranes, pelicans and flamingos haunt the banks of lake and river, in their thousands, and many other interesting birds are common to the territory.

The lion, leopard and hyena are especially famous inhabitants of British East Africa and Nyanza, and the depredations

[1] *Oxford Survey of the British Empire*, "British East Africa," Sir H. Johnston.

NATIVES OF UGANDA MAKING A REED FENCE

of the first-named, during the building of the Uganda railway, might seem matters of romance rather than actual fact. It is recorded that lions frequently carried off the workmen, and work was delayed from this cause.

However, the superb development of savage animal life is a resource of much value to the colony, and this is being realised, by degrees, before too late, and before the enthusiastic but destructive big-game hunter has hunted the great carnivoræ and others to destruction.[1] The remarkable zoology of the protectorate is a source of considerable attraction for travellers and tourists, and more remains to be done in its preservation. The presence of these animals in the wilder regions does not prejudice agriculture, and the locomotive has overcome the difficulties that attended animal transport during its construction, when lions devoured the donkeys and mules, and the tsetse fly destroyed cattle and horses.

The conditions for settlers in British East Africa are not without attractions. There is no free land, but land is offered by the Government under public auction, the price naturally varying according to local conditions. The cost of living is held to be cheaper than in any other part of Africa. The unskilled native labourers will work for ten shillings a month, including food. Labour, however, is not always forthcoming, except in the case of native stock tenders, which are very numerous, and their labour is described as the cheapest and best in the world. In the highlands a lower rate of six shillings and fourpence to ten shillings is paid, and if working near their own homes and allowed to return at night, the men are not given rations, but if they come from considerable distances it is customary to provide them with beans or other similar rations, which cost from two shillings and eightpence upwards per month. Native " boys " for domestic service earn from thirteen shillings upwards per month.[2] The natives are in general of an intelligent nature, and well disposed. Such low rates of pay for native

[1] " Why it should be a matter of national pride, displayed in a yearly Blue book, that from 200 to 300 rhinoceroses have been slain annually is difficult to understand."—Sir H. Johnston, in " British East Africa," *Oxford Survey.*

[2] The above particulars of wages, etc., are from the official pamphlet issued by the Uganda railway.

labour are doubtless valuable for the employer, but as to whether the native can sufficiently support himself thereon does not appear to be the subject of equal information.

The native cattle are small, hardy and yield a limited supply of milk, which is, however, rich in quality. The breed is susceptible of great improvement, but the native owner, whilst expert in feeding and watering, makes no effort to improve the milking capacity, nor the hygienic conditions of the calves. The supply of native cattle for market, though very considerable, is insufficient for the demand, as the native owner prefers not to sell, his cattle being his idea of wealth, of which he probably has more in proportion to the population than in any other country in the world. Some improvement in cross-breeding has been made by the white settlers. Cattle land may be bought at from five shillings to two pounds per acre, according to situation. The sheep industry has been heavily backed by British capital. The native sheep are without wool.

One of the drawbacks to ostrich-farming is the desire of the native to have the feathers for adornment in his own tribal customs, also the destruction of the birds by predatory animals, but the business has claimed considerable attention.

The cultivation of coffee is an important industry of the colony, and may be considerably extended. Sisal-hemp production also offers a good future. Wheat-growing is now a well-established industry, and maize is grown by all native tribes. The experiments in cotton-growing have made good progress.

From British East Africa it is said that the largest elephant tusks ever known since the extinction of the ancient species have come, and the best ivory in the world has been that from the region of Lake Rudolf and the Mountain Nile, the supply finding its way to Mombassa.

In the grandly diversified surface of the Uganda protectorate, with its lofty tablelands, dense forests, snow-capped mountains, beautiful lakes and desolate and arid valleys, we approach that wild region which contains the birthplace of the Nile, described in treating of the Soudan, which joins with Uganda in preserving that mighty stream to British control.

The western part is overlooked by Mount Ruwenzori, within

the Congo border, whose cooling influence pleasingly affects what is otherwise a hot and unhealthy climate on that side and in Baganda. The Rudolf region, terrible in its aridity, borders upon British East Africa. But the eastern part of Uganda is abundantly watered, especially near the Victoria Nyanza and the "noble Debasien mountain," and Mount Elgon, the enormous mass of whose giant crater—ten miles across, with a rim 14,000 feet high—throws out terraces and buttresses down which pour innumerable cascades through the dense forests which clothe the mountain-sides. On the north coast of the Victoria Nyanza we remark the Ripon Falls, descending into the "exquisitely beautiful Napoleon gulf," whence sallies the fully born Nile on its long course through the Soudan to Egypt. The Uganda protectorate is markedly a land of lakes—Victoria Nyanza, Rudolf, Albert (or Edward) Nyanza and others.

The affinities of native races, fauna, flora and climate are largely with West or Central Africa rather than with East Africa, and to mention them would but be to give a list of African quadrupeds and other creatures, and plants. The oil-palm, that valuable product of West Africa, is, however, not found in East Africa.

The protectorate shelters over 3,500,000 inhabitants, of which a few thousands are Indians and Arabs, and a few hundreds British and other Europeans.

Among the negroes the remarkable Baganda people—from whom, by a corruption, the name of the country is taken—call primarily for notice. They had attained to an indigenous civilisation greater than that of any other negro people, unaided by Arab or white man, and their line of monarchs has been traced as far back as A.D. 1400. They are now mostly Christianised, but were "always scrupulously clothed," at first in bark-cloth, but now in calico, and they have an extreme regard for decency, if not for morality. The kingdom of Baganda, ruled by "His Highness the Kabaka," has a thoroughly efficient and recognised native Government, with royal revenue, pensions for the royal family, and a "royal salute of eleven guns."

Agriculture is generally in the hands of the natives, and Baganda chiefs have developed cotton, cocoa and rubber plantations, assisted by the Government Botanical Department. A few plantations are owned by white men, and sugar and coffee are grown. Cotton and cotton seed are increasingly valuable exports, as are ground-nuts and rubber, also ivory.

Steamship service is maintained on the Victoria Nyanza from Port Florence, terminus of the Uganda railway, and with German territory bordering thereon. There are also steamers on the other lakes and navigable rivers ; the Albert Nyanza and the Mountain Nile ; and a railway line runs towards the navigable Nile, upon which stands Gondokoro, once famous for its ivory and slaves, on the border of the Soudan.

Entebbe—"the throne," as the native tongue has it—standing on the Victoria Nyanza shore, is the capital of Uganda, and the town contains well-built official edifices, hospital and churches, and is protected from the waves of the great lake by a breakwater. Twenty miles therefrom is the native Baganda capital of Mengo, with a population of some 70,000, stragglingly built upon seven hills. One of these, Namirembe, or the " Hill of Peace," was struck by lightning a few years ago, and the cathedral standing thereon destroyed. The building was a fine Gothic structure of brick, and was immediately rebuilt. Wadelai and Nimule are other important places, as are Jinja, Hoima, Butiaba and Mbarara, and there are others of less note.

Returning to the coast of British East Africa, we may remark various places of interest, such as the ancient Arab centre of Lamu, buried in the sands, and Melindi, of *Paradise Lost*, where stands Vasco da Gama's pillar, erected in 1498. Opposite Mombassa is the settlement of freed slaves founded by Sir Bartle Frere, and the headquarters of the Church Missionary Society in East Africa, which here is extremely active.

British East Africa and Uganda are animated by the British Imperial spirit, and would seem to be set in their ways towards that enlightened development which follows thereon.

Turning to the south-east, we look towards Zanzibar, the island protectorate of Britain.

Zanzibar and its companion, the " spice island " of Pemba, are singularly interesting portions of the British African dominion, from the point of view of the traveller on whatever purpose bent. The geographical position of Zanzibar, the historic coral island, set twenty miles off the East African

A Native Market in Uganda

coast, and visible therefrom on a clear day, has rendered it of vast importance—the " key " of East Africa from as far south as Delagoa Bay. An old Arab proverb expresses this by saying: " When you play on the flute at Zanzibar all Africa, as far as the lakes, begins to dance."

Here we find an ancient Arab dominion for whose possession many people have striven, an island dowered with some of the most delightful of tropical attributes, and which was the scene of the earliest and latest traffic in negro slaves, a traffic which was struck down by the gallant barques of the liberty-loving British nation, to whose control the isle has fallen, and under whose protection it rests in present security.

The area of Zanzibar and Pemba—the former being almost twice the size of the latter—is 1020 square miles, Zanzibar being nearly fifty miles long. Some adjacent islands also form part of the sultanate and protectorate.

The seaport and capital, Zanzibar, presents, as the steamer enters the roadstead, a scene of constant animation, due to the presence of foreign steamships, men-of-war, Arab dhows and other craft, and as we raise our eyes to the town the white walls of the flat-roofed houses, the quaint mosques, the forts, round towers and other buildings stand out picturesquely, and the " stone ship," into which form is cleverly carved the front of the now disused water-tanks, immediately arrests the attention.

Like many another picturesque Oriental town, Zanzibar embodies in part a labyrinth of dirty, narrow streets and alleys, and here dwell Indians—Banyans and Singalese—negro porters, fishermen, half-castes and outcasts. The Shangani quarter, as one portion of the town is termed, contains the Sultan's residence, the homes of the chief European and Arab merchants, and the Protestant and Roman Catholic missions. The massive, carved doors of black wood of the buildings, which face the street, contrast strongly with the whiteness of the stone walls, to which effect the brilliant red of the acacias in the gardens adds.

Imposing buildings, among them the British Agency in its beautiful grounds, the post office, custom house, English club, cathedrals—the Roman Catholic being one of the finest structures in East Africa—the law courts and the consulates, attest the importance of the place. Most of these institutions he along Main Street, leading from the Sultan's palace to the British Agency. The home of Tippoo Tib stands adjacent. A public park, golf course and good water supply are among the modern amenities of the town, whose aspect has changed considerably since the British occupation.

Under the protectorate slavery has disappeared, with the large old slave market, and sanitary methods are being enforced in the regimen of the town. Before the protectorate was established Zanzibar was described as " a cesspool of wickedness Oriental in its appearance, Mohammedan in its religion, Arabian in its morals, and a fit capital for a Dark Continent." [1] It was, however, at least the starting-point for the exploration of the Dark Continent, whence the civilising streams of explorer, pioneer and missionary took their way to the opposite mainland.

The population of the two islands of Zanzibar and Pemba is over 200,000, two-thirds of which are negroes, with about 11,000 Arabs and over 15,000 Indians—Banyans, Parsees and others. Christian Goanese or Portuguese half-castes, some Madagascar natives, Swahilis and others make up the motley population, whose origin indicates the commercial activity of Zanzibar, the town and port embodying perhaps half the entire inhabitants of the protectorate.

Among the natives are skilful craftsmen and carpenters, and the Swahilis of Tumbatu island furnish excellent sailors for the British merchant marine and men-of-war.

We cannot here enter at length upon the chequered and romantic history of Zanzibar. Negro, early Arab, Malays from Sumatra, perhaps before the beginning of the Christian era, enter into its activities, and trade with China, India and Persia was flowing at the time when, in the sixteenth century, the Portuguese arrived and subjected the island to their cruel rule. Later the Arabs drove them out, and the Iman of Muscat established his capital there in 1840. The Americans, several years earlier, had seen the commercial importance of the place, and established trade relations therewith, and agents of the English Church Missionary Society established themselves there, having been driven from Abyssinia. The sultan ruling in 1844 was tolerant of faiths other than his own, and a more

---

[1] *Tropical Africa*, Drummond, 1888.

enlightened development took place under his rule, when Sir John Kirk, British Consul in 1886, largely influenced the affairs of the sultanate. The great slave trade of the time, in negroes torn from the African mainland and shipped to Persia, Arabia, India, and the United States and Cuba, began to give way, and insistent war was declared on the piratical dhows by the British cruisers, whose operations form a thrilling and romantic part of East African history. The British protectorate dates from 1890, and was brought about partly as a result of German intervention in the affairs of the island.

So varied and profuse is the plant life of Zanzibar and Pemba—plants, trees, flowers and spices introduced by Arab, Portuguese, Indian and British, brought from many parts of the world—that the islands have practically been converted into vast botanical gardens. There are cloves from the Moluccas, mangoes from India, nutmegs from the Banda Islands, cinnamon from Ceylon, guavas from the West Indies, oranges and bananas from India and China, and plants and fruits from Central America, added to the rich, indigenous flora, which includes the beautiful oil-palm of West Africa, and to other magnificent palms, orchids and characteristic native tropical vegetation. In this respect it is noteworthy that the native flora of the islands is that of West as well as of East Africa.

Zanzibar, notwithstanding the rise of Mombassa and Dar-es-Salaam, on the mainland, preserves its supremacy as the great distributing centre of the East African seaboard, and, with Pemba, exports great quantities of cloves, copra, oil-seeds, copal, tortoise-shell, and so forth, with ivory and rubber brought from the mainland, the largest part of its annual trade, worth about £2,000,000 sterling, being done with India and the United Kingdom.

The climate of Zanzibar, warm and moist, with few cool nights, is trying to the European. The hottest period is when the north-east monsoon blows, from the Indian Ocean.

The island of Pemba lies thirty miles north of Zanzibar, and is covered in part with dense vegetation, much of which has been cut away, however, for the clove plantations which form its great industry. The Arab name of Al-huthera—" The Green Island "—is descriptive of its appearance, and the cultivated land is covered with the large plantations of Arab

o

proprietors, who work the soil with Swahili and negro labourers, which, prior to 1897, were all slaves. Some rubber-planting has been done of late. The population of the island is about 60,000, and Shaki-Shaki is its capital. Pemba is governed from Zanzibar, and yields thereto a good revenue from the duty on the clove export.

IVORY SELLERS OF ZANZIBAR

CHAPTER XXI

ABYSSINIA

THE very ancient and classic land of Abyssinia, or Ethiopia, as it was in antiquity termed, and is to-day officially known to its own people, lies entirely within the tropics, but, consisting in large part of high, massive mountains and plateaux, its climate embraces all changes of temperature from the heat of the desert to those cold regions far above the sea, which form the eastern sources of the mighty Nile.

Far back in the known history of mankind Abyssinia takes us, to that remote period characterised by the civilisation of Egypt, a civilisation, moreover, which has continued comparatively little altered in the lapse of those thousands of years since—according to the traditions of Ethiopia—Abyssinia's Queen of Sheba visited King Solomon, and bare, as a result, their son Menelek, from whom the kings of Abyssinia still claim their descent. We hear poetically and menacingly of Ethiopia in Holy Writ :

"Woe to the land shadowing with wings, which is beyond the rivers of Ethiopia. That sendeth ambassadors by the sea, even in vessels of bulrushes upon the waters." [1]

The country is surrounded on every side by the possessions of European powers, and covers an area of 350,000 square miles. It approaches to within about three degrees of the Equator, and in the north reaches to within forty miles of the Red Sea coast. In the south the kingdom is 900 miles wide.

The northern boundary of Abyssinia is formed by the Italian possession of Eritrea, and the western by the Anglo-Egyptian Soudan. On the south lies British East Africa, and on the south-east and east are British, Italian and French Somaliland,

[1] These rush vessels—such as that figured on the tomb of Rameses II;—are similar to those used by the Incas of Peru.

shutting off the ancient kingdom from the Red Sea and the Gulf of Aden.

Abyssinia is unlike every other tropical land in being governed by an autocratic potentate, and so shut off from the modernising influences such as are at work throughout the whole of tropical Africa, America and elsewhere. Wrapped in this cloak, it must be long before Ethiopia acquires the character of a modern state, unless some very radical change occur, to throw off the habit of three or four thousand years.

The character of the Abyssinian reflects the country's history. The frequent executions and other cruelties are largely matters of habit, for cruelty is not necessarily a marked feature of the national disposition. The whole history of Abyssinia has been well described as " one gloomy account of civil wars, barbaric acts and unstable government," amid which occurred the constant usurping of the throne by upstarts who themselves wore the crown for but a brief space, going out in pillage and rapine. It is the history of a Spanish-American republic, translated to Africa, and extending back much farther in time than those turbulent states.

The comparative isolation of Abyssinia tends to perpetuate its mediæval gloom. The country has no harbours. It is cut off entirely from the sea, and its principal cities and their human element occupy a high mountain and upland mass— forming almost an orographical entity—at an elevation of 5000 to 8000 feet above sea-level.

Difficult and remote as are the Abyssinian highlands, the scenery is impressive. A mighty mass of archæn rocks rises in the north to a plateau, in a depression of whose surface reposes the great Lake Tsana, wherein the Blue Nile has its birth. Mountains of weird and fantastic form overlook the enormous fissures which the rivers have carved out in cutting down their valleys on their way to the lowlands. In some cases the walls of these profound gorges rise almost perpendicular for thousands of feet, yet are but 200 or 300 feet from edge to edge. The mountain summits attain elevations up to 12,000 and 15,000 feet,. the highest culminating north-east of Lake Tsana in the snow-covered Daschan Peak.

The traveller in Abyssinia, in the compass of a day's journey, may pass from tropical to almost Alpine conditions of climate, so considerable is the range of elevation, and the plant life of the kingdom follows these changes. On the high plateaux the vision extends for vast distances, and, due to the condition of high visibility, objects on the far-off horizon stand out clear and distinct. The pellucid atmosphere and the cool and bracing mountain air is in sharp distinction from the hot climate of the valleys and the dryness of the deserts, or the hot and swampy districts where fevers are prevalent. Over a great part of Abyssinia the climate is healthy and temperate, but in winter it is cold and bleak in the mountains, conditions which remind the traveller of the inclement uplands of the equatorial Andes, in Ecuador and Peru.

The capital of Abyssinia, Adis Ababa, meaning, in the language of the country, " The New Flower," stands remote from the coast, 450 miles from Jibuti, on the Gulf of Aden, in the centre of the kingdom on the bare, grassy mountain-tops at over 8000 feet above the sea.[1] The town is not old, having been founded in 1892 by Menelek II. Rather than a town, however, the place is an extensive, straggling community, whose buildings possess little architectural merit. The royal enclosure or Gebi covers the top of a hill, having a view of the whole district, and grouped around are the residences of the nobles and foreign representatives, with the military camp a mile beyond. Some trade is carried on by Armenians and Hindus, and there is telegraphic communication with the outside world.

No railway connects the Abyssinian capital with the coast, and to ascend thereto we must traverse the caravan route from Dire Dawa, which itself is reached by the only railway in the country, 188 miles long from Jibuti, on the Gulf of Aden, in French Somaliland.

Some thirty miles from the railway terminus lies Harrar, one of the most important towns of the kingdom, with a population of 40,000. Ascending the mountain slopes to an elevation above sea-level of 5000 feet, we reach a lofty stone wall, pierced by gates and flanked by numerous towers, frowning down upon the hilly road. Passing this barrier we enter narrow streets, dirty and steep, paved, or rather

[1] Its situation and latitude are similar to those of Bogota, the capital of Colombia, on the other side of the globe.

unpaved, with boulders, along which man and beast struggle painfully.

The town of Harrar, it is generally believed, was founded by Arab emigrants in the seventh century, and it was at one time the capital of the country. The first European visitor arrived in 1854—the English traveller Burton—in the guise of an Arab. Rough stone generally forms the material of which the houses are built, except the governor's palace and the residence of the foreign consuls, which are more elaborate. An Abyssinian church, of the customary circular form, and various mosques, all of stone, are other principal structures.

Harrar is a commercial centre of some importance, the trade in coffee, which is extensively grown on the hills surrounding the town, and in other merchandise, being principally carried on by Abyssinians, Armenians and Greeks. Mules, donkeys, pack-horses, and camels in the lower districts, are the means of traffic over the caravan routes, and there is a well-made carriage road from railhead at Dire Dawa to Harrar.

Abyssinia possesses no other towns of any considerable size nor of a population exceeding about 6000, and this somewhat unusual condition is due to the almost continual state of civil war which has marked the history of the country, war between the various large provinces into which it is divided, and to the frequent change of the Court from one to the other. The exhaustion of the fuel supply also influenced the condition. The most ancient capital of Ethiopia was Axum, whose extensive ruins and ancient monuments are of vast interest to the archæologist.

A historic natural stronghold is Magdala, once the scene of operations of a British expedition. It lies 250 miles from Jibuti, and the rocky basaltic plateau upon which it stands rises to over 9000 feet above sea-level, and 1000 feet above the plain at its base. Here Theodore of Abyssinia imprisoned his British captives, but the stronghold was attacked by the English in 1868, under Sir Robert Napier, and its buildings were burned.

Another early capital of Abyssinia was Ankober, as was also Debra-Berhan, or "The Mountain of Light," where once a royal residence stood. Debra-Tabor was the capital of King John, and stands in a strategic position overlooking the fertile plains near Lake Tsana, at over 8600 feet above the sea. A few miles away lies Mahdera-Mariam, or "Mary's Rest," also at one time a royal residence, and now an important centre of pilgrimage, with a large market-place. Its two churches, known as the "Mother" and the "Son," are greatly venerated by the Abyssinians. Another great market centre is Sokota, where important trade routes converge, and here considerable dealing in salt blocks, which are procured from Lake·Alalbed, is carried on. Foreign merchants and traders from the Gulf of Aden also frequent Bonga, a large commercial centre.

The highlands of Ethiopia are of great topographical import-ance in that the Blue Nile rises therein, the Bahr-el-Azrak sweeping in a vast semicircle in deep gorges from its source in Lake Tsana and draining a great part of the Abyssinian plateau. This lake, which forms the principal "reservoir" of the Abai, as the Blue Nile is termed in Abyssinia, lies at a height of 5690 feet above sea-level, and is forty-seven miles long, somewhat pear-shaped in form. High as is the lake, it lies 3000 feet below the encircling plateau, and thus beheld looks like the flooded crater of a volcano, and is 250 feet deep in places. From the stalk of the pear-shaped surface a deep crevasse opens, letting the Blue Nile escape upon its long course, amid crag, rapid and magnificent cascade, down to the plains of Sennar across the Abyssinian border, to where it joins the White Nile at Khartoum.

The Atbara or Black River, another of the Nile's great eastern tributaries, also rises in the Abyssinian mountains and flows into the Nile far below Khartoum. Its headwaters are formed by the Takazze—"The Terrible"—which falls for thousands of feet through the tremendous crevasse along which it sweeps, dividing at flood-times one part of these rocky highlands from the other by an impassable barrier. In the south-west the Sobat, a further great Nile tributary, also descends from the mountains, falling in high cascades, flowing into the Anglo-Egyptian Soudan to join the White Nile.

In the history of exploration of the long-mysterious river it was to a Scotsman, James Bruce, that the discovery of the Blue Nile was due. Staying for two years in Abyssinia, and gaining the favour of the people by reason of his fine presence and ability, he was able to journey unmolested, and reached the long-sought goal in November 1770.

The prosperity of the Eastern Soudan and of Egypt is greatly dependent upon the Abyssinian Nile. A scarcity of rainfall and consequently of water in the great natural reservoirs of the Ethiopian highlands means a low Nile, as the flood-waters of the river, as contrasted with the constant flow, so vital to the land, are practically all derived from the Abyssinian tributaries. The Abyssinian rainy season is caused by the south-west monsoon.

The lowlands of Somaliland and Danakil have a hot, dry climate, with semi-desert conditions, and the country traversed by the Sobat, in its lower part, is swampy, hot and malarious. But as we ascend from lower elevations we mark the great range of plant and animal life. The denser vegetation shelters the elephant, especially upon the Sobat, and the rhinoceros, hippopotamus and the crocodile inhabit the westward-flowing rivers. The glens and ravines, thickly wooded, offer a delightful contrast to the open downs above the line of tree-growth.

In the lower country and in Somaliland the lion is constantly encountered, but in Central Abyssinia it is only seen in the valleys. Great leopards, black or spotted, are found, and the numerous hyenas are hardy and fierce. The zebra, the wild ass and the giraffe roam freely, and monkeys range from the tropic lowlands upwards to the high hills.

Many trees, fruits and flowers familiar to the traveller in tropical lands meet the eye here, such as the date-palm, mimosa, wild olive, sycamore, and laurel, junipers, myrrh and other gum trees whose gnarled forms are found on the eastern foothills. The splendid yellow pine resists the attacks of the white ant. The orange, fig, peach, pomegranate, blackberry and raspberry are among the many fruits. Some sugar-cane, and cotton and indigo, are cultivated, and on the tablelands cereals and vegetables in great variety are grown.

The world owes to Abyssinia a debt of gratitude for its gift of coffee, whose original home and name may have been in the Kaffa highlands, in the south-west part of the kingdom above the source of the Sobat tributary of the Nile. Coffee is among the most important products of the country, and two well-known qualities are cultivated: the " Abyssinian " in the south and the "Harrar-Mocha" in the Harrar highlands, producing this highest-grade coffee, cultivated with great care.

The soil of the Hawash Valley is peculiarly favourable for cotton-growing. The Hawash is a copious stream even in the dry season, 200 feet wide, and at flood-times inundating the plains along its banks. But the river does not reach the coast, being lost and sunk in the lakes of the saline depression of Aussa, several hundred feet below sea-level, some sixty miles from the sea.

Very little land is left uncultivated in the northern part of Abyssinia, the hill-sides being terraced and watered by irrigation channels, often led from the streams for several miles. Under this careful system even poor lands are coaxed to yield crops from eight to ten fold. Indeed, several crops a year are reaped in most districts, the soil being extremely fertile, a condition to which Egypt owes much, for Abyssinian soil forms the fertilising Nile sediment.

Methods of cultivation are generally primitive. The plough is a pole with a couple of iron teeth, hauled by oxen, and the crop is reaped by women and girls. The threshing is mainly done by the ox treading out the corn, and the grain is deposited in clay-lined pits. Barley is very plentifully sown, and sowing and reaping often go on simultaneously. The labourer obtains as his recompense the small sum of threepence a day, to which his food is added, which may cost his master perhaps a penny. The land is subject to the control of the king or the Church, and is not held as freehold.

Great flocks of sheep cover the Abyssinian hills in some districts, not generally wool-bearing, and weighing often only twenty or thirty pounds each. Goats with enormous horns are seen in great quantities, and the meat of both these animals is excellent. The horses are small, but very strong and numerous, and the ass and the mule are both of excellent quality, the latter being bred in large numbers, and preferred to the horse as a transport animal. The traveller on the rocky steeps does well to entrust himself to these sure-footed beasts, which withstand every condition of climate, travelling over the most difficult mountain road and carrying, if need be, a load of 200 pounds. Abyssinia is a good country for stock-raising, and one class of the cattle resembles the Jersey. There is a trade in live stock with Madagascar.

The people who dwell under the ancient and feudal regimen of Ethiopia are estimated as numbering between 3,500,000 and 5,000,000. Abyssinia is a geographical rather than an ethnical expression, and the name itself means "mixed," from the Arabic *Habesh*, a term once applied somewhat derisively by the Arabs to the collection of people of the Abyssinian plateau. From earliest times a Hamitic people occupied the region, and their descendants form the bulk of the population to-day, a race whose prevailing colour is a deep brown in the central provinces, pale olive to fair in the north, and chocolate to sooty black in the south. The majority of the population, however, are described as a mixed Hamito-Semitic people, generally well formed and even handsome, with regular features and dark, straight hair. The official language, spoken by the upper classes, is of Semitic origin.

Upon this admixture of barbaric African peoples there was suddenly imposed, about the middle of the fourth century A.D., a form of Christianity which became the religion of the state. The result has been a strange mixture of savagery with lofty ideas, and laws springing from both. Peaceful occupations have remained undeveloped by constant wars, and the native indolence has been augmented from this cause. "The soldiers live by plunder, the monks by alms." "The haughtiest Abyssinian is not above begging." They are a bright and intelligent people, fond of pleasure, selfish and vain, heavy eaters and drinkers, seizing any excuse for excess. Morals are described as loose, polygamy is common, and marriage easily dissolved, with consequently but little family affection. "Children of the same father but of different mothers are said to be always enemies to each other" in Abyssinia, a statement which reveals family life in an aspect.

The people are beginning to adopt European dress to some extent, but the native costume somewhat resembles the Arabs'. The Christians go generally bare-headed and without footgear, but the Mohammedans among the people wear turbans and sandals. The women wear a smock, having sleeves loose to the wrists. Cotton is the ordinary dress material, but the aristocracy may wear silken robes, and the high-born covers himself to his mouth when conversing with his inferiors. Christians generally carry a small crucifix slung round the neck ; women wear silver ankle rings and bells, necklaces, and ear-rings or rosettes in the ears, with, at times, rings on their toes, and they are fond of the strong perfumes brought from India and Ceylon. The dress of the priests is an elaborate one.

The ancient weapon of the Abyssinian was a sickle-shaped sword, and no man appears without a long curved knife. Great hunters and tamers of wild beasts are these Ethiopians, and it is stated that the nobles hunt antelopes with leopards, and run down the ostrich and the giraffe with horse and grey-hound. The elephant and the lion are hunted, and the skin of the latter belongs always to the emperor.

As may well be expected in this remote and backward land, with its primitive social development, houses are rudimentary and sanitation non-existent. The ordinary dwelling is described as filthy, without ventilation and infested by vermin, the walls inside being plastered with a mixture of cow-dung and chopped straw, or sometimes formed of stone set in mortar. But the general type of house is a circular thatched hut of the roughest kind, sometimes made of straw, although in the north, square, storeyed houses with flat roofs are built. Chimneys are not used, and the smoke soon blackens the interior.

The rock-hewn churches are the most striking buildings in Abyssinia, and for their religious structures the crude art of the people is mostly reserved, often exhibiting itself as frescoes imaginatively and gaudily painted in Byzantine style. The churches are generally circular, with thatched roofs. There are some fine frescoes in the Adowa and Kwarata churches.

Wide powers of government are exercised by the provincial princes or *rases*, a council of whom occasionally assist the *negus negusti*, or " King of Kings." Best known to the European world of these monarchs are the names of King Theodore, King John and Menelek, the last-named the redoubtable sovereign who inflicted defeat upon the Italians. He has been described as having comported himself in this victory with marked dignity and toleration.

The history of Abyssinia is of vast interest, but cannot be more than touched upon here. Ethiopia and Egypt were intimately connected in early times, and indeed were governed by one ruler. Many Jews settled in the country during the captivity. The Greek Auxume state flourished for the first seven centuries of the Christian era, and its port of Adulis was the main outlet for African ivory, gold, perfume and leather. Christianity grew slowly at first, after its establishment in A.D. 330. The Ethiopians conquered and held the richest part of Arabia, but were expelled by the Mohammedans in the seventh century, and Ethiopia, surrounded by the enemies of their religion, was cut off from and forgotten by the world for a thousand years.

In Europe vague knowledge was held of a Christian kingdom in East Africa, with a monarch called Prester John, and search was made therefor, the Portuguese in 1490 being the first to arrive, and for a century thereafter Portugal assisted Ethiopia against the Mohammedans.

In 1805 a British mission was sent to treat with Abyssinia, after which travellers and merchants of various nationalities journeyed there, as also missionaries, which latter were well received, although their secular work was more acceptable to King Theodore than the religious work. The rivalries of British and French factions and missions—religious and scientific —became acute later, the Roman Catholics opposing the Protestants, and vice versa.

King Theodore is described as having been a man of intelligence and education, of noble bearing and untiring frame, but giving way later to intoxication and extreme cruelty. A quarrel with Great Britain arose, due, to some extent, to dilatory British tactics, and the expedition of Lord Napier followed, the British landing near ancient Adulis in January 1868, with the objective at the Magdala fastness. Nearly 32,000 men were involved in this affair, and to march through 400 miles of mountainous country with savage foes on every hand was the task to be performed. But Theodore's army of 100,000 men melted away, and after the capture of Magdala the king's body was found amid the fortifications, where he had died by his own hand.

The struggles between the rivals Johannes or King John and Menelek followed upon this period. Menelek claimed direct descent from Solomon and the Queen of Sheba, as Menelek II. He was, however, obliged to submit until, in 1889, John was killed in battle against the Italians, who were bent upon increasing their possessions upon the Red Sea. Later, after treaties had been made between Italy and Menelek, a desperate situation arose, and the Italians, though they fought with great bravery and gained both success and territory, suffered terrible loss at the battle of Adowa.

Under the "King of Kings," Menelek II., some efforts were made in 1905 to develop the resources of Ethiopia, and the value of its trade reached £1,000,000 per annum. A new dollar coin replaced the old one and the cartridges and bars of rock salt which were used as currency, and the first bank was established, at Adis Ababa. This was founded by the National Bank of Egypt, under a concession from Menelek.

The export trade, consisting mainly of coffee, skins, civet, ivory, gum, pepper, ostrich feathers, kat—a plant of stimulating properties used by the Moslems—live stock and a little gold, is perforce carried on through the ports of the European territories which shut off Abyssinia from the coast. American, English and Indian cotton goods are the chief imports, with arms, ammunition, and food-stuffs. Trade through the Soudan is also of some importance. The output of gold from the placer and quartz mines is worth about £500,000 sterling per annum, but not much is exported. Silver, iron, coal and rock salt are other minerals found in the country.

The rudimentary educational system of Abyssinia is carried on by the clergy, and in 1907 education was made compulsory. In the same year the *negus* decreed the formation of a Cabinet of Ministers, after European models.

CHAPTER XXII

THE SOUDAN AND THE NILE

FEW countries of tropical Africa hold out a stronger appeal than does that enormous belt of territory somewhat vaguely termed the Soudan. The traveller, the tourist, the soldier, administrator, economist and engineer have, since the comparatively recent fall of Mahdism, wrought changes in that portion of the region under British control which it is no exaggeration to designate as little short of marvellous. The names of Gordon, Khartoum and Omdurman alone suffice to shroud it in an atmosphere of interest and romance, and the vast rivers, from the Niger to the Nile, which cross the desolate yet not infertile breadth of the Soudan, are among the most interesting topographical features of Africa.

The Soudan is a geographical expression rather than a well-defined territory, an ethnological rather than a physical unit, whose inhabitants may be defined as the negro people who, north of the Equator, are under the influence of the Mohammedan religion. The Soudan was so named by mediæval Arab historians, the name *Bilad-es-Soudan* meaning " The Country of the Blacks," or Nilotic negroes, although the term is scarcely applicable to the large area inhabited by the Hamitic and Semitic Arab people.

The Soudan stretches across Africa from Cape Verde on the Atlantic to Massawa on the Red Sea, with the various European colonies of West Africa, the Congo, and the great lakes of East Africa upon its southern side. Abyssinia lies on the east, with the Sahara and Egypt to the north. The length of this great belt is 4000 miles, and its breadth 1000 miles, the whole covering more than 2,000,000 square miles of territory, upon which lives or roams a population of 40,000.

This great territory of the Soudan is by no means all desert, as we have seen in French West Africa, Northern Nigeria and elsewhere. From the arid, sandy wilderness of the Sahara

NATIVE SCENE IN THE SOUDAN

border stretches a belt of level, grass-covered plains, giving way to the well-watered and cultivable lands of the Southern Soudan. The waste of broken hills and empty deserts, a landscape with no sign of life save perhaps an occasional vulture, and no vestige of vegetation, which the traveller ascending the Nile passes on his way to Khartoum, might seem to be the veritable abomination of desolation described of old, but this, Nature's most sterile mood, is succeeded by the fertile plains which soon unfold, stretching east, west and south, the basis of the wealth of the Soudan to which the northern desert now takes its place as only an approach. Fertile cotton soil replaces the barren sand; the flocks and herds of the primitive tribesmen graze as they did 5000 years ago, and the green of the trees and the grass refreshes the traveller's eye.

The various states into which the whole Soudan is roughly divided may be named as those of the Niger, within the great curve of that mighty stream, belonging to France; those between the Niger and Lake Chad, under British control; those of the French Congo, and those under the control of Great Britain and Egypt. These first-named have been described in their places.

The last-named group forms the Anglo-Egyptian Soudan, the subject of this chapter. It covers nearly 1,000,000 square miles of territory (950,000), equal to a quarter of Europe; 1000 miles from west to east and 1200 from north to south, a compact territory which ensures—with Egypt—British control of the whole Nile Valley from the great lakes to the Mediterranean. The Red Sea, Eritrea and Abyssinia lie to the east, and Uganda and the Belgian Congo on the south. The Lado Enclave, a piece of territory formerly administered by the Congo Free State, forms the southernmost point of the Soudan, and is important from being traversed by the navigable headwaters of the Nile.

Remote from the coast, in its greater part, and in general less than 1500 feet above the sea, the Soudan—as we shall here term the Anglo-Egyptian Soudan—lies under a beating sun, " its soil like fire, its wind like flame," as the Arabs long ago have said of it, the heat of the climate, however, being least in the desert zone and greatest in the central regions. Whilst the river valleys are malarious, the great dryness of the air renders the Soudan in general healthy. The mean annual temperature at Khartoum, which is centrally situated, is 80° Fahrenheit. The northern desert is almost rainless ; the central belt has an annual fall of ten inches, crowded into perhaps fifteen days of the year, with from twenty to forty inches in the eastern and swampy districts. From June to August terrific sandstorms sweep over the land, and heavy thunderstorms occur in the sudd region.

Such are, roughly, the general conditions of the region into which Great Britain plunged, in the closing years of last century, in the reconquest of the Soudan, with the double purpose of avenging the death of Gordon and of securing for Egypt the control of the Nile. Under a firm but just rule—first military then civil—Britain has rescued this home of the Mahdi and the mad dervishes from savagery, restoring order to a region lost to civilisation, putting an end to slave-raiding, and working to the end that it might yield its increase of the fruits of the earth and of its human inhabitants. These objects have been secured to a remarkable degree, and in a time peculiarly short, a transformation such as never before occurred in the history of the world.

Let us take our stand in Khartoum, the picturesque city, with its handsome buildings, gardens, palm groves and squares, with its streets laid out upon the plan of a Union Jack, the town-planning of a great soldier and Imperialist.[1] It is now a city with a population of about 80,000 people, which has arisen upon the spot where first a small fishing village stood. Khartoum stands at the point where the White Nile, coming from the great lakes of the Equator, mingles its waters with the Blue Nile, born on the rugged roof of Abyssinia ; the promontory where these two mighty streams meet having, by its fancied form of an elephant's trunk, given origin to the name in the Arabic dialect of *Khartoum*.

We reach the capital of the Soudan either by river steamer and railway from Cairo, a journey of 1345 miles, or by rail from Port Soudan on the Red Sea, a line 432 miles long. The town lies 1250 feet above sea-level, and from its geographical position it is admirably adapted as the political and commercial region of this vast British dependency. A heterogeneous population dwells here, the traders being generally Syrians, Greeks and

[1] Kitchener.

Copts, with types of all the negro and Arab inhabitants of the Soudan included, controlled by the British and Egyptian official class.

The Khartoum esplanade, which has been laid out for two miles along the river, contains the governor-general's palace, a handsome building upon whose walls is a tablet marking the spot where Gordon fell, slain by the Mahdists in 1885. In the gardens of the palace is the huge stone figure of the " Lamb of Soba," brought from an ancient Christian city of Ethiopia on the Blue Nile, and in front of the main façade is the bronze statue of Gordon, depicted astride his camel. The Government offices, private villas, zoological gardens, churches, museum, race-course, hospital, great mosque, barracks, fortifications, missions and college—the last-named the educational centre of the Soudan—attest the wealth and strength of Khartoum, which thus has arisen after being entirely destroyed by the Mahdists in 1885. The model villages of the principal tribes, built after native fashions, are a feature worthy of remark.

To reach Khartoum from the coast the railway from the Red Sea branches at Berber ; Port Soudan and Port Suakim, connected by rail, being the principal harbours upon the 400 miles of the Soudan coast.

Up to the time of the Egyptian conquest of the Soudan, in 1820, Christian Nubians and Abyssinians and Mohammedans had fought constantly among themselves for possession. Gordon, in 1877, strove hard to give the country good government, but all the old Egyptian abuses were revived at his departure. The fanatic Mahdi prevented all British schemes of betterment, preaching a species of communism, also denouncing the Turks as unworthy Moslems whom Allah would punish. In 1883 Hicks Pasha's army was massacred, and later followed that grievous page of evacuation by British order, the " too late " expedition of Wolseley and abandonment and death of Gordon. These matters, however, were wiped out by Kitchener at Omdurman in 1898.

Under British administration great attention was at once given to economic affairs, agriculture, the erection of an educated class, missionary work—with full respect to Mohammedan institutions and the Arab language—and the suppression of slave-raiding; and the esteem of all classes of the Soudanese was thus secured.

Let us follow the Nile, the primary topographical feature of the Soudan, from its source.

Escaping from a beautiful arm of the great lake, Victoria Nyanza, in Uganda, over a natural rock weir 500 yards long, at an elevation of 3700 feet above sea-level, this great sluice-way, terminating in cliff abutments on either hand, is the central feature of a landscape striking in its grandeur. Plunging over the cascade, as if rejoicing in its birth, the Victoria or Somerset Nile, as it is here termed, races swiftly between its rocky walls, its surface broken by beautiful islands and innumerable rapids, gathering other streams to swell its volume. Thence through swampy lakes the current takes its way, joined by backwaters, pouring over the great Murchison Falls, raging furiously through rock-bound passes, leaping sheer for more than 100 feet and flinging its spray on either hand about the cleft abyss, through which it passes, a noble stream now.. Flowing thence, amid forest-covered hills, the river enters the north-east end of Lake Albert Nyanza.

Issuing from this lake the Nile has changed its name, and is now the Mountain river, or Bahr-el-Jebel. Lake-like stretches of quiet water form its course, the current 700 yards wide, narrowing again, however, leaving the Fola rapids like a mighty mill race to hiss through a gorge over 300 feet long and only fifty wide, plunging thence into a boiling cauldron, whence the water roars and thunders. Thence passing amid islands and small settlements, including Fort Berkeley, and beneath the rocks of Rejal, looking like the ruins of some stronghold of antiquity, the Nile reaches Gondokoro, near the line between Uganda and the Soudan.

Here we enter upon a very different landscape. From Fort Berkeley, above Gondokoro, the Nile is navigated by regular Government steamers from Khartoum, a distance of 1090 miles, and as we proceed down-stream the view extends over a wide valley, through which the river winds tortuously, with a very gentle slope. This is the region of swamps. There are no banks to be seen. Vast areas of tall reeds and papyrus, with lagoons opening from the stream unfold, and for a long distance the Nile divides into two channels. The view becomes

THE NILE, TRAVERSING THE SOUDAN

depressing as mile after mile of this low, flooded land is passed. The steamer's way is barred at times by the sudd, the great masses of floating vegetation which accumulate on the surface of the river.

The word " sudd " means, in Arabic, " to dam." The material consists of masses of papyrus and " mother of wool "—another reed which stretches along the river in high walls—together with little swimming plants. Loosened by storms, this vegetation clings together, forming veritable dams across the channel, often in solid blocks of the consistency of peat, so much compressed by the force of the water that masses sometimes twenty-five miles long and twenty feet deep are formed, so solid that men can walk upon them, and even elephants have been known to cross without danger. These masses of tropical vegetation are a formidable obstacle to navigation on the Upper Nile, and powerful sudd-cutting machines are necessary to keep the channel open. Due to the sudd blockade, enormous labour was necessary to open the channel to Gondokoro in 1904.

Of recent years an industry has been created in recovering and treating sudd as a source of fuel.

Wearied as we may be in traversing these monotonous stretches of the great river, by reason of the " mother of wool " reeds and papyrus, the vast grasses which form barriers twenty feet high on either hand, by the excessive moisture in the air and the swarm of mosquitoes and flies, we shall not fail to note such beauties as are here displayed. Blue, white and crimson water-lilies adorn the channel's surface ; the curious whale-headed stork stands sentinel among the reeds, and as the steamer passes at night a myriad fire-flies light up the gloomy landscape.

At length we steam into Lake No, where the river is joined by the great western tributary of the Bahr-el-Ghazal, which comes from far away in Darfur and the Belgian Congo. This is also a sudd-obstructed river, with many varieties of the " swimming " sudd plants, and lined with papyrus. It drains an enormous area of the Soudan. We shall remark the interminable array of ant-hills on the shores and plains, their mounds in some places so close together as to resemble " a gigantic graveyard," as one traveller has described it. Steamboats ascend the Bahr-el-Ghazal to Wau—a distance of 930 miles from Khartoum.

Emerging from Lake No, the river again changes its name, and we are afloat upon the White Nile, or Bahr-el-Ghazal. Forty miles below the Sobat enters from the east, flowing from the highlands of Abyssinia and across the vast plain below. This river is navigable as far as Gambela in Abyssinia. At low water the Nile here is the colour of milk; at flood-times a brick-red; the first-named hue giving rise to the name of the White Nile, and its volume is equal to that of the main stream itself.

Again, onward for nearly sixty miles, we reach Fashoda, now renamed Kodok, its earlier designation rendered famous by Marchand and the dispute, now happily forgotten, between France and Britain over the dominion of the region. For several hundred miles thence the scenery again becomes monotonous. But the end of the sudd is passed at last, and 250 miles from the Sobat confluence the steamer approaches Khartoum and the confluences with the famous Blue Nile, whose course from the Ethiopian highland roof we have already marked in Abyssinia.

Enclosed in the triangular-shaped territory between the White and Blue Niles is the territory of El Gezira, the most fertile portion of the Soudan, and here great irrigation works are being carried out for the purpose of cotton cultivation. A sum of £3,000,000 was granted by the British Government for this purpose.

The Nile is navigable only at high water through the cataracts, between Assuan and Khartoum, and at the flood periods steamers traverse the entire river from the Mediterranean to Fort Berkeley, a distance of nearly 3000 miles.

Continuing our journey down-stream from Khartoum we reach the last tributary of the Nile, whose confluence is 200 miles below the city. This is the Atbara, the Bahr-el-Aswad, or Black River, whose source is in Abyssinia, the river flowing across the Nubian plains. Notwithstanding its great volume, the Atbara runs dry after the rain, and numerous creatures— fish, crocodiles, turtles and hippopotami—are imprisoned in the isolated pools it leaves. Between the Atbara and the Blue Nile lies the enclosed land known in ancient history as the Island

WILD ELEPHANTS AT HOME IN THE SOUDAN

of Meroe, whose monuments are of great interest to the archæologist.

Before reaching the Atbara, and about fifty miles below Khartoum, we pass the sixth cataract, where the Nile runs through a narrow and picturesque channel, and 188 miles beyond is the fifth cataract, below the town of Berber, which place is the terminus of the Red Sea caravan route. The fourth and third cataracts are passed in succession.

In places high, barren cliffs shut in the view, and the river, making a mighty curve and doubling upon its course, leaves the fertile land to plunge into the Nubian desert, the wilderness coming up to the brink of the now narrow stream. The waters of the Nile become diminished here from the excessive evaporation caused by the great dryness of the desert air. Dongola is passed, and at Wadi Halfa and the second cataract the border of Egypt is reached. Thence we navigate for more than 200 miles a smooth, broad Nile, over 1600 feet wide, with an almost imperceptible current, noting the spurs of masonry built out by the ancient Egyptians to guide the stream, and, crossing the tropic of Capricorn and passing the island of Philæ, with its famous temple, and other islands, we reach Assuan and the first cataract, which has been greatly changed by the famous barrages and hydraulics works of Assuan. Hence the Nile flows through Egypt to the sea.

The Nile embodies one of the greatest hydrographical romances of the world. The source and regimen of the great river formed a mysterious problem to the ancients. They did not understand the causes of its annual flood, but Aristotle, arguing from the accounts brought back by the explorer sent by Alexander, thought this was due to the heavy rains of Ethiopia. The annual height of the flood has been recorded since 3600 B.C., and the ancient Nilometers at Assuan and elsewhere still exist. Now it is well understood that the Nile obtains its constant supply from the White Nile, coming from the lakes, and its flood-waters from the Blue and other Nile branches of Abyssinia, the remarkable journey of Speke and Grant, in 1862, having solved the problem of the Nile, together with Stanley's discovery of the Albert Nyanza.

The inhabitants of the Soudan, the Arab and negro tribes, cannot, in contrast with the Egyptians, be described as an industrious people.   Many of the Arabs prefer their free desert life, where the climate demands no exertion in the making of houses or garments, and this easy existence is followed even by the owners of great flocks and herds.   Their wealth is mainly in these and their camels and horses.   Yet the Arab people are described as having a real desire for education, and their moral status is improving.   Houses are built of sun-dried bricks, as in the days of the Pharaohs, and are cool and of easy construction, but south of Khartoum huts of grass are used.   The province of Darfur is under native rule, and the various provinces into which the Soudan is divided are administered by English governors generally, with Egyptians or Soudanese in secondary official places.

The education of the people is carried out by the Department of Public Instruction, Arabic being the general language, and there are elementary schools in the villages, and higher branches in the chief towns, also industrial workshops, whilst among the pagan negroes Protestant and Roman Catholic missionaries are at work, wisely giving importance to industrial as well as religious teaching.   The famous Gordon College at Khartoum is the educational centre of the Soudan.   The negroes own large flocks and herds, in some cases.

The agricultural land north of Khartoum consists of narrow belts along the river, and in the Libyan Desert there are cultivated oases.   The valleys and wadis alone are cultivated in the provinces of Kordofan and Durfar, where the durra or millet is grown, the chief food of the Soudanese, who obtain two crops yearly.   Wheat and barley, numerous vegetables, dates, ground-nuts and sesame for their oil, and cotton equal in quality to that of the Delta are produced.   A feature of the native farming is the frequently seen *sakia*, or water-wheel, turned by oxen to raise water for irrigation, and the *shaduf*, or bucket-sweep, for the same purpose.

The Soudan yields many of the forestal products of tropical Africa, such as rubber—especially in the equatorial rain-forests —and gum.   The gum-bearing acacia grows in such quantities on the plains of Darfur and Kordofan, and between the hill and foot-hills of Abyssinia that it constitutes one of the most valuable economic products.   Para rubber has been planted in the Sobat district.   Ostrich farms yield some income to their owners.

CAMEL MEN OF THE SOUDAN

Sunt wood is used for boat-building, and mahogany and fibre are among the exports, as are also senna, wax, cereals, live stock, hides, cotton, cotton-seed, mother-of-pearl and gold. A foreign trade—mainly imports—of nearly £3,000,000 annually is carried on.

The ancient gold mines of Nubia, between the Nile and the Red Sea, are marked by some extensive ruins, and modern mining was inaugurated here some years ago, in the desert at Um Nabartli, a light railway having been built to serve the district. Iron and copper ores are found in other districts of the Soudan.

The plains of the Soudan, under irrigation, may become one of the most important cotton-producing countries in the world, for here vast areas of soil are available for planting. The system of " basin " irrigation is to be employed in the Gezira territory. In this system low banks are thrown up, forming large shallow reservoirs or deposits, into which the turbid flood-waters are led by channels, there depositing their fertilising silt for six weeks or more, until they are allowed to run out. This, which was an early irrigation system, was carried out in perfection in Upper Egypt.

The simple agriculturist of the Soudan has little need for modern or mechanical appliances in tilling the soil, for the holdings are small, and the rude plough and the hoe which served his forbears also serve him well enough. As to manufacture, there is but little, but cloth is woven from native wool and cotton, although the bulk of the textiles used, now imported from Europe, are largely supplying the demand. Native pottery is made in certain special places, and is much esteemed.

The great caravan routes which traversed the Soudan, used by thousands of laden camels, cross the deserts from all points of the compass, taking away native produce and bringing in goods from beyond. Silk, carpets, precious stones, gold, ostrich feathers, ivory and slaves were thus transported, but the advent of the railway and the suppression of slave traffic has reduced the camel to the position of a feeder to the railway lines. Domestic slavery, however, has not yet been abolished.

The desert and forest animals of the Soudan are largely those which the traveller encounters in other parts of Africa, the lion roaming the hot plains and preying upon the cattle of the natives, and in the river-side districts the elephant still abounds. Gazelles and antelopes are numerous and widely distributed.[1]

The multiplication of the Soudanese people increases steadily under British rule. The thousands who perished and the tribes which were completely annihilated by sword, famine and disease throughout the evil fourteen years of Arab domination, which ended in 1900, are being replaced by a steady growth under the sway of the British Empire.

The Soudan is now a solvent country, with a revenue that has increased from a few thousand pounds in 1898, the year of reoccupation, to a sum numbered by millions, and since dervish rule was abolished trade and commerce have shown a natural growth which is remarkable for its rapidity. In the Upper Nile provinces, where a horrible slave trade thrived, the primitive inhabitants are learning to trade in modern ways, and in Dongola, where the mad dervishes destroyed the water-wheels and the date-palms by which the people lived, prosperous market towns have sprung up. El Obeid, ruined and abandoned, is now a thriving town upon the railway, and indeed from east to west and from north to south the same sense of security and progress prevails.

Much, we see, has been done in this tropical land for the benefit of all connected therewith, but even this is but relative, and the scope for greater organisation and more work is very considerable in the great Soudan.

[1] An interesting résumé of the Soudan and its affairs is published by the Soudan Government and Railways Board, for the use of travellers.

NATIVE SOUDANESE WOMEN OF OMDURMAN

CHAPTER XXIII

SOMALILAND AND ERITREA

WE have reached the " Horn of Africa " in approaching Somaliland, where the continent, projecting sharply into the Indian Ocean, terminates at Cape Guardafui, the tip of this mighty horn, a sombre precipice rising in a sheer wall 900 feet from the sea as the steamer rounds it to enter the Gulf of Aden, across which arises the Arabian coast.

Three European powers—Italy, Britain and France—are the masters of this wild and generally barren part of Africa, whose total area may be taken at 356,000 square miles. The coastline upon the Indian Ocean is 1000 miles long, formed by the shore of Italian Somaliland to the Juba river, which marks the frontier of British East Africa. The Gulf of Aden shore-line is 600 miles long, westward to Tajura Bay and Jibuti, and thence to the Straits of Bab-el-Mandeb. Upon this northern shore of the great triangular-shaped territory of Somaliland the British and French spheres lie.

Somaliland was the last portion of Africa to be explored, and many a stubborn fight with the " mullah " forms part of its history of but a few years since. Englishmen, French, Italian and American explorers have suffered thirst, wounds and death in the exploration or conquest of this great territory, some of it of so poor a character that a British Government a few years since, declaring it useless to pursue natives whom it was impossible to overtake, and that the economic resources were not worth development, ordered withdrawal of the British forces to the coast.

A stony and thorny landscape, waterless generally, is the characteristic scenery which unfolds to the traveller in this part of Africa. Yet Somaliland is the ancient land of frankincense and myrrh, a land which its early inhabitants termed *Regio Aromatifera*, by reason of the growth, amid the desert scrub, of these sweet-perfumed trees.

233

Steep mountain slopes and escarpments arise beyond the narrow, coastal strip, and ascending these we reach a high plateau stretching for vast distances at an elevation of 3000 feet above the sea. This is the Ogaden plateau, the mcnotony of whose surface is only relieved by a few scrubby plants and broken here and there by the dry beds of streams—the waterways of a region where scarcely any rain falls. Yet the climate of this upland is dry and bracing, the heat appearing much less than that actually marked by the thermometer.

Tall, coarse grasses cover a part of the high plains in places, with isolated trees arising therefrom. Mimosas, acacias, aloes, with the highly fragrant balsam, gums and resins before mentioned, and the " balm of Gilead " have their place with the briars and thorns. Again, as on the Golis range, there are cedar forests to gladden the eye, the trees sometimes reaching 100 feet to their tops, and thus from the *ban*, or fertile plain, to *aros*, the bare desert, transition constantly occurs. Indeed, a park-like aspect prevails in certain parts, where the escarpment is approached, and dates and figs grow in the fertile vales.

The elephant has been driven from the coastal districts of the Gulf of Aden, and his retirement has followed upon the advent of the explorer, who, in his pursuit of big game, has often added to our knowledge of the region. The lion and other great carnivora prey upon the Somali cattle, and the rhinoceros roams freely in his particular districts of the interior. In general terms it may be said that Somaliland is rich in the great and varied African fauna, whether of beautiful birds, graceful antelopes and gazelles or ferocious felines and reptiles.

The Somali people, who, according to very approximate estimates, number over 1,000,000, are of the Hamitic family, although they claim Arab origin. They show traces of Arab, Abyssinian and negro, and vary in colour from brown to black. Nomads who breed cattle, sheep and camels, following the rains for pasture, settled people near the coast, and hunters or workers in iron and leather, or collectors of gum and resin— such are the varied classes of the Somali. Their main riches are their flocks and herds. They go armed, as befits a fighting and cattle-raiding race, and he who has killed a man may wear an ostrich feather in his hair. The women occupy a very low position in the social scale among this vain and avaricious people, doing most of the work. Yet the Somalis must also be described as intelligent and bright, and loyal as soldiers. The Arab influence has been strong, and the Mussulman religion is strongly stamped upon the Somali. They have comparatively little political cohesion, however. The national dress is the *tobe*, a cotton robe falling to the ankles.

Italian Somaliland embodies an area of perhaps 146,000 square miles, and upon its coast-line, 1200 miles long, not a single good harbour is found. The winds and currents which beat upon the dangerous Horn of Africa render navigation almost impossible at certain seasons, and "Beware" is the interpretation of its name in the lingua franca of the Levant.

The towns are rarely of more than 5000 inhabitants. Mukdishu, facing the Indian Ocean, is an ancient place, described by Marco Polo and others in the fourteenth century as an immense city. It is mostly in ruins at the present time, but when captured by the Portuguese in 1510 it had many mosques and was a flourishing port. A massive square tower is a remainder of the Portuguese period, since when the town has had a chequered history. The Italian governor's palace stands hard by. Many houses are of Moorish type. Arab, Swahili and Indian settlements are interspersed with the few coastal towns, whilst in the interior no town exists.

British Somaliland occupies a much smaller area, covering some 68,000 square miles, facing upon the Gulf of Aden for 400 miles, and running inland, in the widest part, for over 200 miles. It embodies the more mountainous part of the Somali region, and here, upon the slopes, the aromatic flora giving the frankincense and myrrh especially flourishes.

A strip of narrow, sandy plain skirts the gulf, crossing which the rugged cliffs and their steep, intersecting valleys give access to the Harrar plateau and Abyssinia. To the south of Berbera two ranges of mountains parallel the coast, culminating in the lofty, cedar-covered Golis mountains, south of which is the waterless Haud plateau.

The chief seaport and town of British Somaliland is Berbera, whose haven is the only sheltered one upon the coast of the Aden Gulf. A normal population of 10,000 is swollen from October to April by the arrival of caravans from the interior to three or four times that number, vast trains of camels arriving, the owners' tents on their backs, and bearing merchandise worth hundreds of thousands of pounds, in gums, resins, ostrich feathers, sheep and goats. This merchant trade is destined for Aden, upon which place Berbera might be regarded as commercially dependent. The fort, hospital, barracks and residence of the commissioner are the centre of British control, Berbera having come into the possession of Britain in 1884.

No line of railway exists in British Somaliland, nor yet are there any roads, and camel traffic on the caravan routes is practically the only means of transport. The camels may cover fifteen to twenty-five miles a day, of seven hours, marching so as to avoid the greatest heat, and carrying 200 to 300 or more pounds. To and from Aden a busy trade by sailing boats and steamers is carried on.

Near the French line stands Zaila, upon the edge of a desert fifty miles wide, its buildings of white coral stone. Here is a good sheltered harbour, with Arab sailing craft. A picturesque procession of old Somali women, with camels bearing water in goatskins from the springs three miles away, comes into the town every morning. Ivory, coffee, skins, mother-of-pearl and cattle are exported.

The French portion of Somaliland lies at the entrance of the Red Sea and extends from the Straits of Bab-el-Mandeb to beyond the Gulf of Tajura, and runs inland for about 130 miles at one point, the total area covered being but 12,000 square miles. Here we encounter the same waterless plains, and a coast with a single seaport, that of Jibuti in the Tajura gulf. Back of this lies a chaotic mass of volcanic rocks destitute of vegetation, giving way to a chain of salt lakes. The climate is hot but not necessarily unhealthy.

Jibuti is the terminus of the railway to Abyssinia, and the governor's palace and other buildings mark the seat of administration. Stone, flat-roofed houses form the dwellings of the Europeans. The harbour is a capacious one, and trade is chiefly in coal for vessels. Jibuti was founded by the French in 1888, and the population of the port and native town embodies types of many races.

Dracyoli photo

A BAZAAR IN SOMALILAND. WOMAN TRYING ON A SWAHILI ROBE, AND BOOTHS OF INDIAN TRADERS

The Italian possession upon the Red Sea, of Eritrea, extending from the Strait of Bab-el-Mandeb for 650 miles to its boundary with the Anglo-Egyptian Soudan, is a land diversified in its physical formation and of ancient historic interest, having been the home of the old civilisation of the Axumites, whose obelisks and monuments of sculptured stone—standing or fallen—bear witness to the culture of its early people.

The early history of this place belongs rather to Abyssinia, which kingdom bounds Eritrea on the south. French Somaliland is its south-eastern neighbour, and the area covered is about 60,000 square miles, inhabited by some 450,000 people, which includes 3000 white settlers.

The hills and highlands of Abyssinia, approaching the Red Sea in the north, become Italian soil, entering Eritrea. We reach this high and fairly fertile plateau, lying at over 6000 feet above the Red Sea, by the railway which runs for sixty-five miles from Massawa, the principal seaport, to Asmara, the capital, at a still higher elevation, surrounded by rich, agricultural lands.

A coral-built coast and thick group of islands face the Red Sea, the coast indented by the Bay of Annesley, south of Massawa, and by the Assab Bay, the littoral plain widening out for nearly 200 miles towards Bab-el-Mandeb, as the Ethiopian hills recede from the coast, the broad plains belonging in great part to Abyssinia. This is the Afar country: arid volcanic plains covered in places with the mimosa, and crossed by streams which, descending from the mountains, die out before they reach the sea. These torrent beds are generally bordered by dense vegetation, and water can be obtained therefrom in the dry season by digging wells. Far inland, upon these plains in Abyssinian territory, the rivers drain into salt-pans and basins, such as those of Asali and Aussa, which are 200 feet below the level of the sea, for the region is part of the East African rift-valley. In places the salt lies thick, like snow or ice, upon the plains, and only the fierce heat of the sun undeceives the traveller. Volcanoes with vast craters terminate this singular region, the crater of Alid, half-a-mile in diameter, with walls 350 feet deep, rising therefrom, with its companion Dubbi in Eritrea. The lion, the elephant, the panther and the antelope all have their home in the wilds.

The excessive heat and humidity of the coastal zone continues throughout the greater part of the year, and malarial fevers prevail. At Massawa the average temperature is 88° Fahrenheit, rising at times to 120° in the shade. We encounter more favourable conditions, however, as we ascend towards the plateau.

Massawa is built partly upon coral islands and partly on the mainland, and is the natural port for Abyssinia, were it not an Italian possession; and it formed part of the Abyssinian dominion for centuries. It was occupied by Italy in 1885. The harbour affords good anchorage. The coral-built houses of the European and Arab merchants, the public buildings and the forts arrest the eye as we enter. An ancient aqueduct supplies the town with water. Trade has developed considerably under the Italian regimen, and coffee from Abyssinia, butter, hides and civet are exported, with Indian and European cotton goods as imports.

To reach Asmara the railway ascends to an elevation of 7800 feet above the sea, and we pass from the hot coastal temperature through two zones of climate as we rise to the plateau—a temperate to a cool climate. The highest belt is of a subAlpine character, rising from 8000 feet to its summit of 10,000 feet. The name of the Eritrean capital is derived from an ancient word meaning the "place of good pasture," and the surrounding plain, whose fertility bears out the designation, was known to the Abyssinian as the plain of a thousand villages. The population of about 10,000 includes, perhaps, 1000 Italians and Europeans, soldiers and civilians, and Italians cultivate its rich lands and carry on a busy commerce. There is also some gold-mining near by.

We shall remark the enormous sycamore-trees that flourish on these high uplands, and the dark-hued olive groves which clothe the slopes of the hills, extending downwards to within 3000 feet above sea-level, where the vegetation gives place to desert flora, and the maize and millet plantations of the native farmers. This low country, however, is little suited for crops, although in the middle zone and on the highlands the cereals of Europe flourish, the lack of agriculture on the lowlands being compensated for by the pastoral industries. The nomadic people are masters of large herds of cattle and camels, whilst sheep and goats are also plentiful in the uplands.

The Abyssinian native, as we have seen in Ethiopia, is a good agriculturist, and the science of irrigation is ably understood. The Italian farmers of the plateau raise excellent crops, and in the middle zone the experimental cultivation of such staple products as cotton, coffee and tobacco has given good results. Thus we remark that this Italian colony, whilst entirely within the tropics, is dowered with areas of land and a climate suited to the white settler, a fortunate economic circumstance. The growing external trade includes a considerable export of coffee and ivory, wax and gum, and the pearl fisheries of the islands are worthy of note, as is the native salt industry—salt being a valuable article of commerce in these regions.   Butter, meat and hides, articles of local commerce, attest the powers of a pastoral community.

The camel and mule roads that cross Eritrea in every direction are in some cases suitable for vehicles, or at least those from Massawa to the capital and those to Abyssinia, Kassala and across the Soudan to Khartoum are of this character. Camel roads lead through all the valleys running to the Red Sea, wherever Nature has been sufficiently beneficent in a water-supply.   Among the interesting towns of Eritrea is Zula, the old centre of the Axumites, and an early slave and ivory centre, also the point of departure of the British expedition to Magdala.

The Abyssinian people of the plateau here have pacifically accepted Italian governance, notwithstanding their martial nature.   The Arabic or Hamitic people of the plains are largely nomadic shepherds, and there are various negro tribes.   Among these varied people it is interesting to note the native industries of cotton-weaving and silver and iron working, as well as leather-working and mat-making.   A civil governor from Rome administers the colony, and commissioners control the provinces in conjunction with the village headmen and councils of native elders.   A grant-in-aid is made by Italy towards the revenue, largely consumed by military expenses.

Here we leave tropical Africa to take our way eastwardly through the torrid zone of Asia.

CHAPTER XXIV

ARABIA

FROM tropical Africa to tropical Asia we pass almost insensibly, for, across the narrow Straits of Bab-el-Mandeb, Arabia is topographically similar to Somaliland, and the two formed a whole in earlier geological ages.

The origin of the name of this famous strait—" The Gate of Tears "—which connects the Indian Ocean with the Red Sea, and is twenty miles wide, may have been in an Arabic tradition that it was formed by an earthquake which swallowed up a multitude of people in the act, opening a chasm which separated Africa from Asia. Probably, however, it is to the danger of its navigation that its poetic designation is to be ascribed. The passage is divided and commanded by the small island of Perim, over which the flag of Britain waves.

Arabia is a world of its own, a closed world as regards much of its desert interior, although one of the oldest of lands. Only half the country lies within the torrid zone, the tropic of Capricorn crossing it nearly midway. But Arabia, as a whole, is one of the hottest regions on the earth's surface, a condition arising both from its latitude and generally arid surface. Yet its great plateau, sloping from the great escarpment of the Red Sea to the Euphrates and the Persian Gulf, lies at a considerable elevation, the south-western escarpment rising to from 4000 to 8000 feet. Mexico, in the same latitude, offers somewhat similar topographical conditions, but is more fertile and temperate. One-third of the whole country is desert, and throughout the remainder but a relatively small part lends itself to the life of a settled and progressive people, due to the absence or scantiness of rainfall and water. The mountains are not high enough to intercept the monsoon rains, such as make of Abyssinia, partly in the same latitude, across the Red Sea, a mighty upland reservoir for the Nile.

From Bab-el-Mandeb along the Arabian Sea the coast-line

runs for 1300 miles, whilst the Red Sea coast—or from Port Said to Aden—is 1500 miles long. The northern side, from Port Said to the Euphrates—with which we are not here concerned—is 600 miles in length, and the great trapezium-shaped peninsula thus enclosed covers approximately 1,200,000 square miles.

The northern portion of Arabia is, in the main, desert tract, with some pasturage during a part of the year, the central region a dry, stony steppe, whose few wells afford the means of life to a nomad people, and in addition the great wadis or ancient river-beds which traverse the deserts are fertile enough to support the population of some large towns and villages. A vast empty space upon the map of Arabia denotes where lies the great uninhabitable southern desert of Dahna or Ruba el Khali—*Arabia Deserta*.

It is in the highland plateaus, where, due to the high elevation of the country and its proximity to the sea on the south and west, that the settled, fertile and cultivated *Arabia Felix* is found, with a sufficient rainfall and a temperate climate. But these pleasing areas are little more than a fringe to the country. Asir and Yemen in the west, facing the Red Sea, and Jebel Akhdar and the Batina coast of Oman in the east, lying under the tropic of Capricorn, irrigated from the mountain streams, are perhaps the most fertile parts of Arabia, and Hadramut, lying along the southern coast, is also a region of villages where man dwells in sufficient prosperity.

But Yemen, in the jebel or mountain land, was the real *Arabia Felix* of the ancients. Let us ascend the deep valleys, which, winding through the foothills, lead to the higher regions and the mountains, the scenery changing as we rise. The trees are covered with trailing plants, the swift-flowing streams are dammed at intervals with primitive barrages, whence channels take their way to irrigate the fields on either hand. Every height is crowned by stone towers, the *dars* of the inhabitants, and the roads, paved with large slabs of stone, conduct us to substantially built villages, offering a pleasing contrast with the mud or wattle huts of the lower country.

From base to summit the hills are covered with terraced fields, rising one above the other, and reveal both the fertility and value of the soil and the enterprise of the inhabitants.

Q

Striking and varied is the scenery. The bare, rocky slopes and dry ravines which scar them suddenly give place to the verdant terraces, extending upwards for thousands of feet, and perhaps carrying back the recollection of the traveller to the equally or more extensive terracing upon the Andes of Peru, the work of the ancient Incas. Here in Arabia, as in Peru, a vast amount of labour has been performed in building up rough stone walls—from five to ten feet high each—tier after tier, forming small terraces whose width is little over twice their height.

The jebel or highland towns which we reach in our ascent from the coast sometimes convey the impression of clusters of forts rather than groups of dwellings. The houses are high and of three storeys their walls of quarried stone, and the small windows with which the façades are pierced are surrounded with Arabesque ornament and a whitewashed border. But we are here in the midst of what is practically a vast coffee-garden, which covers this ancient, fertile and interesting western point of Arabia, overlooking the " Gate of Tears," the Red Sea and the African coast.

We may have entered the country through Mocha—or Mokha—a name which falls gratefully on the ear of the lover of coffee. Mocha is the town and seaport a little north of Perim and Bab-el-Mandeb, and earlier was the principal point of export for Arabian coffee. As we approach Mocha, from the Red Sea, our first impression is of an imposing city, but upon traversing its streets we remark that the houses, stone-built and of ample proportions, are half ruined. An ancient aqueduct supplies the place with water, and an arid plain surrounds it. The fame of Mocha has diminished, and, due to the poor anchorage, none but native vessels now use the port whilst the receding sea has left other ports upon the coast inland and remote.

The coast of Yemen, upon which Mocha lies, has now as its main seaport Hodeda, the only harbour of any note since the days of steamship traffic began. It is but a few miles away from Mocha. The coast is obstructed by islands and shoals. We traverse one of the typical coffee regions in ascending from Hodeda to Sana, the first plantations being reached at an elevation of 4300 feet above the sea, and this important

CASTLES OF TROPICAL ARABIA

cultivation extends to over 7000 feet up the western slope of the Arabian highland.

The mountains here reach some considerable elevation, the highest points rising to 11,000 feet. Upon this south-western point of Arabia, overlooking Bab-el-Mandeb, with its narrow straits, arises Jebel Sabur, a rocky mass and landmark from far away, nearly 10,000 feet above the sea.

From the crest of the range we descend some thousands of feet to the broad and open valleys of the plateau, and in such an environment lies Sana, the capital of the Yemen province. In this region the old cultivation is falling away—some of it has gone long ago—due to the drying up of the land or the washing away of the soil, the rocky ribs of the mountains having been laid bare and naked by these agencies, above the line of the wadis or water-courses. However, there are fertile vales where dates grow in profusion, with wheat, millet and other cereals, and fruits, and the slopes are overgrown with the juniper.

The Arabian coffee shrub does not grow wild, and it is believed to have been brought in from Abyssinia in the sixth century. In the valleys of Surad, the slopes of Haraz and Sabur, at Yafa and Zubed, the principal centres of its production are found, the coffee being planted in terraces upon the sloping hills, and shaded from the sun by the fig and the tamarind. Led from neighbouring streams, irrigation channels traverse the slopes to perform their function of watering these terraced gardens. The coffee plants raised from seedlings are set out when about six weeks old, in rows some five feet apart, and watered every two or three weeks, and the dried berries are sent down to Hodeda or Aden.

The husk which remains from the cleaning process is used by the people of Yemen in making their favourite drink. We may also remark the khat, a small bush grown hereabouts for the purpose of a stimulant. The leaves and shoots of this plant are chewed and have the same sustaining properties which the *coca* leaves of Peru and Bolivia possess. We must not fail to remark in some districts the rose-gardens, whose fragrant blooms are destined for the manufacture of atr, or attar of roses. Aromatic plants and flowers are much appreciated by the Arabians—jasmine, mignonette, lavender, thyme and others—who are fond of placing a bunch of such in their hair.

Very different from the jebel or mountain land is the appearance of the lowlands, which in south-western Arabia, in the districts of Hejaz, Yemen and Asir, consists of a narrow, coastal strip not more than twenty miles wide, lying at the foot of the mountain region we have described. This lowland strip of Tehama is partly sandy desert, partly pasturage or cultivated, especially where the torrents descending from above have deposited patches of alluvial soil. So fertile are these deposits under the warm, equable climate, and under the irrigation system—in which every available stream of water is controlled by dams and channels—that two or three crops a year are obtained from the soil, and a large settled population and important towns exist along the seaboard of the Red Sea. The date is here a principal article of food.

The great gem for which the south-western coast of Arabia is the setting is the sacred and famous Mecca, which lies about 150 miles below the tropic of Cancer. Mecca lies inland from the port of Jidda, on the Red Sea, a distance of " two camel marches," or about forty-five miles, in a sterile valley, forming the Haram or sacred territory around the city, as described in the Koran. Jidda was attacked unsuccessfully by the Portuguese in 1516. The district forms the heart of a mass of rugged hills, traversed by narrow defiles, backed by the huge mountain wall which cuts off the Tehama lowlands of the coast from the high plateau of desert Arabia, to which we ascended in passing through the coffee districts farther south.

Outside this—purposely—sterile valley, wells, gardens, date-palms, springs and green pastures stretch, but the agricultural possibilities of Mecca could never have supported a large population, and from what was at first, probably, a halting-point upon the great " incense route," the gateway between the lowlands of the coast and inner Arabia—and thus having a commercial *raison d'être*—Mecca grew to be a privileged and sacred place, and now plays the twofold part dictated by commerce and religion. But commerce, in Arabia, could very generally only be carried on with the sanction of religion, in a country so unsettled, with Bedouin hordes whose occupation was to waylay caravans, or receive blackmail for refraining from molesting them. The people of Mecca have become foremost in the international commerce of Arabia, but this is much less important now than it was in the Middle Ages, when merchandise from all parts of the Moslem world, transported over the long caravan routes, arrived for the great fairs. The great pilgrimages have retained their importance, and in attending to the wants of the pilgrims the inhabitants of Mecca live.

The normal population approaches perhaps 100,000, but the city is overcrowded in the season of pilgrimage, and the lack of sanitary conveniences is gravely felt, though the water is plentiful, and famous by reason of its virtues. " The fanaticism of the Meccan is an affair of the purse ; the mongrel population has exchanged the virtues of the Bedouin for the worst corruptions of Eastern town life, without casting off the ferocity of the desert. The unspeakable vices of Mecca are a scandal to all Islam, and a constant source of wonder to all pious pilgrims. Under cover of the pilgrimage a great deal of slave traffic goes on," says one authority.[1]

The city presents a somewhat picturesque appearance, with high stone houses lining ample streets, which last, however, are dirty and untended, and the enormous mosque, with its courtyard and arcades, is the principal architectural attraction, the streets ascending the mountain-sides. The mosque is used as a lecture centre on Mohammedan law and sciences associated therewith in the intervals of the pilgrim seasons. The low part of the city is subject to the severe floods of the tropical rains from the hills, to avert which the great dam of the Caliph Omar was built, ineffectually, however.

Medina, the " City of the Apostle of God " and of the Hejira and the resting-place of Mahommed, surrounded by its massive stone walls and well-built gateways, lies about eighty miles north of the tropic of Cancer, on the slopes descending to the Red Sea, a camel's march therefrom, and is reached by the railway which, running for over 800 miles from Damascus, extends towards Mecca. It has a population of over 20,000, and the mosque of the Prophet is its chief glory, together with the tomb of Fatima. The railway, completed in 1908, did much to consolidate the Turkish dominion over Medina.

Along the mere camel-track which forms the caravan route from Damascus to Medina and Mecca, some 6000 pilgrims and nearly twice as many pack-animals annually take their way

[1] William R. Smith in *Encyclopædia Britannica*.

across the torrid desert, paying tribute to the Bedouins to secure immunity from attack of these predatory tribesmen, drinking from the reservoirs on the way. This and other caravan routes are the principal means of travel in Arabia, with the railway and the sea routes between its port, served by the steamships and the native craft. The pilgrim traffic increases. At Jidda, for Mecca, 75,000 persons land annually, a quarter of them from British India and a lesser proportion from Java and the Straits Settlement, British steamships bearing the majority of these multitudes who so persistently set their faces towards Mecca, and who bring into Arabia somewhat under £1,000,000 per annum in money.

The coffee trade of Hodeda falls off from its earlier value of nearly a third of a million pounds a year, the decrease being due to the extension of coffee-planting in other of the world's tropical regions.

East of Bab-el-Mandeb, flanked by high mountains, the maritime plain runs for 200 miles. We enter the Bay of Aden and are upon British territory.

Aden was also part of the *Arabia Felix* of the Romans, who captured it about 24 B.C. The Portuguese attacked Aden in 1513, and the Turks later occupied it. The British occupation has a curious and romantic origin. A vessel under the British flag was wrecked upon the coast in 1837, and its crew and passengers were grievously treated and the ship plundered by the Arabs. As part of the compensation which Britain demanded for this cruel and unfriendly act the Sultan offered to sell the town and port to England, and an agreement was entered into. This transfer was effected, however, only under the pressure of a naval and military force, and Aden was then annexed to India by the British.

Since the Suez Canal was opened the importance of Aden as a seaport and coaling-station has greatly increased, and there is also a large trade with Somaliland. The population is somewhat under 50,000. The town is built in what perhaps is the crater of an extinct volcano, and surrounded by natural fortifications of mighty precipitous rocks, the hilly mass upon which it is situated reaching over 1700 feet above sea-level, and, joined to the mainland by a sandy neck of land over which the sea almost washes, the place is practically impregnable. Aden is a fairly healthy place, but suffers from its limited water supply, which is obtained from old rock reservoirs, wells, aqueducts and by condensation. The area of Aden is but fifteen square miles, but a total of eighty square miles of territory are held by Britain in the vicinity, obtained by purchase or cession. With its admirable strategic and convenient position Aden is a place of much importance, and one of the most remarkable of Britain's tropical seaports.

Shabwa, an ancient capital, lies farther east, half buried in the encroaching sand of the high desert, but the long, fertile valley of fields and date groves is occupied by towns and villages.

Continuing along the Arabian coast, we reach a moderately fertile strip of land at Gara. This was an ancient centre for the trade in frankincense, the trees which bear the spice growing on the hills which descend to the shore. Behind the escarpment is reached, 4000 feet above, beyond which the great desert extends as far as the eye can reach.

Off the Arabian coast lie the small rocky, lofty Kuria Muria Islands, belonging to Britain, and used as a cable station.

In the eastern part of Arabia, cut by the northern tropic, we reach the district of Oman, with its lofty mountains and narrow, fertile Batina coast, dotted with populous towns and villages. Here stands Muscat, the principal seaport and capital of the province, lying almost upon the tropic of Cancer. Muscat has been called the hottest place in Arabia. It is encircled by bare, rocky hills, which reflect the heat of the sun. Muscat commands the entrance to the Persian Gulf, and is a valuable naval base. Its history is an interesting one, and the numerous ruins of Portuguese origin date from the time of Portugal's occupation of the Gulf, from 1508. The place is isolated from the mainland, the mountains rising behind like a frowning wall, beyond the low, sandy isthmus which connects Muscat and its lofty ruined fortress of Jalali with the Arabian coast. Walls with towers further defend it. The British Residency is situated here—an Indian political resident—and the Sultan's palace is near the centre of the town. Muscat exports large quantities of dates.

The fertility of the highland region behind Muscat is in remarkable contrast with the rocky seaport itself, luscious fruits and verdant vegetation rewarding the husbandman who by his ingenuity has made so much of the perennial springs issuing from the rocky clefts. We remark the underground channels in which the water is conducted, reminiscent of such in Persia. Irrigated fields and palm groves surround the well-built villages, and beyond are great oases, supporting a considerable population, giving way to saline marshes, sand and the desert mirage.

The opening of the overland route to India brought the west coast of Arabia into greater prominence. The province of Hejaz, in which Mecca lies, fronting the Red Sea, and part of Yemen, came under Ottoman control, mainly by conquest in 1872, but the whole of Arabia's southern coast, east of a diagonal north-east line which cuts inland from near Bab-el-Mandeb to the desert, lies under the British sphere of influence, including the Aden protectorate, with Hadramut and Oman, in the east. From Oman to Bab-el-Mandeb the tribes have agreed not to bargain away any territory without British consent.

Turkish rule over the great provinces of Hejaz and Yemen—with their administrative centres at Taif and Sana respectively—has never been accepted by the Arabian people, except inasmuch as it is imposed by force, and it is only by reason of large subsidies of money that the powerful Bedouins of Hejaz have remained acquiescent. In the northern districts the dilapidated Turkish soldiery live and gather their salaries as best they can. Constant revolt against Turkish rule occurs, and in 1905 Saria was surrounded by the Arabs, and only the most strenuous fighting held it for the Ottoman Empire. Thus is the land held under conditions which cannot endure.[1]

The interesting lands of Midian, of the Sinai peninsula, Palestine and the Euphrates, so interwoven with the familiar history of the great Scriptural characters, heroes and pilgrimages, lie far beyond the tropic of Cancer, and consequently, being outside the torrid zone, do not come within the scope of our present description.

The population of Arabia cannot be stated except by mere approximation, and may reach a total of about 5,000,000.

[1] Considerable fighting took place in various parts of Arabia, especially near Aden, in the Great War—1915—between British and Turkish forces.

ARABIA DESERTA

The Arabs claim descent from two sources—from Shem and Ishmael, those of the first being the more settled and agricultural people, generally of Southern Arabia, those of the latter the pastoral and Bedouin tribes, although these distinctions are merged into each other. The privileged religious class of the south—the Sharifs—claim descent from Mohammed. There are many negroes, slaves or freed, and in Southern Arabia a large number of Jews. British Indians are largely represented in the traders of the coast towns.

Into the archæology or ethnology of Arabia we cannot enter here, but the tropical part of the country is held to have been the cradle of the Arabian people and their civilisation, which has so enormously influenced much of mankind.

The finest type of Arab is found in the southern part of Arabia—whose deserts gave origin to the race—among the roaming Bedouin tribes and mountaineers. Anything in the nature of monarchial control or political organisation is repellent to them. It is held that the highest Arab civilisation flourished before the time of Mohammed; art and poetry and the condition of woman were superior then than under the Koran, with the harem and the veil. Arabia, however, has never formed a real nation, and its one-time brilliant kingdoms are now levelled.

Physically the Arabs are regarded as more perfectly constituted than the European races, and with the simple habits which characterise many of the tribes they often reach an extreme and healthy old age. Mentally they have been kept back by their lack of organising power, to a large extent. Dignified, unsmiling, but not unkindly, is the typical Arab's countenance, and his demeanour is courteous, brave and manly, but he is also cruel, superstitious and untruthful. He " defends his guest at his own peril and submits to the reverses of fortune with resignation; he pities and aids the wretched and never forgets the generosity of an enemy. The Turk is cruel, the Arab grateful." His tranquillity, however, may cover meditations of revenge.

Arab towns are generally surrounded by walls and a dry moat, and many of the shops are kept by women. The lack of sanitary methods is somewhat offset by the dry climate. We remark the use of colour and whitewash on the walls of the houses, a custom which has filtered, through Arab and Spaniard, to Mexico and Peru, along with the use of sun-dried bricks, of which Arab houses are exclusively formed.

The hospitality of the Arab has become a byword, and the *kawah*, or coffee-room, to which guests are conducted, is one of the principal apartments of an Arab house—often elaborately adorned. Tea is now largely consumed also. One solid meal a day is the Arab custom, generally eaten after sunset. Sandals are more often worn than shoes, and a long silk gown over the long shirt distinguishes the richer folk, with loose cotton trousers in some cases. The domestic servants are often slaves, as slavery is a recognised custom, negroes being brought in, although under illegal trafficking. There is no particular " colour line " between the Arab and the negro, and the two races intermix. The Arab women are as brave as their husbands, and in history have taken part in the wars. Education is almost unknown as understood by Europeans, but reading, writing, grammar and the eloquent conversation and courtesy that distinguishes Arab children are acquired from the home teaching of their parents. Arabia, it is to be remarked, is not famous for its horses, contrary to the popular illusion.

Let us turn from Arabia seaward, to where the island of Sokotra lies in the Indian Ocean, 130 miles east from Guardafui, the Horn of Africa, on the route to India by the Suez Canal, and about 200 miles south-west of the Arabian coast. Imposing in appearance as we behold it from the sea, culminating in rugged pinnacles rising from a terrace plateau, is this ancient and interesting isle, which lays claim to be one of the oldest incense-producing countries in the world. Egyptian galleys must have visited it during the twenty-five centuries of the Pharaohs' reign, probably finding here one of those "harbours of incense " which they exploited. Here the precious " auta " trees, which furnished the ' 'divine dew " for the service of the gods, grow in great abundance and variety, and frankincense and myrrh, aloes and pomegranates are among its principal products, as well as ambergris.

The island has an area of about 2000 to 3000 square miles, and a population, unnumbered, of perhaps 12,000. People of a mixture of Arab, negro, European and Indian blood inhabit the coast, and the ancient Sokotri—probably from Arabia in times unknown—the hilly interior. The capital, Hadibu, is a small place of less than 500 inhabitants, picturesquely situated and surrounded by groves of date-palms.

The people live an extremely simple life, in round, flat-topped houses or caves, and some of them are as white as Europeans, and as tall. The principal food is in dates and milk. Cotton, tobacco and millet are grown, but the aloes and dragon's blood trees are the most valuable products.

This " island abode of bliss "—for such its name originally may have meant—was early a meeting-place of Indian and Arabian ships, and in the tenth century was a haunt of pirates and corsairs, who were described by some as " Christians and pirates." The Portuguese seized the island in 1507, in their desire to control the trade routes to the East, but Arab dominance followed. On the opening of the Suez Canal Sokotra was secured by Britain, as a dependency of Aden, but native rule is maintained.

CHAPTER XXV

INDIA, CEYLON, THE LACCADIVE AND MALDIVE ISLANDS, ETC.

OF that vast, populous and diversified portion of the Asiatic continent, to which Europeans have given the name of India, but for which its own people cannot be said to have any comprehensive nomenclature, it would be impossible here to enter into any description beyond that necessary to the scope of this work. India is cut midway by the tropic of Cancer, and consequently its southern half lies within the tropics, and similar conditions obtain in Burma and Indo-China, or " Further India," of which the independent kingdom of Siam is a part, with Cambodia, Cochin China and other regions adjacent.

The outstanding sociological feature of India is its enormous population, embodying, with Burma, nearly 320,000,000 souls, all but a fifth dependent upon the single industry of agriculture. The total area of India and Burma is over 1,766,000 square miles, and of this nearly two-thirds consists of British territory directly administered by British officers, the remainder being formed of native states acknowledging the supremacy of the paramount power. There are about 3,000,000 native Christians.

The great provinces of Madras, Mysore, Hyderabad, Bombay, the Central Provinces and some smaller ones, lie wholly or mainly within the tropical half of the country, covering the peninsula, the tropic of Cancer cutting this near the Asiatic mainland. To the south lies the island of Ceylon ; to the west and south the Laccadive and Maldive Islands ; to the east the Andamans and Nicobars, all British tropical possessions.

The great central state of Hyderabad contains over 11,000,000 people, and is in parts mountainous and wooded and elsewhere flat or undulating. Much of the land is fertile and rich, and its scenery often picturesque, but there is still a large quantity of good land not yet brought under cultivation, whilst other portions are sterile. It is generally well watered, with numerous artificial "tanks," and the climate good, with an absence of hot winds. Rice and millet, wheat, cotton, oil-seeds, tobacco, sugar-cane and a great variety of fruits and garden produce are raised. Native silk is largely used, and hides are articles of commerce. Cotton, oil-seeds, country cloths and hides are exported. Iron, coal, copper, diamonds and gold are chief among its minerals. The Nizam's State railway yields a handsome profit from the working of the coal mines, and in the rich cotton country many cotton presses and mills have been erected. The capital city is the fourth largest of India, with nearly 500,000 inhabitants, and its picturesque bazaars are thronged with natives from all parts of India.

The vast province of Madras, facing the Bay of Bengal, has a population of nearly 42,000,000. Irrigation cannot be practised generally here, and agriculture is more hazardous than in Hyderabad, but the customary tropical products of India are raised. The greater part of the soil is held by the cultivators direct from the Government, about a quarter being owned as private estates. There are but few staple manufactures in Madras, and hides, coffee and raw cotton are exported.

The Bombay presidency, facing the west, has a population of over 25,000,000. It is partly mountainous and partly formed of rich, highly cultivated alluvial plains with extensive cotton tracts. The great River Indus enters the Arabian Sea in this state, slightly north of the tropic of Cancer. Bombay, the great tropical city on India's western shore, has an area of twenty-two square miles and a population of over 1,000,000. It is the second city of India, having fallen slightly behind Calcutta. Its position on the side nearest to Europe and its railway and port advantages, together with its monopoly of the cotton industry, are great advantages, but it has no great river serving it, such as the Ganges affords to the Calcutta region. The water front is perhaps the finest of any city in the world, and the well-built and handsome native bazaars and spacious streets devoted to commerce with Europe attest the importance of this great city.

The docks are very important, and the cotton-milling industry comes second, followed by flour and other mills. Plague and over-production brought something of a temporary decline in the closing years of last century. The city owes its modern foundation to the British, since 1668. A famous industry is that of its black-wood furniture and inlaying.

The province of Mysore embodies a hill country and an open country, the last covering the greater part of the state. The valleys and wide-spreading plains are covered with populous towns and villages, and the whole population is over 6,000,000. The streams are everywhere embanked to form reservoirs to irrigate the fertile soil, the water being used again and again at successive levels. There are 20,000 of such tanks in the country. Mysore may perhaps be described as the most prosperous native state in India, and is generally immune from famine, although one-fourth of its people perished from that cause in 1876. Large profits from railways and gold mines are made. Coffee, gold, sandalwood and ivory are its chief exports.

The Central Provinces have a population of over 11,000,000, and consists of upland and plain, and in the fertile areas wheat and other grains are grown, and cotton is the staple crop. There are several hundreds of cotton factories in the state, and these, with coal, form the only important non-agricultural industries. Raw cotton, wheat, rice and oil-seeds are exported. The export of wheat fluctuates during periods of famine.

Only the southern or coastal portion of the province of Bengal lies within the tropics, as do the mouths of the Hugli and Ganges rivers, which traverse it. Calcutta stands a few miles within the tropics. The valley of the Ganges is one of the world's most thickly populated and fertile districts, and it abounds with every product of nature, from cereals and spices to cotton, silk and all tropical products.

Almost everything necessary to the life of man, or for foreign trade, abounds in this vast and wonderful valley. The delta of the Ganges, where the irrigated rice-fields lie, forms a scene of singular interest and novelty. Calcutta, the vast tropical capital of British India and of Bengal, has a population of nearly 1,000,000. It lies eighty miles from the sea, on the River Hugli, and was called by Macaulay the "City of Palaces." Its public buildings, however, do not compare with those of Bombay. The city owes its prosperity to its riverine position, and the port is one of the busiest in the world, and the shipping rivals that of the port of London in its profusion.

In India the land furnishes the main source of revenue, and the assessment of the land tax is the principal work of Indian administration. The land system is based on the ancient Hindu village community system, in which the land was not held by private owners, but by occupiers, and the revenue was payable by the community represented by its head man. The proceeds of the harvest were pooled and the share of the State set aside before general distribution, a system similar in some respects to that obtaining under the Incas of Peru. This system obtains throughout large portions of India up to the present day (but in some districts the organisation has decayed) and here the land, which was regarded primarily as a source of livelihood and State revenue, is not subject to commercial speculation. British prejudice introduced the landlord system.

Under the community system [1] the cultivation of the soil is the basis of Imperial revenue and the unit of the social system, and the village organisation includes the craftsmen in other industries, who exist for its benefit and who raise grain for their own families. Agricultural life and work is partaken in by every class in these village communities and is surrounded with ceremonious and religious observances. Cultivation is very assiduously carried out, and the experience of centuries has taught the husbandman the arts of irrigation and manuring, and these, with the great native patience and ingenuity which are applied to the soil, give results which not even the most scientific or colossal engineering projects of the Government can always exceed. One of the most important crops is wheat, of late years, but mainly in the non-tropical states, and oil-seeds are universally grown. The oil was used as an illuminant in the native lamps, but there is now also a great demand for it from Europe. Mustard or rape, linseed, til and castor are the principal, the oil being expressed in India in mills and the cake used as food for cattle, and in some cases for the people.

---

[1] These communities of India, which have existed for centuries, and which in some respects have been wonderfully productive of local economic rule and order (without any State interference), although opposed to European ideas of property, are worthy of close study by the social reformer to-day. An account may be found in *Handbooks of Hindu Law*, Cornish, London, 1915.

The indigo planting and export, which was formerly the most important business in India financed by British capital, has been ruined by the German synthetic dye, notwithstanding its superiority over the chemical compound. Tea to a large extent has taken the place of this industry. Assam may have been the original home of the tea shrub. Tea-planting is the result of Government initiative; that of coffee was introduced by the natives; cinchona or quinine was also introduced by Government work. Sugar-growing is important, but does not meet the native demand. The forests of India are an important source of natural wealth, and further devastation is being checked by Government control. Teak is the most valuable timber.

The manufacturing industries of India are developing but slowly, and do not aid so far in reducing the pressure on the land. The most important is weaving, in the cotton mills of Bombay and the jute mills of Bengal, and these are increasing. Bombay, indeed, is likely more and more to rival Manchester. Its cotton mills, moreover, have developed earlier Manchester defects : in the abuse of child life in the factories, and "inhumanly long hours," the last due largely to the introduction of electric light, when, perforce, earlier, they closed at sunset.

One of the most regrettable industrial conditions of India is the decline of the old indigenous industries. Formerly, up to the middle of the nineteenth century, there existed in various parts of India colonies of hand-workers and skilled craftsmen in cotton, silk and carpet making, and in brass and silver ware, as well as in wood and ivory, but both the output and artistic character of these have fallen off. Indian cotton goods found a market even in Europe before the English industry arose, which now swamps the beautiful old native work, and cotton manufactures and yarns, imported exclusively from the United Kingdom, form forty per cent. of the Indian import trade. The modern Indian mills supply some of the demand. The original importance of the hand-loom in the country may be judged from the fact that the census returns at the beginning of this century gave nearly 6,000,000 hand-loom weavers in India.

The export of cotton from India, which rose to a value of 37,000,000 sterling at the time of the American Civil War, now reaches only about 8,000,000, the cotton being inferior to other varieties. It has been shown that Indian labour is inefficient in comparison with the Lancashire mill hand, one of which can do the work of six Indian power-loom weavers and nine hand-loom weavers. Thus it is argued that India cannot compete with Europe in this, notwithstanding its cheap labour and home-grown cotton.[1]

The metal manufactures of India are, from an artistic point of view, among the most important. Special alloys and famous bells are made, and the village brazier and blacksmith manufacture many articles of daily use. The deposits of iron ore of India, found everywhere, are, as regards the antiquity of their exploitation and the purity of the material, perhaps the first in the world. The native methods of smelting, a legacy of countless generations, yield metal of a fine and easily worked quality. The wrought-iron column of Kirlab, 3000 years old, is an instance of Indian metallurgical skill. This column weighs eight tons, and is a puzzle to the archæologist. European ventures in this field have, however, generally been failures, though one very important iron-works has been established. Copper is a further mineral found freely, and the gold of the Mysore mines has become famous, and has produced for its British owners a score of millions sterling.

India, as a whole, is a land teeming with natural wealth and advantages, and within it stand some of the most beautiful temples, palaces and other constructive works that the mind of man has been able to conceive. Over it many princes hold sway—potentates whose rupees and diamonds are not unworthy of the fabled stories popularly associated therewith. But India is a land teeming with millions of the poorest and most backward people in the world, notwithstanding its thousands of years of civilisation and its Aryan origin. The Indian in his abject poverty, ignorance and heathenism strikes a note both of pity and reproach in the observer's mind. Half fed, half naked, subject to the ravages of famine, and with 250,000,000 of its people unable to read and write, is India to-day.

With all its vast population, it cannot be said that India is necessarily over-populated, as considerable portions of the

---

[1] The question to the student of " human geography " might well be as to whether any competition is desirable, and whether the Indian should not supply his own demand irrespective thereof.

R

country are but thinly inhabited. The future of its people will undoubtedly depend upon a better organisation of the natural and industrial resources.

Perhaps the most remarkable economic feature at the present time in India is the lack of manufacturing industries, for, as we have seen, practically the whole population depends upon agriculture. In this connection the disinterested economist will remark what appears to be grave evils in the "human geography" of the country. The industries of India can scarcely be stimulated and increased whilst the country is kept as a "preserve for the manufacturing output of Lancashire" and other goods ; articles, be it said, the right and ability to manufacture which are India's heritage.[1]

The famines of India form a subject of terrible import, and to a large extent they are allied with economic causes, and so are a subject falling within the scope of "human geography." [2] Primarily they result upon failure of the monsoon rains, upon which India depends for its normal crops. It has been stated —principally by Indian economists—that famines have been more prevalent in India since British rule began, 140 years ago, during which period there have been twenty such and various times of severe scarcity. Some of these famines have been terrible. In the famine of 1876 to 1878, for example, 5,000,000 people perished, and, as late as 1899 to 1901, 1,000,000 died from the same cause, with an estimated loss to India of £50,000,000.

But it is difficult to believe that these have been caused by British rule, which, on the contrary, has taken the most energetic measures in relief works and in matters of a permanent character, such as the extension of irrigation and railways, whilst under the native rulers no systematic efforts were made. Yet, on the other hand, the economic situation of India, maintained by Britain, is probably a contributing factor. The Indian people, to the extent of four-fifths of the population, being dependent upon agriculture for their living, are, when this fails, thrown out of work and have no money to buy food. Thus it is not necessarily to lack of food in the country that famine is due, but to lack of means to purchase it, and so it

[1] The author is not concerned with old-fashioned questions of " Free Trade," etc., but looks to a wider economic science.
[2] " Human geography " as meant in this book (vide last chapter).

CEYLON · TWO TYPES OF "CULTURE": MODERN VILLAGE FOLK AND COLOSSAL STATUES (NOTE HUMAN FIGURES STANDING THEREBY) CARVED OUT OF THE ROCK, OVER 1000 YEARS OLD

becomes a question of employment largely. If,, therefore, there were more manufacturing industries and factories, the dependence upon the land would be decreased; but, as we see, the Indian is to some extent deprived of manufacturing occupations, partly by Britain's fiscal policy and exigencies of oversea trade. This brings us again to the almost un-considered science of " industry-planning," [1] under which, in the endeavour to make agriculture and manufacture com-plementary to each other, economic equilibrium might be more nearly approached. There are, of course, difficulties in the way of obtaining capital for Indian manufacturing industries, and even where such have been established by native capitalists the wage paid and general conditions of labour therein are exceedingly poor, and sometimes oppressive.

It has been stated by economists that there is always enough grain in India to feed the people, but the problem is to trans-port it from one part of the country to another in famine times. The railways can scarcely cope with such sudden traffic. Further, the existence of railways, whilst of vast benefit, has also induced the native to sell his wheat stock, when, from the point of view of the community, it might have been more beneficial to store or retain it for home consumption.

To express these conditions perhaps crudely, Britain appears to take vast quantities of wheat—and other raw material—from India, under fiscal conditions satisfactory to herself, and to send out vast quantities of manufactured goods, also under fiscal conditions satisfactory to herself, but conditions not necessarily the most beneficial to India. Doubtless the future will be constrained to find more equitable economic methods than these, and in seeking them Britain would add to her great services to this tropical and sub-tropical land.

Here we leave the Indian mainland and look southwards to the island of Ceylon, one of the most picturesque and valuable in the world, though it would be impossible here to do more than touch upon the infinite variety of its resources and history.

As we approach this " utmost Indian isle," as the old geographers termed Ceylon, the truly beautiful aspect of the

---

[1] Advanced by the author in his lecture, 1913, before the Royal Society of Arts in London.

land unfolds as seen from the steamer's deck : the shores, clothed with an unmatched covering of luxuriant coco-palms, the hills, monarch of which arises the striking " Adam's Peak," the whole forming a picture which does not soon fade from the traveller's mind.

Ceylon is a British Crown colony. It covers an area of over 25,000 square miles, reaching to within 6° of the Equator, and its highest point attains an elevation of over 8000 feet above the sea. Against the lofty mountains the monsoon clouds are driven, and their waters descend in copious but intermittent rivers to the coast, the hydrography varying greatly, however, in different districts. The island seems to have been slowly rising for ages from the sea, according to the indications of its geology. The vertical zones of climate, and in consequence of vegetation, are strongly marked—that frequent characteristic of elevated tropical lands.

Dense masses of tall forest cover some of the hills, and timber trees are found from the beach to the loftiest peak ; and masses of beautiful flowering rhododendron, sometimes seventy feet high, beautify the landscape. Trees, plants and flowers of both tropical and European varieties are widely encountered. Most valuable of the native trees is the coco-nut palm, its fruit and other products yielding oil, fibre, food and drink, and timber for native utensils. Cinnamon is one of the abundant trees, and rubber and quinine are exceedingly important here.

Coffee, one of the earlier famous products of Ceylon, has been largely replaced by tea, and in the growing and curing of this many millions of pounds sterling are invested, the industry having raised the island from its former depression.

We remark how in this beautiful land every valley and plain capable of cultivation yields its crops of grain, and how the system of terrace cultivation has covered the hill slopes with stepped plantation, where verdant rice crops are watered by one of the most ingenious native irrigation systems in the world— conduits carried around the faces of precipices in earthen and bamboo aqueducts.

Indeed the irrigation system of Ceylon, the tanks, aque- ducts and channels built to store or convey water, date in some cases from 500 years or more before the Christian era, and are of the utmost interest to the engineer to-day. The irrigation methods bear witness to the great patience and skill of the native agriculturists, and the ancient reservoirs, which through the centuries had often fallen to ruins, have in many cases been restored under the British. The native rulers covered the whole face of the land with their hydraulic works, and some of the old reservoirs form magnificent artificial lakes.

The population of Ceylon is about 4,000,000, composed of people of various races and religions, the Sinhalese largely predominating, and the Buddhist faith being that of over half the community, with Hindus and Mohammedans following, and about 400,000 Christians of all sects. Ceylon may be called a Buddhist country. A great advance in education has been made since the last quarter of the past century.

The people of Ceylon are skilful craftsmen in almost all trades and manufactures, from jewellery to carpentry and masonry, but comparatively little manufacturing is carried on. The principal article of export is tea, and the chief imports are Manchester goods and other foreign manufactures.

Among one of the valuable adjuncts of labour here is the famous and intelligent Ceylon elephant, which, wild, exists in considerable numbers.

The early history of Ceylon is wrapped in fable, but rock inscriptions in the sacred city of Anuradhupura show ancient Buddhist dominance. The Portuguese arrived in 1505, and established a fort at Colombo, but they did not extend their sway far inland. In 1658 the Dutch drove them out, and entered upon a wise economic policy, which, however, ended a century and a half afterwards in weakness and cowardice. The British East India Company occupied all the Dutch forts at the close of the eighteenth century, since when England has controlled the island uninterruptedly.

Ceylon, like India and some other " heathen " lands, is an example of a community wherein an ancient culture and a wonderful applied knowledge in the development of natural resources and material arts is accompanied by those singular beliefs and practices which show how slowly the world approaches a common-sense spirituality and habit of life. To-day pilgrims from all parts of the East ascend Adam's Peak, where in the hollow of the lofty rock a depression is variously claimed as the footstep of Siva by the Brahmins, of Buddha by the Buddhists, of Adam by the Mohammedans, and of St Thomas by the Portuguese Christians. This "footstep" is protected by a finely built roof, and by the jealous guardianship of a rich monastery.

The cultivation of tea, as remarked, has worked profound changes in Ceylon. In 1877 there were well over 25,000 acres under coffee, but this "monoculture" was ruined by a plant disease, which brought on a severe economic crisis and almost total decline of cultivation of the berry. The planters then turned their attention to tea, and at the beginning of this century nearly 400,000 acres were under tea cultivation.

Millions of acres of land still await more general cultivation, and the growing of quinine, or cinchona—which earlier was so widespread as to bring down the cost of the drug to a tenth of its former price—of cocoa, rubber, cardamom and so forth, is increasing. Rice, protected by import duties, is being more freely grown, but there are still imports of millions of bushels annually. Local tobacco is esteemed and some exported, as are coco-nut products and cinnamon. Articles of food and drink are largely imported. The minerals of commercial value for export are principally gems and plumbago, and in the obtaining of the latter Cornish and Italian miners are superseding the natives.

The whole hill country of Ceylon is traversed by splendid roads, and every village has the advantage of excellent secondary roads, the highway construction policy being a gift of British rule. Colombo, the capital of this fertile island, is a city with many fine institutions and modern utilities, and it stands upon one of the finest artificial harbours in the world. It is the converging point of the island's railway system, which now gives through access to India by means of the viaducts across the dividing strait and island of Adam's Bridge. The population of Colombo is somewhat under 200,000.

West of the Indian coast lie the small Laccadive Islands, which, discovered and fortified by the Portuguese in 1498, fell under British sway in 1877. The total area is about eighty square miles, with a population of over 10,000, people of mixed Arab and Hindu descent. So low does this archipelago lie upon the surface of the Arabian Sea that, were it not for the coco-palms which thickly clothe the islands, they would scarcely be visible. Enormous excavations in the removal of coral have been made here in unknown ages, with the purpose of exposing the sand for cultivation—vast " gardens " which, according to local tradition, were made by giants in olden times. The making of coir, or coco-nut fibre, is the principal industry of the people, largely carried on by women.

Southward lies the Maldive Archipelago, an immense number of small islands of which 300 are inhabited by a Mohammedan population numbering about 30,000. These people are described as industrious and intelligent, and they grow their own food and make their own clothes, mats, boats and other articles. The excellence of the island weaving is noteworthy. Coir and cowries, tortoise-shell, coco-nut products and dried fish are exported. Upon the Minikoi atoll stands a lighthouse familiar to the traveller in these waters, as are the numerous wrecks that mark its dangerous reefs.

The Andaman and Nicobar Islands, nearer to Burma and Malaysia than to India, deserve more than the passing notice here accorded to them. The total area of the first, 240 islands, is 2500 square miles, and of the second, of nineteen islands, 635 square miles. The natives of the Andamans are remarkable in having remained savage—with a reputation terribly but exaggeratedly savage—notwithstanding their position relatively near lands where great civilisations and empires have come to being and near the track of a great seafaring commerce which has gone on for 2000 years. It is as a penal settlement, however, that the Andamans are most interesting: the " scientific " treatment of convicts. Both archipelagos yield most of the customary tropical products, except that, peculiarly, coco-palms do not exist in the Andamans.

CHAPTER XXVI

INDO-CHINA, BURMA, SIAM, CAMBODIA, COCHIN-CHINA, FORMOSA

THE Asiatic continent dips deeply into the torrid zone in its south-easterly portion, and below the tropic of Cancer here lies the southern extremity of China and the Indo-Chinese peninsula, with Burma, Siam, Annam, Tongking, Cambodia, Cochin-China forming " Further India," terminating in the long, narrow Malay Peninsula.

Burma, a province of British India, is a peculiarly shaped region extending for a great distance along the eastern side of the Bay of Bengal, from latitude 10° to far beyond the tropic of Cancer, a length of some 1200 miles. On the east, in its upper portion, Burma is bounded by the great Mekong river, and, below, by Siam and French Indo-China, whilst on the narrow southern boundary are the Siamese Malay States.

The total area of Burma is about 239,000 square miles, but the upper portion lies beyond the tropics. The population is about 12,000,000. The Burmese, who predominate, are a brown people, active and well proportioned, and from their lively, gay but somewhat lazy temperament they have earned the sobriquet of " the Irish of the East." The women are more active in business matters than the men. As to the Burmese family relations, it is said that the people adore their children, who are described as the happiest and merriest in the world.

Burma is a land of varied physical characteristics. There are rich alluvial plains, barren, hilly tracts, fertile and well-populated, open plateaux, and masses of rugged hills cut deep with narrow gorges in which great rivers are confined. Burma proper is enclosed by a wall of mountains on three sides, some of the peaks rising to nearly 10,000 feet. It is traversed by huge rivers running from north to south, the principal of which is the Irawadi, which rises in unexplored regions far to the north, where India, Tibet and China meet. This great river is navigable from the sea for 900 miles to Bhamo, and its tributaries for lesser distances.

The Irawadi forms an enormous delta, a vast expanse of level plain, and here, twenty miles from the sea, stands Rangoon, the capital of Burma. This fine city draws much of its life from the river and its rich delta, and has risen within half-a-century from comparative insignificance to the position of a city of more than 250,000 inhabitants, surpassed as a seaport and in the volume of its trade only by Calcutta and Bombay. The port at times presents almost as busy a scene as the Hugli at Calcutta. Rice-mills and teak wharves line the river front, and the export of rice is valued at many million pounds sterling annually. Rangoon is the centre of Burmese religious life, and the magnificent shrine, Dagar Pagoda, a great pile covered with gold from base to summit, and 368 feet high, dominates the city architecturally.

Indeed religious buildings, which include various Christian cathedrals and churches, monasteries, temples, pagodas, mosques, synagogues and so forth, cover an area out of all proportion to the size of the city, and cemeteries are equally numerous. The death-rate is high. The markets are good, and the native industries of wood and ivory carving, silver-work and so forth are noteworthy. Many improvements have been effected in public works and utilities of late years.

Burma is divided into Upper and Lower Burma. The old province of Tenasserim, a narrow strip between the Bay of Bengal and the range of hills on the border of Siam, is the territory from which British control extended. The capital is Moulmein. The Shan States are intersected longitudinally by the great Salween river, rising, like the Irawadi, in the almost unknown territory of the Tibetan mountains.

The principal industry of Burma is agriculture, only ten per cent. of the people being town-dwellers. Men, women and children labour in the fields, their food consisting principally of boiled rice and fish, chillies, onions, oil and various vegetables. Elephant flesh is also eaten, as indeed is almost every animal except snakes. The principal crop is rice.

One of the chief assets of Burma are the forests, which are the finest in British India, and the existence of vast supplies of teak—largely used for shipbuilding—was one of the reasons for annexation of the province. Teak may be used advantageously in conjunction with iron in shipbuilding, on account of the nature of its essential oil, which, unlike that of oak, does not corrode the metal. The teak forests are very carefully preserved by the Government, under a special forestry department.

Many of the Burmese farmers or land-workers work also as artisans, practising various trades in the off season, such as timber-felling, fishing, rice-husking, silk weaving and dyeing. The native industries of Burma, however, suffer by reason of modern commercialism. A blow has been dealt at hand-weaving by the import of cheap foreign cottons and silks, and the fine native vegetable dyes have been driven out by the imported aniline preparations.

However, large numbers of people—over 420,000—are engaged in the production of textile fabrics, of silk and cotton—mainly girls and women. The most expert male silk-weavers are found at Amarapura. Brown dye, known as *cutch*, obtained from the *sha* wood, is greatly used, and cutch-boiling is an important occupation for a large number of the poorer class in parts of Lower Burma. The bright-coloured fabrics of which the Burmese are fond, such as yellow and pink, harmonise well with their dark olive skin, but these fashions are changing towards the dingier colours of Western modes.

The wood-carving and silver-work of the Burmese is characterised by much breadth and freedom, and such work constitutes the principal native arts. But it has to be recorded that these national arts are losing their special character through European influence—a very marked condition in Burma. Many thousands of people are engaged in the fishing industry, and minerals—especially tin, jade, ruby, oil, coal and gold—are worked. Petroleum production is becoming important. Amber and salt are other products. Iron is found in many districts and worked by the people, who make native implements therewith.

The chief Burmese exports are rice and timber, followed by cutch dye, rubber, cotton, petroleum and jade ; and the imports are largely cotton, silk and woollen goods, hardware, and so forth.

Among Burmese cities the name of Mandalay is famous from

its subjection by Britain in 1885, after King Thebaw had issued a manifesto commanding his subjects " to drive the British into the sea." However, the king, the day after the occupation, was himself taken down the river to Rangoon, and independent Burma became part of the British Empire. Mandalay has a dry, healthy climate, and there is much cultivated land along the Irawadi here.

Burma takes a high place as regards the capabilities of its people for education, ranking over other Indian provinces and even over some European countries. The literate males number nearly forty per cent. of the male population, females, however, being in a much lower proportion. The credit is given to the primary education of indigenous schools, taught by Buddhist monks. The standard of education, however, is a low one.

The south-eastern part of China, which falls below the tropic of Cancer, is but a fraction of the total area of the country. Nevertheless it embodies a long stretch of seaboard. Canton and the British port of Hong-Kong are but little below the tropic, and the French possession of Kwangchow and the large island of Hainan lie to the south of these.

Hong-Kong, or the " place of sweet lagoons " in its native meaning, is an island, one of a cluster lying close upon the coast. It has an area of less than thirty square miles, but a portion of the adjacent peninsula forms part of this colony, an extremely important seaport of the British Empire. The chief centre is Victoria, whose handsome suburbs enjoy a climate generally cooler than that of the colony in general, which is very hot.

As we approach the magnificent harbour from the sea the aspect of Hong-Kong is singularly beautiful, but the interior of the island is treeless, bare, wild and dreary. In the alluvial valleys, however, food-stuffs are grown, and the population numbers about 100,000, over three-quarters of which are accounted for at Victoria. The harbour presents always an animated appearance, with the arrival and departure of steamers, and there is a continual firing of salutes. Beautiful public gardens have been laid out, and fine roads traverse the place.

Cut by the tropic of Cancer, and lying somewhat over 100 miles from the Chinese coast, is the large island of Formosa, belonging to Japan. Its area is about 13,500 square miles, over a third of this being in the tropics.

Formosa, "the beautiful," is so called from its majestic scenery, which appealed to the early Spanish navigators who bestowed its name upon the island. Mountainous, and in large part covered with virgin forest, there are, nevertheless, valleys whose rich alluvial soil produces rice in such quantities as to have earned for the island in earlier times the name of the " granary of China." The beauty of the wild flowers, the orchids and other ornamental plants are sources of delight to the traveller. In the forests the famous camphor-tree abounds, and camphor, whose production is a Government monopoly, is —with tea—the principal article of export, and is under strict control.

The magnificent cliff scenery of Formosa is strikingly displayed in the mighty walls which arise for over 2000 feet sheer from the sea, and inland Mount Morrison rears its head to over 14,000 feet, being the highest peak in the Japanese Empire.

The climate of Formosa is hot and malarious. The population is over 3,000,000, mainly Chinese, with over 50,000 Japanese, and there are many semi-civilised or savage aboriginal tribes in the wilder parts of the island. The Japanese took possession of Formosa at the time of the war with China.

French Indo-China covers a large area of tropical territory of about 290,000 square miles, and includes Annam, Tongking, Cambodia, Cochin-China, Kwangchow Bay and part of the Laos country. To the west lies the great Mekong river, with Siam and Burma ; on the east and south is the China Sea. The population of these French possessions numbers about 18,000,000, of whom three-quarters are Annamese. There are about 13,000 French in the region.

Tongking has an area of 46,000 square miles, with about 6,000,000 people. Its surface embodies the great Song-Koi or Red River delta, with low-lying fertile plains facing the China Sea, and the interior mountainous tracts and intermediate plateaux region. In the wooded districts the elephant roams freely, and the tiger, panther, buffalo and monkey. In the Mekong river crocodiles abound, and there are many poisonous snakes. Savages inhabit the northern districts. The climate is more agreeable to Europeans than that of Indo-China generally.

Rice is the most important crop here, grown mainly on the delta lands; and coffee, tobacco, cotton, sugar-cane, pepper and other tropical products are raised, and silkworm cultivation is important. Some gold and tin ores are worked, and a large quantity of inferior coal.

Cotton-spinning is carried on in some districts, and fine carved and inlaid furniture is manufactured, well known to traders. The chasing and enamelling of gold is a noteworthy native art, and paper-making from the fibrous bark of the paper-tree is an extensive industry, as is the distillation of rice-spirit. Rice, rubber, hemp, lacquer, coal, mats and gamboge are among the principal exports.

Annam is a curving strip of territory about 800 miles long, forming the eastern side of the Indo-Chinese peninsula, facing the China Sea, in the direction of the Philippine Islands, hundreds of miles to the east. The population is over 6,000,000, the Annamese being the predominating people throughout French Indo-China. The Annamese have been described as the least pleasing of all the Mongolian Indo-Chinese people. As to their customs, both sexes wear wide trousers and a tunic. The houses are without fronts and open to the view except at night, when a bamboo curtain is let down. The principal apartment contains an altar to the family ancestors. The wife enjoys a good position in the social system.

Although fond of ease, the people are industrious and are attached to their native soil. Rice-growing, fishing, silk spinning and weaving are the main occupations, and the customary tropical products are raised. The interior of the country is formed of plateaux and wooded mountains, and savages dwell in certain districts here. The midday heat is compensated for in some degree by the cooler nights.

The country is ruled by an emperor, under the French, the centre of administration being at Hué, a town of about 42,000 people.

In Cambodia, lying towards the southern end of the

peninsula upon the basin of the Lower Mekong, we enter the land of the famous royal city of Angkor-Thom, with its marvellous old temple of Angkor Vat and other stone structures, the relic of the ancient Khmer civilisation.[1]

The area of Cambodia is about 65,000 square miles, and its population perhaps 1,500,000, three-quarters of which are Cambodians.

The hydrography of the country is somewhat peculiar, the annual overflow of the Mekong river, whose waters come from the far-away Tibetan snows, and the remarkable Tonlé Sap Lake, affecting the conditions of life to a considerable degree. The lake supports a very large fishing population, and great quantities of fish are caught in large nets after the inundations. A famous native sauce is made from fermented fish here.

Many of the inhabitants live on the shores of the lake and river, in huts on piles or floating rafts. The men are good hunters and woodsmen, but do not love continued labour. The people are described as clean, moderately intelligent, proud and courageous, and the wife enjoys a good position in the household, polygamy being confined to the wealthy. There is no hereditary nobility, a condition commonly found in Indo-China, but Brahmin and mandarin castes exist.

The Cambodians are skilful workers in gold and silver ; also iron ; and earthenware, bricks, mats, and silk and cotton textiles are made ; and although fishing and agriculture are really the only industries worthy the name, nevertheless the people are their own artificers, and are self-sufficing as regards their own wants. They care little for foreign commerce, like other people of the peninsula, and this is here maintained by the avid and business-like Chinaman. Rice, dried fish and pepper are largely exported, mainly through Saigon, in Cochin-China. Gamboge, the pigment from the gum-resin of a laurel-like tree, takes its name and is exported from Cambodia, and elsewhere in Indo-China.

The remarkable Angkor ruins are situated in the forest region north of the lake, and date from the ninth century A.D. Towers, galleries, columns, carved faces of Brahma and temples to Buddha—for the two faiths in succession were reverenced here—and wonderfully carved and moulded walls, whose stones

[1] Wallace, *Geographical Studies.*

are cut from huge blocks fitted closely together without mortar, cover a wide area, and are among the archæological wonders of the world.

Cochin-China covers the south of the peninsula, with an area of about 22,000 square miles, and a population of 3,000,000. This immense, flat and monotonous plain—for of such the country consists—is traversed by the Mekong river, which through its numerous delta channels empties into the China Sea.

The country owes its fertility to the annual inundation of the river, the cultivation of rice-fields being the principal industry. Pepper, sugar-cane, tobacco, coffee, cotton and jute are also raised. The warm, moist climate is trying to Europeans.

Savages to the number of about 50,000 are found in the interior of the country, and the typical fauna of the peninsula—elephant, rhinoceros, buffalo, tiger, panther, leopard, wild boar, ichneumon, monkey, crocodile and so forth—abound in their respective haunts.

The general religion of the Annamese is a tolerant but somewhat vague Buddhism, largely ancestor-worship. The people are intelligent, with ancient artistic and intellectual traditions, and, as has been seen, are adepts in various crafts. Fair activity in educational teaching marks the French regimen, and public works are not neglected.

The ancient kingdom of Siam is one of the very few monarchies in the tropics. It is, however, by no means the backward or effete state which the European may often picture it, economic and sociological activities being noteworthy, as we shall see upon a brief examination of its life and customs.

The total area of the country is about 220,000 square miles, with a population of about 6,250,000, half of which are Siamese, the balance being Laos, Chinese, Malay, Cambodians, Burmese and so forth.

The country is watered by a number of rivers, and in the north there are parallel ranges of hills. The Mekong river sweeps around its eastern boundary. On the west is Lower Burma, Siam extending partly down the narrow Malay Peninsula to British Malaya.

In Eastern Siam, an inhospitable, barren, sandy and hilly region, about 2,000,000 of perhaps the poorest peasantry in the world support life. In Central Siam, however, the heart of the kingdom, we reach a vast, fertile plain which produces all but a fraction of the national wealth.

The three rivers and their tributaries, which traverse this portion of Siam, and the series of canals which cross it, give a complete and extensive system of water communications throughout this broad district with Bangkok, the capital. These waterways serve instead of roads, which in land so flat, low and soft would be extremely expensive. Some of the canals are ancient, early rulers having recognised the value of this system of transport in view of the peculiar topography. The busy boat traffic, under a small toll, yields a revenue more than sufficient for all expenses of maintenance.

Southern Siam faces the Gulf of Siam, with Bangkok lying twenty miles from the mouth of its navigable river. This, the capital of Siam, was earlier built on piles and floating pontoons on the margins of the many canals and waterways which served in place of streets. Now, however, well-made streets with brick houses have taken their place, with parks, warehouses, rice-mills, electric tramways and many other utilities, and with four railway stations. Some of the buildings are very handsome, especially the public offices and institutions, and in the centre of all stands the royal palace. The beauty of the Buddhist temples with their gilded spires and coloured roofs is noteworthy. The population is nearly 500,000, and the annual foreign trade of the port is valued at over 14,000,000 sterling, three-quarters of it being with the British Empire. The climate has improved, but the death-rate is high. This prosperous city, before 1769, was but a small village on the banks of the river.

The climate of Siam, though in general enervating, does not suffer from extremes, but its humidity is trying to Europeans. There is, however, a cool, dry season. The country is famous for its big game, large numbers of wild elephants roaming free, and these furnish valuable transport animals when caught and tamed, which is extensively done. Tigers, bears, rhinoceroses, many kinds of monkeys, leopards and crocodiles are among the abundant tropic fauna here.

As regards the plant life, the chief economic products are rice, very abundantly cultivated in the central plain, and the coco-nut and acacia palms, with the customary tropical plants, trees and fruits. In the cooler uplands oaks, pines, chestnuts, great apple-trees, vines, honeysuckle, raspberries and so forth abound, with forests of teak-trees in the north. A good deal of the " Burma " teak exported from Moulmein is in reality of Siamese origin. Sago, sugar-cane, bananas and other tropical products abound in the lowlands. Betel-nut is an important product and is chewed by everyone.

The main occupation of the Siamese, excepting the people of Bangkok, is that of growing paddi, or rice. The efficient cultivation yields good harvests, although methods are primitive and irrigation rudimentary, as the water is not raised from the low-lying canals.

The teak-cutting occupies many workers, Siam being the largest teak-producing country in the world. The industry is mainly in the hands of the British and other Europeans, but the forests are cultivated by a special department, with experienced officers recruited from the India Forest Service.

Mining is carried on for silver, gold, rubies and tin, the Government Mining Department being highly organised, with expert British advisers. Silk culture is also carefully organised, with the aid of Japanese experts. As regards the land, the Government give careful attention to its survey, with a Land Registration and Settlement Department, which has performed excellent work. Under the ministry of Lands and Agriculture instruction is given to the Siamese youth in surveying, irrigation and other branches of rural engineering, and good pupils are also sent abroad to the best engineering and technical schools.

The manufactures of the country are but few, all articles of daily use being imported. But the Siamese are, nevertheless, skilled in arts and crafts; architecture, carving, goldsmith work and so forth following upon Indo-Chinese lines, the gold working being famous for its execution, as is the repoussé silver-work. This industry dates from very ancient times, and each district has its special processes and patterns.

The Siamese are a light-hearted but apathetic people, of medium height and olive complexion, ugly from the European standpoint. They are submissive to authority and courteous to strangers, and quarrelling and crime are rare. They are good cultivators of the soil, but do not love sustained labour of other kinds, and they lack the business faculty, as a consequence of which trades and industries are largely in the hands of the alert Chinese—that stolid, shopkeeping people whom we encounter throughout tropical (and other) lands, literally from China to Peru.

In this absolute monarchy, for such is Siam, there is no system of caste, and the lowest-born may attain to the highest dignities. Nor are there any hereditary titles, such dying with the individual upon whom they may have been conferred. The upper class, however, have the " privilege " of practising polygamy, but the position of women is a good one generally. There is now no slavery. Ten Ministers of State, like those of the departments of a European country or American republic, assist the monarch, who, however, is an autocrat, both in theory and practice. The Civil Service supplies candidates for many governmental posts. An advanced railway policy has been followed, and there is keen competition abroad for trade with Siam.

CHAPTER XXVII

AUSTRALASIA—THE MALAY PENINSULA

ALTHOUGH a vastly scattered area, of islands and peninsulas, to describe all of which in detail would take not part of one but the whole of more than one volume, there is a geographical excuse for treating Australasia as a unit, in that its scattered lands together form the south-eastern extension of Asia.

Within this geographical expression are embodied Malaysia and the Dutch East Indies, with the Philippine Islands; Australia with Tasmania and New Zealand; Melanesia, formed of New Guinea and adjacent islands, including the Solomons, New Hebrides, Santa Cruz, Admiralty, Fiji, Loyalties and New Caledonia; Micronesia, which is made up of the Ladrones, Pelew, Carolines, Marshall and Gilbert Islands; and lastly Polynesia, including Samoa, Tonga, Cook, Tahiti, the Marquesas, Ellice, Hawaii, and the various clusters of islets between them.[1] Not all of these widely separated lands, however, lie within the torrid zone. The southern half of Australia, with New Zealand, lies beyond the tropic of Capricorn and therefore does not come under our survey.

On the east is the Pacific Ocean, extending to the shores of America; on the west the Indian Ocean, to the coast of Africa. The Equator cuts Australasia almost midway.

The Malay Peninsula is a continuation of the mainland from the kingdom of Siam, a lozenge-shaped tongue of land 750 miles long and sixty to seventy wide. It embodies several political divisions: the British colony of the Straits Settlement, the Federated Malay States, the independent Malay state of Johor (under British influence), the non-federated states, also under British protection, and those states to the north of Perak and Pahang, which lie within the sphere of Siam. Its area is estimated at about 70,000 square miles. On the south is the

[1] Australasia thus includes Oceania, the term used by some geographers.

strait and island of Singapore, on the east stretches the China Sea, towards the shore of Borneo, and on the west lies the Strait of Malacca, the narrow waters of which wash the coast of Sumatra.

The peninsula is divided by a backbone of granite mountains, and between the base of the range and the Strait of Malacca lies the mineral wealth of the region, containing the great alluvial deposits of tin, among the most important in the world. The highest mountain reaches 8000 feet. Mangrove swamps cover almost the whole of the west coast, and their sombre foliage and the lugubrious mud-flats form generally an uninviting approach. On the east coast the land is free from mangroves and their muddy shores, a change due to the force of the monsoon, which sweeps in from the China Sea, and here beautiful sandy beaches alternate with forest-covered, rocky headlands.

The Malay Peninsula is largely covered by forest, through which innumerable rivers and streams thread their way, forming one of the most characteristic hydrographic systems of any land; but these remote forests, except in a comparatively few places, are untrodden by man, nor indeed have ever been traversed. Even the Malays and the aborigines upon the river-banks rarely leave the water-side in their journeys, save to pass from one stream to another.

Some of the rivers are navigable by native craft and steamers for considerable distances. The Malay travels by boat in preference to going on foot, and the rivers are almost the only highways in some districts. They are rich in scenic beauty.

But modern means of communication have rapidly increased of late years and Malaya has over 735 miles of railway between the principal towns, ports and mining and planting districts, and over 2250 miles of excellent roads, practicable for motor cars.

The climate of the peninsula was formerly considered as unhealthy and was disparagingly compared with that of the west coast of Africa, but it is now regarded more favourably. Equable temperature, bright sunshine, with cool breezes and nights may be said to characterise it. It would appear that malaria has often been the result of clearing and digging, connected with newly turned soil, and disappears after a time.

Labourers in deep mines suffer much less from it than those engaged in the surface alluvial mines.

To Europeans the climate is enervating, and a change every five or six years is imperative. But a temperate life, a sufficiency of exertion and the avoidance of chills ensures for the white man as good a state of health as at home. The Malays are in general a remarkably healthy people, but Chinese immigrants suffer often from malaria and beri-beri, and other diseases. At Singapore, the hottest district of the peninsula, the temperature seldom rises above 90° in the shade, with a mean below 80°. On the eastern slope the maximum is about 93°, falling to 70° at night. The Sumatras, the violent gusts of rain-bringing wind, are common, and the rainfall varies from 75 to 100 inches on the west coast, to about 155 on the monsoon-beaten east coast.

Countless ages of decaying vegetable matter have produced an extremely fertile soil on the plains and mountains of the peninsula, and in the vast dense forests this gives birth to a marvellous profusion of matted vegetation, consisting of huge trees, climbing plants, ferns, mosses and trailers. Without cutting through, a way cannot be made among the jungle growth, and even the wild beasts roam along their own well-trodden paths. Miles of cane-brakes line the river banks of the interior, and good hard-wood timber abounds, whilst the flora includes many varieties of orchids. Of fruits, the coco-palm, pomegranate, custard apple, banana, sugar-palm, areca-nut and others are common, and coffee, rice, tobacco, sugar, pepper, cotton, sago and gambier are successfully cultivated. The great development in rubber-planting is perhaps the most remarkable feature of agriculture in the peninsula, as described elsewhere.

In these wild lands a profuse and varied animal life exists : the elephant, bison, rhinoceros, tapir, deer, Asiatic tiger, panther, leopard, wild cat, wild boar, wild dog, flying squirrel and flying fox, alligator, python, cobra and a host of smaller animals and reptiles. Birds of many kinds, including pheasants, partridges, herons, ducks, pigeons abound, with many small varieties. Anthropoid apes, including the gibbon and twenty kinds of monkey, are found, among the latter being the coco-nut monkey, which the Malays train to gather coco-nuts

from the trees. The profusion of insect life—still partly unstudied—is greater than in any part of the world.

The dominant race of the peninsula and archipelago are the
Malays, although there are other people who have had their
home there for long periods, such as the Samang Negritos, a
small, very dark folk, who attained to but a low culture, also
the Sakai, and the Siamese. The Chinese immigrants are
extremely numerous. The distribution and origin of the
various races here is one of much interest and some obscurity.

The Malayan family, broadly, includes all the brown races
inhabiting that part of Asia south of Siam and Indo-China,
and the islands. They have occupied the land for a thousand
years, but still have a tradition that they are not the original
inhabitants, but came " out of the sea," and they would appear
to have attained to some state of civilisation before their
arrival. They are, on the coast, a maritime people, and have
been daring pirates, with, earlier, an evil reputation among
Europeans. Formerly they were regarded as one of the
cruellest and most treacherous of the world's people, but this
opinion about them has changed, and certain traits of loyalty
and courtesy which they possess are now acknowledged. They
differ from other Orientals in that friendliness displayed
towards them does not necessarily breed familiarity or lack of
respect. They are, however, indolent and improvident, fond
of pleasure and gay costumes, with an aversion from toil.
Without much idea of the value of money, they have little
notion of honesty where it is concerned, and would borrow
rather than earn. Left to their own resources, they are often
unjust and cruel, as well as improvident. The term " to run
amok " has its origin among these people : outbursts of homi-
cidal mania ascribed to them, a state which, however, is probably
the counterpart of suicide among the white races.

The Malays are Mohammedans, but religion enters lightly
into their lives. In some of the islands they profess a form of
Hinduism, especially in their superstitions. Their piratical
habits gone, they have become peaceable fisher-folk, building
houses generally on piles on the banks of the rivers and planting
coco-nut groves, sugar-palm, sugar-cane, maize, tapioca and
rice—the latter their main food.

Much ingenuity is exercised in their industries, and irrigation is carefully practised. Jungle produce is worked, including gutta and valuable woods, rattan, etc., and cotton and silk cloths, earthenware, mats, weapons and silver vessels are made. Silver-ware exhibited in London in 1886 from Malaysia was pronounced the most artistic of any native craftsmanship of this nature. The Malays had learned to make gunpowder and forge cannon at the time when the first Europeans visited them, and they would appear to have invented an art of writing. They are also expert boat-builders. Their chief weapon is the *kris*, a short dagger with wooden or ivory handle, and they also make swords and scimitars.

The Malay language is used over a considerable area throughout the region, and it has been held that the Malays possessed an ancient alphabet, although there are neither manuscripts nor monuments extant prior to the advent of Islam in the archipelago, about the thirteenth century. The Perso-Arabic alphabet, then introduced, is in use for most purposes, but since the Dutch advent Roman characters have come to be largely used also. The number of foreign elements in the language is a result of the various human ingredients which from time to time have entered into the make-up of the people. The Malays are held to be allied ethnologically, though remotely, with the Polynesians, and it has been thought possible that the two peoples once shared a common land—some continent whose sinking and disappearance separated them.[1]

The Crown colony of the Straits Settlements is composed of the island of Singapore, the town and territory of Malacca, the islands and territory of the Dindings, the island of Penang, and Province Wellesley. These possessions lie scattered along the south and west sides of the peninsula, connected by the Federated Malay States.

The Federated Malay States are the British protectorates—but not possessions—of Perak, Selangor, Negri Sembilan, along the west coast, and the state of Pahang, on the east coast, each ruled by a sultan.

The " City of the Lion," as the word Singapore, translated from Malay, means, is the most important part of the Straits Settlements and one of the most valuable possessions of Britain

[1] By Dr Alfred Russel Wallace, and discussed in the author's book, *The Secret of the Pacific*. London, Unwin, 1913, second edition.

in its smaller colonies, the seaport standing midway between China and India on the great ocean trade route to the Far East and a half-way house for steamers thereon. It is both a commercial and a strategic centre, and is well fortified.

As we enter the excellent harbour we observe the numerous vessels, ocean steamers from various quarters of the globe, and native craft from all parts of the Malay Archipelago. There are extensive coaling wharfs and docks, and in the neighbourhood stands the largest tin-smelting works in existence, its furnaces annually producing over half the total amount of tin consumed in the world's industries. Singapore is also famous for its tinned pine-apple, familiar to the consumer abroad.

The island, which is twenty-seven miles long, is separated from the mainland—the native state of Johor—by a narrow strait, and to the south are the Dutch Islands. The highest hill is Bukit Timah. The climate is generally humid and hot, and the green vegetation contrasts strikingly with the red stone of which the roads are formed. The jungle has largely given place to plantations of tropical products, established by Chinamen.

The island of Singapore was first occupied by Sir Stamford Raffles, who, in 1819, persuaded the Sultan of Johor to cede it to him in the interests of the East India Company. Of its population of about 230,000 the Chinese number much more than half, and the excess of men over women, and the low birth increase, is due to the absence of women among the Chinese. The women of China do not accompany the men emigrants. Malays and Indians are numerous, and of Europeans there are less than 5000. The churches, temples and mosques of all faiths and denominations demonstrate the customary religious toleration under British rule, and the banks, public buildings, clubs, race-course, polo, golf, cricket and other grounds indicate British activity in matters of business, administration and sport. At Singapore is the residence of the Governor of the Straits Settlements.

Malacca, lying to the north-west upon the mainland, giving its name to the straits which separate the peninsula from Sumatra, is an ancient, dreamy little town and district. But it has a remarkable history, in which the Portuguese, the Dutch and the British have figured, and fierce fighting between the adventurers of the first two nations has reddened with blood the waters of Malacca Straits, waters at both ends of which the Union Jack now securely floats. The British acquired the place from the Dutch in 1824.

The population of Malacca is about 100,000. A once flourishing port, it has declined, due to its shallow harbour and to the growth of Singapore and Penang. The area held is about 700 square miles, and the cultivation of rubber and coco-nuts is extending. Good roads have been made, and the land is almost all in the hands of the natives, with 50,000 acres under tapioca, and perhaps a fifth of that under rubber. A railway runs inland to Seremban.

Dindings, to the north, and Province Wellesley occupy islands and territory on the same coast, the former near the Perak river. Fertile plains thickly populated by Malays form part of the districts, and sugar-planting and other agricultural industries give employment to Chinese and Tamil labour. Railways run to Perak, Selangor and to Malacca. The islands are but sparsely inhabited. The fine natural harbour has not developed into a commercial centre, such as had been hoped.

The beautiful island of Penang, or " areca-nut island," as its native name has it, is, after Singapore, the most important part of the Straits Settlements. As we enter its excellent harbour we remark the coco-nut groves along the shore of the island upon which the colony is situated. Above, Penang Hill rises high from the promontory upon which the town is built, with government and private bungalows upon its slope. The churches, colleges, race-course, golf links, and so forth, are indication of its European life. Much of the land, however, is still covered with jungle, the total area of the island being somewhat over 100 miles. The shipping trade, though important, is falling off somewhat, due to the rise of Singapore. The population is about 130,000, mainly Chinese, followed by Malays and Tamils and other Indians, with perhaps 3000 Europeans and Americans.

The total population of the Straits Settlements is about 600,000, nearly half of which are Chinese and a third Malays. Men greatly predominate among the population.

The independent state of Johor stretches across the southern end of the peninsula and has an area of about 9000 square miles. It is covered from side to side by one vast forest, whose continuity only the smallest clearings and cultivations break. Facing Singapore is the capital, Johor Bharu, whose fine sultan's palace is the most noteworthy building. Described as a sort of "Oriental Monte Carlo," this place is much visited from Singapore. Notwithstanding its Malay rule, Chinese greatly predominate among the population, forming three-fourths of the total of about 200,000. For a country so near the Equator, Johor has an equable and healthy climate, cooler than Singapore. The cultivation of tropical products is the principal industry, and revenue is derived from taxes upon exports, and upon gambling. Moderately well controlled, and subject to European influence to some extent, Johor has so far escaped the need for British administration. A recent sovereign of Johor, a man of intelligence and interested in problems of Government, was made Maharaja by Queen Victoria, on his visit to England, with supreme rule of the state.

The state of Selangor covers an area of about 3000 square miles, a narrow strip of territory between the mountains and the Straits of Malacca. At the mouth of the Klang river is Port Swettenham, the principal entrepot, where ocean steamers lie alongside the wharf. It is connected with the capital, Kuala Lumper, by railway, and the main trunk line of the Federated Malay States passes through the country. The scenery of the capital is very attractive, and a great deal of money has been spent in the town upon public buildings, waterworks and so forth. Kuala Lumper is the seat of the Resident-General, and in a sense the capital of the whole Federation. It has the finest public buildings of any town in this region.

The population of the state exceeds 200,000, more than half of which are Chinese, and as usual on the peninsula males greatly predominate. The easy approach to Selangor by rail from north and south is a factor in the increase of the population, and a score of different nationalities are included in the census. Deaths, as elsewhere in the peninsula, outnumber births, and immigration is responsible for increase.

Selangor is one of the richest tin-producing Malay States. The extraction of the metal from the alluvial soil forms the chief industry of the people here, and nearly 100,000 acres of tin-bearing land are worked, a great body of Chinese being employed thereon. The tin export reaches many thousands of tons annually.

The rubber-planting industry of Selangor is very important and many Tamil coolies have been brought in to provide labour for the estates.

The Negri Sembilan, or Nine States, a small native federation lying east of Selangor, also produces large quantities of tin for export. But agriculture is the mainstay of the people. The area of this federation is 3000 square miles, and the population, about 120,000, has largely increased by immigration of late years. The seaport is Port Dickson or Arang-Arang, connected with Serenam, the capital, by rail, the trunk line traversing the town. The beautiful watering-place of Magnolia Bay is a pleasure resort for this part of the peninsula.

Perak lies to the north of Selangor, with Province Wellesley on its northern side, and covers about 10,000 square miles. The population is over 400,000, half of whom are Chinese, with the disproportion of a lack of females as elsewhere on the peninsula, the rapid increase being due to immigration. The capital of the state is Taiping, connected with the port of Teluk Ansac, and with Selangor and other places by rail. Hundreds of miles of excellent metalled roads for vehicles have been made in this state. The residence of the sultan is Kuala Kangsar, a place of some note. The Perak river is navigable for long distances by native craft, and near its mouth for steamers.

In the granite and other formations of the state lie important deposits of tin, and the output of the metal has increased to a value of many millions of dollars each year. This value fluctuates greatly, due to labour conditions and market prices. The labour is generally recruited from Southern China.

The rubber-planting industry is also very important here, and the area under plantation of Para rubber has increased immensely. Copra and other produce is exported, and the agricultural possibilities of the state are very considerable, resulting upon a fertile soil, regular rainfall, and good means of communication. The general conditions of Perak seem to point to a prosperous future.

The state of Pahang is the largest in the peninsula, covering over 15,000 square miles, with a coast-line over 200 miles long. It lies upon the eastern side. The seaports are difficult of access when the monsoon blows. The Pahang river is of some service for navigation, but is shallow. A vast forest covers almost the entire estate, but a broad, grass-covered plain—almost the only land of the character in the peninsula—covers the Lipis Valley. The population is estimated at over 100,000, three-quarters of which are Malays. The country was earlier much more thickly populated, but civil wars reduced the number of the inhabitants.

The administrative capital of Pahang is Kuala Lipis, and the town is served by a magnificent road which crosses the mountains for eighty miles to Kuala Kubu, a station on the Selangor railway. The sultan's residence is at Pekan, the former capital.

The important output of tin here is partly obtained from lode mines, as contrasted with the alluvial deposits in which the metal occurs elsewhere in the peninsula, and the export reaches a value of several millions of dollars annually. Gutta-percha is also exported. The fertile soil and steady rainfall are conducive to the yield of many tropical products. The natives are engaged in agriculture, mining being largely carried out by Chinese. The financial affairs of Pahang are less advanced than those of the other states of the Federation, but are improving.

To sum up the general conditions of the Federated Malay States, it may be said that, under British protection and sound administration, this rich, tropical region has developed in an amazing manner. An energetic policy in expenditure upon public work has been amply rewarded, and 400 miles of railway and 2500 miles of the " best-metalled cart roads in the world " have been built, with other improvements, all out of revenue, without any debt being incurred. This has, of course, been possible because of the extraordinary wealth of the tin deposits, which, under British protection, could be economically exported, in what, previously, was a land of bandits and savages.

The alluvial tin mines are worked almost entirely by Chinese capital and labour. The rubber estates are mainly in European hands, largely worked by Tamil labour.

In the administration of these states great use is made of the native aristocracy and chiefs, under British supervision. Each sultan is consulted, and the Mohammedan religion is respected. The territory in general has never been ceded to Britain, and it is recognised that the Malays are the owners of the soil and entitled to first consideration, and under this just and benevolent autocracy an extraordinarily rapid amelioration in the life of the community has taken place.

The non-federated Malay States were given over to British suzerainty in 1900 by treaty with Siam, of which kingdom they were very ancient dependencies. They include Kelantan, Trengganu, Kedah and Perlis. They lie north of the federated states, on the east and the west sides of the peninsula. The areas of the first and second are respectively 5000 square miles, of the third, 4000, and the last is very small.

The Kelantan river is the largest on the east coast and is navigable for eighty miles. The climate of the Kelantan state is remarkably healthy in the cultivated and open districts. In the population of 300,000 the Malays greatly predominate, but Chinese and Indians are increasing. The only town, Kola Bharu, upon the river, has been laid out with excellent roads, and has many good modern buildings.

In this state coco-nuts, rice and betel-nuts are largely grown, and much live stock is raised. A good deal of gold has been dredged from the river, and 20,000 expert Kelantanese fishermen find occupation in its waters in fishing and fish-curing. The industry of silk-weaving is growing. The people, formerly of doubtful and often evil character, are becoming very law-abiding. The ancient town sites and mines point to an early history of which little is known.

In Trengganu we encounter somewhat analogous conditions of climate, people and products. the population being under 200,000, fishing and agriculture, pepper and gambier being the principal industries, with some tin-mining, although the mineral wealth is so far little known.

The capital and its suburbs, at the mouth of the river—from which the state takes its name—has a population of 30,000, but the interior, of very difficult access, is sparsely populated. The garments woven from imported silk by the natives are celebrated for their quality and design, and are largely exported, and native weapons and brass utensils are made. The carrying trade is done largely by native-built ships.

Under British influence very superior conditions of life have taken the place of those under which the misery of the down-trodden peasantry and rampant crime were the features of the land, where "holy" men exercised a pretentious religion. The evil Malay rule is past.

Kedah has a population of about 250,000, Malays, Siamese and Chinese. The capital, Alor Star, lies on the Kedah river, and is well laid out with good streets and handsome buildings. It has a population of 20,000, half of which are servants of the local aristocracy and half Chinese. There is good wharfage for vessels, the town lying ten miles from the coast. The surrounding country is flat, fertile and unhealthy. Rice, cattle and tin are exported, and coco-nut, betel-nut and rubber plantations have been successfully established. The centre of the tin mines is at Kulim, and there are some good roads in the state. The population is largely distributed in small villages. A sound Siamese government already existed when the British assumed control of Kedah, and the sultan rules by reason of long hereditary right.

The little state of Pérlis, with its chief town of the same name, upon the river, has a population of about 10,000, Malays and and Chinese. Rice, pepper and tin are the principal exports.

The implanting and growth of the rubber industry in Malaya might be described as one of the commercial romances of the tropics. The Para rubber seed was first brought from the remote interior of the Amazon forests,[1] many thousands of seeds being selected on the Tapajos and Madeira rivers and sent to Kew, in 1876. The first plants sent out thence to Malaya perished on the way, but of more which followed a few flourished when planted, and from these have descended the innumerable trees of the Malaya plantations at the present time, except a few of Ceylon origin.

The area of land under rubber in British Malaya in 1905 was about 38,000 acres. In 1912 it had risen to nearly 700,000 acres, with a production of nearly 43,000,000 pounds of rubber for that year, figures which are truly remarkable.[2]

The exports of tin amounted in 1912 to a value at 107,000,000

[1] By Mr H. A. Wickham.
[2] The average yield was about 250 pounds per acre, and the value of the export in 1913 was £8,500,000.

dollars. The alluvial tin is found both in the lowest swamps and on the mountain-tops in the western Malay States, from the grass roots to depths of perhaps 200 feet.

The growing of coco-nuts and export of copra gives promise of great extension. By far the greater part of the area under cultivation is held by natives in small parcels of land, the total being perhaps 160,000 acres.

Whilst from the commercial side this industrial development is of the utmost value, it is to be recollected that single vast products of this nature are subject to great fluctuation in market price, with great depressions when these fall unduly. They also tend towards the neglect of smaller industries and the production of food for local wants. The condition of a vast excess of men over women, and low birth-rate, cannot be regarded as satisfactory, and is due to the inrush of immigrant labour.

The Siamese Malay States form two groups, a northern and a southern, numbering nearly a score. They are of small area, except that Patani, the largest, covers 5000 square miles. It has a population of about 400,000, and possesses the only town of importance in the group. The mining of tin is a growing industry, and fishing and salt-drying are other occupations furnishing articles of foreign trade. Elephants are among the exports. British consuls are stationed in these states, but other countries are unrepresented.

CHAPTER XXVIII

AUSTRALASIA : MALAY ARCHIPELAGO, JAVA, SUMATRA, BORNEO, NEW GUINEA, ETC., PHILIPPINES, TROPICAL AUSTRALIA

In the Malay Archipelago we approach the group of largest islands in the world, a region curious and remote in some respects. The Dutch East Indies, Java, Sumatra, Borneo, Celebes and others of these islands stir the traveller to interest. They have long been rendered famous by colonist, naturalist and explorer. In the Java forests lie some of the most remarkable ruined temples in the world, and here was discovered the assumed link between man and his animal ancestors, the higher apes. In Borneo, " Great Britain might be set down and still be surrounded by a sea of forests." [1] It was at Ternate, in the Moluccas here, that the evolutive idea of the " survival of the fittest " suddenly flashed upon Alfred Russel Wallace, as he lay wrapped in a blanket sick of a fever in 1858.

The tropical products of these islands are of great value and variety, and the vegetation, whether wild or cultivated, is of much interest. The islands in general belong to the great equatorial forest belt. Food plants in large number grow here, many of the luscious fruits being little known abroad. The orange, the mango, guava, shaddock, mangosteen, durian and others are common. Coffee, sugar, rice, maize are widely cultivated. The coco-nut, the bread-fruit, the banana, plantain, sugar-palm, tea-plant, sago-palm, yam, ground-nut, cassava, do not exhaust the list. The Dutch greatly encouraged coco-nut cultivation among the natives, and its oil is largely used in native cookery. Sugar and coffee growing also owe much to the Dutch. Spices—pepper, nutmegs and cloves— were for a long period matters of Dutch commerce here. Camphor, gutta-percha, benzoin, damma and other produce of the forests are exported.

Borneo, one of the world's largest islands, is cut midway by

[1] Dr A. R. Wallace.

the Equator, and covers an area of nearly 290,000 square miles. Politically, it is divided into four states : British North Borneo, administered by a chartered company ; Brunei, a Malay sultanate under British protection ; Sarawak, ruled by a raja under British protection ; and Dutch Borneo, the largest and best part of the island.

Borneo is generally mountainous, with low, alluvial coast-lands, and bold headlands in places, with banks of black, mangrove-covered mud at the river mouths. Some of the rivers form highways, and along their valleys the population principally lies. Steamers and native prahus ply upon the navigable rivers.

The principal and in fact the only really good natural harbour in general use in the island is that of Sandakan, in British North Borneo, or Sabah. This territory has an area of 31,000 square miles, and a population of about 200,000. Sandakan is the most important settlement in the British territory, and has a population of about 9500. A railway traverses the western coast region from Jessellon to Weston, but the roads in general are only bridle-paths. The climate is damp, hot and generally enervating. The natural resources are undoubtedly great, but have been very partially explored so far, and the sparse population does not warrant the extension of railways. Traces of mineral wealth are found, including petroleum, iron and copper, also coal and gold, but it is upon the abundance of excellent timber and other forestal products that business at present depends here. Rubber and gutta, edible nuts, mangrove bark, edible birds' nests—used by the Chinese—guano and rattan are exported. Tobacco, coco-nuts, sago, pepper, gambier, sugar-cane, coffee are grown. In the interior there is much broken forest-clad country, over which means of communication are difficult. Native " villages "—often consisting of a single long community hut or house—are dotted through this wild land, and upon the verandah of such a dwelling, where the various families who inhabit it congregate at times, in some cases hang festoons of their enemies' skulls. A very low state of civilisation exists among some of the tribes, who drink to intoxication their native liquors.

Dutch Borneo has a population of nearly 1,250,000, but vast areas remain unexplored, and the population is sparse. Wars, piracy, poverty, head-hunting, drunken orgies and other matters have maintained a low stage of civilisation among the interior tribes, who are often of a wandering nature. In the more fertile districts, and where mines have been opened and vessels ply upon the rivers, there is much more industry.

The chief centres of government are Pontianak, with about 10,000 inhabitants, and Samba and others in the western division, and in the eastern is Banjermassin, with 50,000 people. Few of the Europeans in Dutch Borneo are planters, the whites generally holding official posts. Chinese and Arabs carry on mining, agriculture, trading and fishing, and the bulk of the native population is engaged in its primitive agriculture. Native industries are also carried on, such as spinning, weaving and dyeing, the making of boats and of iron weapons and implements. In the south-east shipbuilding, iron-forging and gold and silver work, with diamond-polishing, are highly developed industries among the larger settled communities.

Sarawak covers about 55,000 square miles, and has a population of over 500,000. The chief town is Kuching, with about 30,000 people, and is reached by steamers, sailing twenty miles up the river. The state is governed by the Raja Brooke.

The population embodies the few European officials, and a number of Chinese and Indian traders, the semi-civilised Malays forming the great bulk. It has been said that " nothing could be done without the Chinese " here. Some of the Malay tribes are industrious and contented people. Head-hunting has been suppressed, except that the Dyaks break out at times into their raids. The animal life is rich, and the jungle swarms with insect life, whilst a wide variety of ferns and orchids is found.

The principal exports are gold, pepper, sago, rubber, gutta and gambier. The mineral resources may prove to be of great value, and excellent timber abounds. Coal is mined and shipped, and trade in general has increased considerably of late years.

Brunei was a once powerful sultanate, but now is reduced in area to about 1700 square miles. It is controlled by a sultan under a British resident, and has a population of about 25,000, Malays and others. The pretty town of Brunei stands on the

A DYAK GIRL OF BORNEO IN DANCE COSTUME OF CANE, BRASS AND SILVER

river, cleared hills and magnificent forest-clad heights forming its background. Most of the houses are built upon piles. This interesting and historic place is also known to the natives by its Arab name of Dar-ul-Salam, or the "City of Peace," and its earlier sultan assumed the name of Mohammed. It had a "golden age," but after the Spaniards captured the place in 1580 this came to an end, and decadence set in. The stone wall between the two small islands off the coast, built prior to that time, still remains, its foundation resting—history records—upon forty rock-filled junks, which were sunk there for that purpose. The handsome stone tomb of the great warrior of Brunei, who kept splendid court in his time, still remains, a shrine for the Malays.

The mineral, forestal and agricultural wealth of Borneo in general is very considerable. Diamonds, quicksilver, gold, copper, iron, petroleum, sulphur, rock-salt, coal, tin, marble, are all embraced. The oil-fields may have a valuable future and Dutch Borneo already produces considerable quantities of oil for foreign markets.

The greater part of the island is clothed with luxuriant vegetation, amid which the forest monarch, the *tapan*, rises as a splendid column to where it is crowned by a beautiful dome of foliage. The palms, nuts, fruits and vegetables furnish an infinite variety of food-stuffs, and the spices are famous. Cotton, sugar-cane and tobacco are very important. Beautiful rhododendrons, orchids and other aerial plants display their hues in the tropic wilds, the flora being exceedingly rich. The climate is hot and damp, practically unchanging throughout the year, and tremendous rainstorms occur. The conditions of health naturally vary over so large an area, but fever is not more to be feared than elsewhere in Malaysia.

The great island of Sumatra, under the Dutch, is nearly all explored and under control. It covers over 178,000 square miles and is narrow, and 1100 miles long, paralleling in part the Malay Peninsula, from which it is divided by the Malacca Straits. A range of lofty mountains forms a continuous backbone, facing the Indian Ocean, with vast alluvial plains upon the east side. Balanced on the Equator, as is the island, the climate varies little throughout the year. Two of the rivers—

the Musi and Jambi—are navigable for nearly 400 and 500 'miles respectively. The volcanoes of the great backbone pour forth ash and scoria upon the land, and large lakes occupy extinct craters.

The capital and residence of the governor of the western region is Padang, with over 35,000 inhabitants, and on this side, under the influence of the few thousand Europeans, the fertile soil, good roads, timber supply, coalfields and agriculture, well-populated districts have grown up. On the east coast conditions differ, and the people are much less concentrated in the towns, and live by fishing, navigation, cattle-raising and forestal products. Medan, where the high officials for the eastern division dwell, has a fine Government House. Patembang has a population of 53,000, and possesses the finest mosque in the Dutch East Indies, and a " traditional tomb of Alexander the Great." A number of regencies make up the domain of the island. The capital of the Padang highlands is Bukit Tinggi.

Much of the surface is forest-covered, but agricultural wealth is marked. Pepper, rum, rubber, copra, nutmegs, mace, gambier, bamboo or rattan and coffee are all exported, and the numerous tobacco plantations which flourish in some districts are of much importance, being worked in some cases by British capital with alien coolie labour, largely Chinese. Gold, tin, copper, iron, coal and petroleum all exist, and in some cases are profitably worked. The production of petroleum is especially important, the crude oil being conveyed in pipe lines to the east coast, and there refined.

Various railways and good roads traverse the western and some other districts, and the coffee grown on the first-named side of the island thus finds means of transport.

The population of Sumatra is over 3,000,000, all Malays, with the exception of perhaps 100,000 Chinese and some thousands of Europeans and Arabs.

Between Sumatra and Java but a narrow strait exists, which, however, marks a wide change in the plant and animal life of the two islands.[1]

The island of Java is narrow, but over 600 miles long, covering an area of over 48,000 square miles. It is the centre of

[1] Wallace's Line.

Dutch Colonial Government in the East Indies, and has a population of over 30,000,000. Other long, narrow islands continue the singular topographical formation—Flores, Timor and others—leading on towards Australia.

With its companion islands in the Indian Ocean Java forms a sort of mighty breakwater, as if protecting the rest of the archipelago, and it is part of the rim of the ancient submerged continent here. Volcanic forces made the island and continue to fertilise or devastate it. It was in the east of Java, in a dense forest, in the riverine deposit of the late Phocene age, that the famous skeleton of the anthropoid ape-like man, the *Pithecanthropus erectus* of Dubois, was found, remains of vast interest to the evolutionist.

The famous old seaport of Batavia stands upon the north coast, and is the capital of the entire Dutch dominion in the East. It grew from a Dutch "factory" in 1610, and was taken by Britain in 1811, but restored to the Dutch later. Very important harbour works were carried out during the latter half of last century, towards the close of which the population had increased to 120,000. The modern part of the place is higher and healthier than the old, and here the European people dwell, the houses being generally of single storeys and separated by rows of trees. The governor-general's palace and other civil and military buildings attest the importance of the city, which has a number of noteworthy educational and scientific institutions, including a Society of Arts, with a museum, a royal physical society, and one for the promotion of agriculture and industry. These have been established in some cases for over a century. There are theatres, club-houses and fine hotels.

The Dutch residents of Batavia are relatively few, and are intermixed with Portuguese and Malays. Arabs are engaged in trade and navigation, Malays as boatmen and sailors, Javanese as agriculturists, and there is a numerous and industrious Chinese contingent. These people were earlier greatly oppressed, and even ruthlessly massacred. Batavia exports a wide variety of the agricultural, mineral and forestal products of the archipelago, from gold-dust and diamonds to coffee, sugar, tobacco, timber and edible birds' nests, bees'-wax, tortoise-shell, dyewoods, camphor and drugs, and tin, pepper, spices and fine clothes—all from her own and adjacent isles.

Were it left to itself Java would soon become clothed everywhere, even where cleared, with luxuriant vegetation, and indeed it is remarkable that with its great and growing population the island should still have such extensive tracts of almost primæval forest. This last circumstance is explained by the simplicity of the native life. Bamboo is sufficient for house and furniture construction, and little fuel is required. However, forest fires made to clear the land for rice have already devastated considerable areas, and the Government have now assumed control of the forests to a large extent.

The variety of trees, flowers, ferns, mosses—" the absolute dominance of the mosses," as one traveller has put it—and fungi is remarkable. The flowers include violets and rhododendrons. The tall and noble liquid-amber tree rises with a clean column to over 100 feet, spreading out thence with a beautiful plume, the highest tree of the forests, and wonderful combinations of colour are formed by the climbing plants around it. Most valuable from a commercial point of view is the teak, which is largely cut for export. The upas-tree is another inhabitant of the forests.

The bulk of the population of Java, indeed all of it save about 1,000,000, are distributed in small villages forming independent communities. Family life is well ordered. The Javanese are nominally Mohammedans, but with the great mass of the people this faith is overlaid with rank superstition. Thousands annually make the pilgrimage to Mecca. The work of Christian missionaries is relatively restricted.

Descriptions of all the interesting towns, places and objects of Java cannot be entered upon here. The palaces and tombs of sultans are found in various places. Sugar factories and fisheries give employment to a number of people. Railways, harbours and canals have been built, or their building attempted, in many parts of the island. In some of the residencies large areas of land are under cultivation. The roads and railways generally follow the fertile plains along the coast or between the areas of volcanic formation, and their total length, including state, private and light railways, reaches 2500 miles.

In Java we encounter those remarkable old Hindu temples

and ruins, of which Boro-Budur is the largest and most magnificent, ranking among the architectural marvels of the world. A stone-encased hill, rather than a temple, the structure rises 150 feet above the plain, its imposing terraces made of sculptured lava blocks. This extraordinary structure of Boro-Budur—a kind of " picture-bible in stone "—is but one of the many remarkable groups of temples on the island. Indeed, the archæology of Java forms an extensive and fascinating subject by itself alone. It is impossible here to describe these ruins. Those at Brambanam have been described as precious relics of the Golden Age of Hinduism, and " in a relatively small space are crowded together some of the most stupendous and best-finished specimens of human labour and the science and taste of ages long since forgotten."

The climate of Java is generally healthy and has been compared with that of the Riviera. The richness and fertility of the soil are unsurpassed, and it has been said that nowhere on the face of the globe can be found such fine beauty in foliage, such magnitude in leaf and flower, such delicacy and such diversity and brilliance in colour. Some of the flowers are so soft and delicate in texture that to breathe upon them is to destroy their beauty. Some bloom by day and others by night, and their delightful perfumes are unforgettable. The alternation of cool and fragrant air from the mountains, with the refreshing breezes of the sea, the perpetual sunshine and light rains, all tend to conserve the beauty of this island, so pleasingly dowered by Nature in these respects, a land of " perpetual life and beauty," as its enthusiastic admirers are pleased to term it.

South of Java, 150 miles, we remark the tiny but interesting Christmas Island, visited by the British buccaneer Dampier—whose activities in the tropics were so widespread—in 1688. The island is the flat summit of a submarine mountain which rises 15,000 feet from the ocean floor, and it has been the subject of various scientific expeditions. It is the only tropical island (excepting perhaps the Galapagos, off Ecuador) which had never been inhabited by human beings before the white man approached it ; the first settlers arriving in 1897, at which time a thick forest of giant trees covered it.

Hundreds of miles farther west are the lonely Cocos-Keeling

Islands, of coral, covered in part with coco-nut palms, the home of enormous crabs which devour the nuts, and of myriads of sea-fowl, undisturbed by man. Some of the islands, however, are inhabited, and the great guano deposits and the coco-nuts are profitable matters of trade. The people form a patriarchal colony, descendants of British and Dutch adventurers and their slaves, now governed from the Straits Settlement. The northern island has an excellent climate, and might be utilised for sanatorium purposes. The south-east trade wind blows almost continuously, and terrific storms break over the atolls. Scientifically the islands are interesting from Darwin's upheaval and subsidence theory, later disproved.

The island of Celebes now claims us, Dutch in its governance, like Java and Sumatra, beautiful in its scenery, rich in its vegetation. "Nowhere in the Archipelago are seen such gorges, chasms and precipices, perhaps 500 feet high, yet completely clothed with a tapestry of vegetation." [1]

This irregular-shaped island has an area of about 69,000 square miles, and a population of 1,250,000, people of Malayan stock. The men of some of the native tribes are described as brave, ambitious and revengeful, but not treacherous, and the women as clever, amiable and vivacious. Bodily exercises of running, hunting, wrestling, dancing are much indulged in, and although intoxication is rare " running amok " is frequent. Mohammedans by name, the people are overloaded with superstition, and worship animals and spirits. Houses are of wood or bamboo ; cotton cloth is woven by the women, and but little attention is given to agriculture. Gold, tin, iron and copper are all found and used in native industries. There are no large beasts of prey, nor yet the elephant or rhinoceros. Coffee, coco-nuts, sago, bread-fruit, pepper, melons, sugar-cane, cotton, tobacco, indigo, oranges, rice, millet, ebony, teak, sandalwood and a host of other foods, fruits, timber and products make up the lavish gifts of Nature here. The climate is a maritime tropical one. Many of the rivers are navigable by native prahus and rafts, and by canoes formed of hollowed trunks. The coasts are fertile and well populated, but there are few good harbours. The seat of government and centre of

[1] Dr A. R. Wallace.

trade is Macassar, with about 18,000 people, and there are other towns of some size.

Large sums of money have been voted in Holland for schools in the Dutch Empire of the East Indies. Entire religious liberty is upheld and missionary work progresses among the natives.

It is impossible here to consider the host of smaller islands which make up the Archipelago of Malay, but New Guinea, from its size, now claims our attention.

New Guinea is the largest island in the world, excepting Australia, from the northern point of which it is divided by the narrow Torres Strait. It was[1] politically divided between Britain, Germany and Holland, in relative proportions of about twenty-three, twenty-eight and forty-nine per cent. of its area, the total of which is somewhat over 312,000 square miles.

High ranges of mountains traverse the island, reaching in one or more places above the perpetual snow-line to nearly 16,000 feet. Some of the rivers are of great length, the Kaiserin Augusta, falling into the Pacific, being navigable for 180 miles by ocean steamers. Fever is prevalent on the coast, but higher up the climate is cool and healthy.

The shores of New Guinea, as we behold them from the sea, display broad grass-lands dotted with eucalyptus, and over those great areas which fires have devastated high grasses cover the land. The successive elevations of the country reveal the varied and peculiar plant life, from palm to pine. The animal life reflects the comparative poverty of this Australian sub-region, but New Guinea is famous for its beautiful birds, chief among them the brilliant bird of paradise. Large areas of the island are unexplored, and it is a land of many surprises.

The inhabitants, as a whole, are of the Melanesian race; Indo-Pacific, the Papuans, a negro tribe of fine physique, predominating; with stunted, low-type people in some districts. The population, however, can only be estimated very approximately in so wild a region, and is placed at about

---

[1] Prior to the Great War, at the beginning of which Australia captured the German portion with her own fleet and Expeditionary Force.

6oo,ooo. A semi-civilisation has come to being on the west coast under Malay influence and Mohammedan teaching, but head-hunters and cannibals occupy other regions. Slave raids and bad treatment by traders have been responsible for distrust and hostility among some of the tribes. In the east a comparatively fair people cultivate and irrigate the plateaux to a high degree. Wild tribes, naked and painted, swarm like monkeys in the trees of the submerged lands upon the Princess Marianne Straits, but the natives of Torres Strait are orderly and furnish labour for the white man, and there are some Christian communities. In the south-west exist a courteous people whose children, educated in mission schools, show marked intelligence.

Papua, the British section of the island, is controlled as a Crown colony [1] by a lieutenant-governor and council, the centre of administration being at Port Moresby, on the south coast. The natives make fishing-nets, mats, canoes, pottery, decorated gourds and so forth, which they barter. The white man's industries are in gold-mining, the planting of coco-palms and export of copra, and in rubber-planting. Gutta-percha, sugar-cane and cotton are likely to yield good results under proper attention. The population is about 400,000, a few hundred being white men.

The German Protectorate included the Bismarck Archipelago—that is, New Pomerania, New Mecklenburg, New Hanover and the Admiralty and Solomon Islands, with 200 islets scattered among them. As late as 1884 not a single white man existed in German New Guinea, but after annexation enterprise was awakened and tobacco and cotton successfully produced. The seat of government is at Herbertshoe, in New Pomerania. The chief harbours on the "German" mainland are Friedrich, Wilhelmshafen and Konstantinhafen. Missionary societies in some districts are active among the natives, of whose number no estimate has been made. The Germans made considerable progress in the elimination of tropical diseases under their administration. The whole protectorate was lost to Germany early in the Great War, having been dexterously taken, with but little resistance, by a small Australian military and naval force, although fighting

[1] The territory is now governed under Australia.

MOUNTAIN NATIVES OF PAPUA

took place at Herbertshoe for possession of the wireless station. The chief "language" spoken between the native and the European is the "pidgin English" [1] so frequently found wherever the British trader has established himself.

Dutch New Guinea occupies all the western part of the island, with a population of possibly 200,000. The possession should be of future value, but so far the Dutch have done little in its development. It forms part of the residency of Ternate, in the Moluccas.

These, the "Spice Islands" of Halmahera, Ceram, Buru, Timor, Aru and many other smaller ones, lie between New Guinea and Celebes. Encompassed by the sea, with a heavy rainfall, their vegetation is generally luxuriant. The Banda Islands, part of the group, are famed for their nutmegs, as is Ternate. But this last-named island and town suffers, like its companions, from isolation, and is of less importance to-day than under its old sultanate of centuries past. The sultan of the Moluccas is still nominally a ruler, but the Dutch broke up the Molucca government long ago in the formation of their East Indian Empire, although a subsidy is paid to the sultan, and some compensation was made for the purposeful destruction of the nutmeg gardens, which was carried out for the purpose of Dutch monopoly of this famous spice.

We now journey northwards to the large group of the Philippine Islands, which is included under the heading of Australasia. Here we enter a remarkably interesting region, a colony of Spain for centuries, wrested from that Empire by the United States in times still fresh to the memory.

The largest islands are Luzon—in which is the port of Manila —and Mindanao, the northernmost and southernmost respec-

[1] We may here give an example of this singular idiom. The official British proclamation on planting the flag in New Pomerania, issued by the British commander, began : "Whereas the forces under my command, etc.," but was supplemented as follows, in order to render it intelligible to the native :—" All boys belongina all place, you savvy big feller Master he come now "—ending with : " Me been talk along with you now, now you give three feller cheers belongina new feller Master." A further edict of the proclamation was : " You no steal Mary belongina other feller man " !—in reference to the need for the sanctity of marital relations under the British regime.

tively, nine lesser and a large number of smaller occupying the space between, the whole extending over 13° of latitude between the Equator and the tropic of Cancer, and covering a total land area of 115,000 square miles. There are, in all, 3141 islands, but all save eight per cent. of the area is covered by the eleven largest.

The Philippines are lands of earthquakes and volcanoes. Indeed, earthquakes are frequent and familiar, though greatly feared, incidents of life here, and at times are exceedingly violent. The plant and tree life is rich and varied, and the greater part of the surface of the islands is forest-covered, broken by many high hills, and watered by many streams and rivers. The coast-line is fringed with coral reefs and broken by innumerable bays and river channels.

The population of this great archipelago numbers nearly 8,000,000—7,000,000 of which are classed as civilised. About a similar number are Malays, a brown race, and the blacks are believed to have been the original people. They number about 25,000. The brown race came from the south, immigrating in successive waves from prehistoric times. Among the wild tribes there is some head-hunting, but most of them are expert agriculturists, who have laid out their lands under the system of terrace farming, with irrigation channels, such as we remark in Peru and elsewhere.

Among the foreign-born population the majority are Chinese. Spaniard and Malay, and Chinese and Malay are responsible for the principal mixtures. Of the white race, numbering about 16,000, half are Americans, the remainder principally Spaniards.

As the steamer enters from the China Sea the harbour of Manila, which since the American occupation has been made accessible to large vessels, we find ourselves within the broad expanse of a nearly land-locked bay, so large that if the typhoon be blowing the water is as turbulent as the open ocean. But the climate is in general excessively hot. To the west arises a rugged, mountainous range. Vessels from all parts of the world lie at their moorings, and a wide variety of tropical produce is carried hence in exchange for foreign merchandise brought in. Manila is famous for its hemp, and is the greatest hemp market in the world.

The city itself covers an area of over twenty square miles of the low ground upon the mouth of the Pasig river, and extends towards a second lagoon, nearly as large as that of Manila Bay. On the south is Intramuros, or the " Ancient City," surrounded by over two miles of high walls, built in 1590 by the Spaniards after an attack by Chinese pirates. Formerly a moat and drawbridge formed part of its defences. The site of Manila was first occupied by a Mohammedan chieftain. Early in the seventeenth century it became the commercial metropolis of the Far East, and fleets from India, China, Japan, Malacca, Mexico and elsewhere entered its bay. The plate-ships from Panama and Acapulco regularly performed their Transpacific voyage thereto. The city was taken by the British in 1762, but was restored to Spain.

The population of Manila, of some 225,000, is mainly of brown people, with over 25,000 yellow and 8000 white, and a few hundred black. More than half the whites are Americans. The public buildings are not striking in appearance, but some of the churches have fine façades and towers, and there is a monument to Magellan, the discoverer of the islands. The white inhabitants possess some handsome villas, two-storeyed houses, of brick or stone below and wood above, roofed with red tile —a form of construction which, with other details, is precautionary against the earthquake danger. The greater number of the houses are, however, mere shanties. Manila possesses a fine cathedral, and there are churches, convents, schools and other institutions, botanical gardens, a long boulevard and the well-known Luneta or principal pleasure-ground. In the time of the Spaniards social life was gay, but subject to the restrictions customary to that regimen, and Spain ruled her colonists with an iron hand to the very last. The Escota is a busy thoroughfare, the shopping and financial centre of the city, and the Rosario is noted for its Chinese shops. There are attractive residential districts at Ermita and Malate, along the bay shore. Over thirty newspapers are published in the capital. The Americans have carried out a number of improvements.

The American fleet under Dewey took possession of Manila, after destroying the somewhat weak Spanish warships in the bay, in 1898. A feature of this conflict was the friendly attitude of the British squadron and the scarcely veiled antagonism of the Germans. Previous to this the Filipinos had revolted against the yoke of Spain, which undoubtedly pressed heavily upon them. Nor did they accept American control without considerable struggle and bloodshed. The Spanish autocracy was replaced by a popular assembly, and the Americans have established a good educational system, and their rule must be regarded as a beneficial one.

It is not to be supposed, however, that the long Spanish regimen in the Philippines was altogether oppressive, or stationary as regards betterment of the population. However much Spain may have failed in the administration of these islands, the Spanish ideal was certainly not merely one of the exaltation of commercialism and the exploitation of the native, and in comparison with Dutch or even British colonisation in that part of the world, the Philippines developed during the 300 years after their discovery far more than other parts of the European-controlled East Indies. On the other hand business was neglected, as far as the native was concerned, and to make converts by ruthless religious methods rather than creating citizens was the main defect of Spanish rule in this thickly populated tropical land, as in tropical America.

The principal industry of the islands is agriculture, but it is generally in a primitive condition. The native farmers very slowly adopt the scientific teaching of the Americans in this respect, and are indolent. A mere fraction of the land in the archipelago is cultivated, less than two per cent. in the great Mindanao Island. Around Manila activity is more marked. Of the nearly 900,000 small "farms," many are mere gardens of less than an acre, with an average of about eight acres.

Hemp, tobacco, sugar, rice and coco-nuts are raised. Betel-nuts are chewed by the natives instead of tobacco. Coco-nut cultivation has extended under American control, and sugar has declined. The main food is rice, and in some districts maize, and some cotton and coffee are produced. The cattle have suffered at times from rinderpest.

Gold, coal and copper are worked to some small extent, and sugar and cigar factories are among the principal industries. The native women manufacture nearly all the material for their garments, from the abundant and varied fibre—hemp, pine-apple and cotton, and the natives carry on other small industries to supply their primitive wants. The foreign commerce is mainly in the export of Manila hemp, copra or dried coco-nut, sugar, and tobacco in leaf, cigars and cigarettes.

The edible birds' nests, so much prized by the Chinese, are found in large numbers, and another noteworthy natural product is in the remarkable molluscs, among them the giant clams, five feet in diameter, the shells of which are used in place of window-glass often. There are about 500 miles of railway in operation, and the Americans have improved the roads. American trade predominates somewhat over British.

Following upon our survey of Northern Australasia we now reach the great island-continent of Australia.

It is perhaps scarcely surprising that the Australians, in view of the heterogeneous, coloured, backward and more or less savage races which inhabit the northern part of Australasia, of which they themselves are the southern inhabitants; races whose modes of life we have remarked in Malaysia; should strive to conserve for their own land the ideal of being and remaining a "white man's land," notwithstanding that half of it lies within the torrid zone, and so would seem to demand the presence of a population which could advantageously till the soil under the equatorial sun.

Australia is divided midway by the tropic of Capricorn, and consequently it is the northern half of the continent which we have here to consider. The respective areas of the tropical and non-tropical divisions are 1,145,000 and 1,801,700 square miles approximately. Thus, within the tropics, lie half of Queensland, three-quarters of the vast region known as the Northern Territory, and a third of Western Australia, a zone about 2200 miles long upon the tropic of Capricorn and about 900 miles from south to north, reaching to within about 10° of the Equator.

Australia does not contain a teeming native or coloured population such as we generally find in tropical lands. The aborigines have been estimated as about 180,000 in number for the whole country, and even these are steadily diminishing, although even at the time of the European advent it is estimated that they did not number more than 200,000. These scanty people, moreover, have been classed as among the most backward and primitive of the whole human family.

The white population of the whole of Australia is somewhat under 4,500,000, but the bulk naturally inhabits the non-tropical half. There are over 30,000 Chinese, numbers of which are in Queensland and the Northern Territory, under 4000 Japanese and under 5000 Indians. There are also a number of South Sea Islanders. All the white Australians, with the exception of three and a quarter per cent., are of British origin.

Australia is one of the oldest lands—geologically—in the world, and must have been dry land in large part when Europe and Asia were submerged. It has few navigable rivers from its interior, and no snow-capped mountains or volcanoes, whilst its outline is singularly regular, the enormous Gulf of Carpentaria, in the north, being the greatest indentation. Enormous deserts occupy a great part of the interior, and a vast alluvial plain, covering 500,000 square miles, stretches right across the continent from the Gulf of Carpentaria to the Murray river, and is the most notable topographical feature of Australia. This plain lies east of the central meridian, 135° E., and west of this are the Australian steppes, entirely different in character, formed of flat-topped and terraced hills, and the stone-covered flats of desert sandstone. Some of the land is below sea-level, but the average elevation is 1500 to 3000 above.

The water-courses shown on the map here disappear in the dry season, leaving nothing but a row of scanty gum-trees to mark their courses. Over these vast steppes, covering 400,000 square miles, the rainfall is scanty, and not a drop of it ever reaches the sea. These dreary areas, however, are by no means confined to tropical Australia, for they are well distributed, and a great belt across Australia, from south to north, has an average of less than ten inches of rainfall.

The north-west coast and the shores of the Carpentaria Gulf are subject to the annual monsoon, which penetrates inland for 500 miles. The climate of Australia, as a whole, is milder than in lands in similar latitudes north of the Equator.

According to temperature, a considerable area of Australia would not seem to be adapted for European colonisation. The interior of the Northern Territory—within the tropic—has a mean summer temperature exceeding 95° F. None of the large cities lies within the tropical half of the continent.

Queensland has many ports on its eastern or Pacific side. Of these Rockhampton, the largest town in the tropics, with about 16,000 inhabitants, lies almost on the tropic of Capricorn, and there are others of importance upon the tropical coast. There are no very high ranges of mountains, the loftiest in tropical Australia being Bellenden Ker, 5200 feet. Palms and tree ferns are plentiful, and there are many kinds of woods suitable for carpentry, or having spice or medicinal barks. The most famous is the well-known eucalyptus. No country is better furnished with fodder-plants, and excellent pastures result. There are many flowers and fruits, some of them very peculiar. Snakes, sharks, crocodiles, turtles, some beautiful and curious birds—among them the laughing jackass—game birds and multitudinous forms of insects, among which the mosquito is abundant. The scanty land fauna is that of Australia in general.

Sugar-cane cultivation is an important industry in Queensland, and nearly 150,000 acres are under plantation. It is of interest in this connection that work in the cane-fields is carried on by white labour, the importation of Pacific Islanders and Kanakas now being prohibited. The cultivation of tobacco also increases, and the soil is well suited for cotton. Some coffee is grown, and all the cereals are produced in one or the other part of the state. The principal industry, however, is stock-raising, and sheep are only secondary. Dairying is also important. There is but little manufacture, except in those matters arising out of agriculture. Of the nearly 3000 miles of railway the greater part is in the non-tropical portion of the state. Brisbane, the capital, lies in this southern portion.

The Northern Territory is much isolated from the rest of Australia, but the purpose has been advancement of building or continuing the railway from Adelaide, in South Australia, to Port Darwin. The line at present is nearly 700 miles long, the total distance which separates Adelaide from Port Darwin, on the north-west corner of the territory, being 1896 miles.

Various attempts have been made to colonise this vast region. Much of its soil is rich and fertile. Sugar, coffee and rubber have been experimented with, as well as sheep-farming.

U

Tin and gold mining have a considerable annual output.   The renting of pastoral land is exceedingly low, and it has been stated that land to support 30,000,000 sheep and vast numbers of cattle, forming one of the most important centres of meat supply for Britain, could be created here.[1]   Agriculture at present cannot pay the high wages for clearing and cultivating at present obtaining, and the " White Australia " ideal and laws prevents the import of low-paid or indentured coloured labour.

It is considered that the growing of wheat in this remote part of Australia may help to solve the problem of development. Experiments with Indian wheat have been successful, but have not yet been made on a large scale.   The Australian understands wheat-growing, and likes it, it is stated, because it does not entail hard and continuous work all the year round.   If wheat can be successfully grown on the uplands the products of the lowlands would also be in demand.

The Northern Territory covers more than 325,000,000 acres —four and a half times the size of the United Kingdom.   It is 900 miles long and 560 wide, with 1200 miles of seaboard, and various navigable rivers intersect its northern portion, with some spacious harbours.   This large country is practically uninhabited, the population numbering less than 3600, over half of which are Europeans and about 20,000 aboriginals.   The coastal climate is hot and humid, but is regarded as much better than that of other equatorial lands of its latitude in the northern hemisphere.   It is stated[2] that there is nothing to prevent white races from thriving here, and that the region is surprisingly healthy.   The dry inland plateau is also suitable for the white man.   In winter the temperature falls to 56°, and in places below freezing point.   The maximum shade temperature on the tropical coast is given at 96°, and the night minimum at 65°.   The rainfall varies from ten to forty inches over vast areas—213,000 square miles having ten to twenty inches.

The land may roughly be divided into three almost equal parts, of coastal land, well watered, with good soil along the rivers, covered with tropical vegetation ; good pastoral country

[1] D. Lindsay, *Northern Territory*.   Royal Society of Arts, 1915.
[2] *Ibid.*

of wide grassed plains and river flats ; and fairly good to inferior pastoral and sandy land, these divisions having 75,000,000, 80,000,000 and 87,000,000 acres respectively. There are large gold and other mineral-bearing areas. A great deal of land is susceptible to irrigation.

The conditions in the tropical part of Western Australia do not differ greatly from those of the Northern Territory.

It is thus seen that in this remote part of the British tropical empire lies a land of very considerable possibilities, at present lying fallow. The great economic questions are in labour and railways.[1] The emptiness of this great territory has given rise to some apprehension that Japan might cast covetous eyes upon it, as an outlet for her own people.

We now leave the great land masses of Australasia to voyage among the remote Pacific isles.

[1] It cannot be held to be beyond the powers of British Imperial statesmanship to make use of this great property and its resources for the benefit of the teeming millions of the United Kingdom, for whom the difficulties of economic life are constantly increasing. The lack of organisation and endeavour in this respect, whether here or in other parts of the tropical empire, is at the same time a reproach and a danger. A more energetic and intelligent stewardship is necessary.

CHAPTER XXIX

THE ISLANDS OF THE PACIFIC

THE isles of the Pacific Ocean are perhaps more associated in the popular idea with "the tropics" than other parts of the torrid zone. Here lie the typical coral shores, fringed by the crisp foam of a blue and boundless sea, overarched by an azure sky, with a background of palms, and the human accompaniment of the brown or black man, among whom trader or missionary has striven each to develop his special activities—predatory, peaceful or proselytising.

We have here the home of the "beach-comber," the land of dusky maidens and island potentates and warriors, the most typical land of the coco-nut and of early barter and cannibalism, and these tropic isles—coral atolls or volcanic mountain tops protruding above the bosom of the vast Pacific—have always been shrouded in an atmosphere of romance, savagery and *dolce far niente*. They have ever appealed to our imagination, and the "desert island" of our youth must have been situated somewhere in these remote, beautiful and lonely archipelagos.

Here, moreover, we find traditions of sunken continents and the long journeys of prehistoric canoe voyages to distant lands, and upon some of these sea-girt isles there are ruins of stone structures of which history can say nothing, among the archæological mysteries of the world.[1]

Portions of the Pacific Ocean, whose enormous extent rolls unhindered by any continent through the tropical belt between Australasia and America, for over a third of the circumference of the globe, are strewn in places with small islands like stars in the sky, in groups and singly, often the least accessible places on the face of the earth. But they are practically all charted and under the possession or protection of one or the other of the maritime nations.

[1] Described in the author's book, *The Secret of the Pacific*, 2nd edition.

GATHERING COCO-NUTS, GUADELUPE.

At the time of the great European War of 1914 the total area of these numerous small colonies was divided as follows :— Great Britain held about 22,500 square miles ; Germany about 26,000 ; France about 9000 ; and the United States about 6700, with some under joint control of Britain and France. The total area of the Pacific Islands is very approximately given at somewhat under 70,000 square miles, and their population somewhat under 1,000,000. Of this about 354,000 were upon British and about 323,000 on German territory. There are more than 2600 islands, with numerous very small islets in addition.

We cannot do more than examine very briefly these numerous islands. Journeying eastwardly from Australia we reach in succession the Pelew,[1] Marianas,[1] Caroline,[1] Marshall,[1] Gilbert, Bismarck,[2] all north of the Equator ; Solomon,[2] Santa Cruz, New Hebrides and Ellice groups and archipelagos, south, which brings us as far as Fiji, near the world's dividing degree of longitude, the 180th meridian. Beyond these lie the Phœnix, Samoan, Tonga, Hawaii, America, Manihiki, Cook, Society, Marquesas, Panmote and other groups, the most eastwardly being the lonely Easter Island (below the tropics), whence we approach the Pacific coast of the New World.

The Pelew Islands, whose total area is less than 200 square miles, are well wooded and possess a healthy climate, with a population of over 3000, Melanesians. Among the native institutions may be mentioned the species of " Mutual Aid Society " that obtains here, and which possesses considerable influence. The government was that of a native kingdom, under Germany, which country bought the group from Spain in 1899.

The Carolines, to the east, were also purchased by Germany from Spain. They cover somewhat less than 400 square miles, the largest island being Ponape. Heavy storms sweep the archipelago, but the climate is healthy, though moist. The inhabitants, who number somewhat less than 40,000, are a comparatively intelligent people, expert boat-builders and sailors and careful agriculturists.

In Ponape we encounter some of the remarkable ancient

[1] Taken by Japan from Germany in October, 1914.
[2] Taken by the Australian forces from Germany in 1914.

stone buildings, the origin of whose existence in various of the Pacific Isles is unaccounted for : Cyclopean ramparts and walls built of basaltic blocks ; canals, artificial harbours and huge breakwaters, without any key to the personality of their constructors.

In the Marianas or Ladrones, to the north, there are also curious monolithic remains, with rows of massive columns. The largest island of this group is Guam, belonging to the United States. The remainder were German, and cover about 245 square miles, with about 3000 inhabitants, mostly the descendants of Tagal immigrants from the Philippines—the nearest land to the west.

Dense woods and luxuriant vegetation mark the scenery of these islands. Coco-nut and areca-palms flourish, and sugar, coffee, cocoa, cotton, tobacco and copra are exported, with mother-of-pearl. But although fertile and well watered agriculture is here neglected. The climate is damp but healthy. Since the early Spanish times of Magellan—who discovered the islands—the population has decreased, it is said, from perhaps 60,000.

Guam covers over 200 square miles, with over 12,000 inhabitants. It was taken from Spain by the United States in 1899, which country it serves as a naval station and port of call on the Philippine route. It is also a cable station. The climate is healthy and agreeable, the lowlands are fertile, and parts of the mountains and plateaux are well wooded, and tropical food-stuffs common to the group are raised. The United States have established an agricultural experimental station here. The Chamorro or Indonesian people who inhabit Guam are of a dark chocolate colour, and are mostly poverty-stricken, small farmers, though their status has improved under American Government, which abolished peonage and provided for education. The island is visited at times by destructive earthquakes and typhoons.

The Marshall Islands, formerly belonging to Germany, are a coral group which rises little above the sea. They cover about 160 square miles, with about 15,000 inhabitants. These people are expert navigators and boat-builders, their canoes and schooners carrying sails. The usual coco-palms fringe the coral shores, and bread-fruit, bananas, yams and so forth furnish food, as does the abundant fish. The people are pagans, but there is some Christian missionary work among them.

The Bismarck Islands, also previously German, include New Pomerania, New Mecklenburg, New Hanover and the Admiralty and other islands. The first-named covers 9500 square miles, and is largely unexplored. Its rich, tropical forests and jungles are traversed by many streams. At Herbertshoe, the chief centre, is the seat of the governor of German New Guinea, whose coast lies less than 100 miles away.

The Solomon Islands, to the east, were divided between Germany and Britain, and are well watered, with some good harbours. Forest-clad mountains and undulating grass-lands characterise the scenery of these islands, and the fertility of the soil is attested by the numerous villages and plantations. The total area is 17,000 square miles, of which Bougainville—German—the largest, named after its French discoverer in 1768, covers more than a quarter. The population, of about 180,000, is of Melanesian races, black-brown or copperish, a people who have been described as treacherous, who practised infanticide and indulged in head-hunting, and whose women are almost slaves. However, they are good agriculturists, and their finely made canoes, inlaid with mother-of-pearl, and high, carved prow, are objects of native art. In the magnificent forests here many kinds of palms are found, with valuable timber trees. Mountains rise to over 10,000 feet on Bougainville, with active volcanoes. Coco-nuts, copra, ivory-nuts, cocoa, coffee, bananas and other fruits are grown or exported, the total cultivated area being small, however.

The Santa Cruz Islands are administered by Britain. They cover 380 square miles, with about 5000 people. The islands are densely wooded, and the sea-fisheries are of great value to the natives. These people, having been kidnapped during last century as labourers and carried overseas, learned to hate the foreigner, and Bishop Patteson was murdered here. Partly as a result, steps were taken in England to supervise the traffic in natives. Copra is exported. Cannibalism has not yet died out, but missionary teaching has made good progress. The British took possession in 1898.

The Gilbert Islands, threaded by the Equator, are low, coral, sandy atolls or reefs strung out in a crescent shape for hundreds of miles. They are noteworthy in that, although not of more than 166 square miles in total area, and scarcely producing anything for human life beyond coco-nuts and the fish of the seas, they support a large population of over 30,000. These prolific and hardy people are of the Polynesian race, tall and stout, vigorous and active, naked, fierce fighters with their sharks'-tooth-studded swords. They make canoes of boards from the coco-palms, sewed on to a framework. Many of them were taken to Hawaii as labourers in 1878 and later, but the unfortunate "blackbirds," as they were termed, were not suitable for the work, and abuses were committed upon them. The archipelago was annexed by Britain in 1892.

The Ellice Islands, south of the Equator, a group also annexed by Britain in 1892, are similarly low, coral formations, coco-nut yielding, covering only fourteen square miles, with a population of perhaps 2400. The people have a tradition that they came from Samoa, "thirty generations ago." They are now all christianised, due to the work of Protestant missionaries.

New Caledonia and the New Hebrides, lying east of Australia, approach the tropic of Capricorn. The last-named are under joint British and French control, the former entirely French. As we reach these elevated, volcanic islands we remark their beautiful aspect. The rugged scenery and rich vegetation is well set in the azure sea, from whose deep bosom the shores rise abruptly. The largest of the New Hebrides is Espiritu Santo, with an area of nearly 900 square miles. New Caledonia, with its grand rocky scenery and beautiful landscape, covers, with its adjacent islets, nearly 6500 square miles. The streams are ingeniously used by the natives for irrigating their plantations. Bread-fruit and bananas grow wild. Maize, coco-nut, sugar-cane, coffee, cotton, tobacco, rice and so forth are produced, as well as indigo and lucerne. Fine timber trees abound, and turtles are numerous on the coast.

The Kanakas—which race inhabits New Caledonia—are regarded as the best agriculturists in the Pacific Isles. They are tattooed and wear ear-rings, and both men and women are naked, or with occasionally a loin-cloth. Their standard of morality is accounted low, and women are almost slaves, but infanticide is being stamped out. This practice, it will be recollected, resulted in some cases from a native fear of over-population. The climate is healthy and the trade winds invigorate the temperature. The population of this fine island numbers about 55,000, of which about a quarter are European colonists, officials and soldiers. The capital, Noumea, stands upon a fine harbour.

The climate of the New Hebrides is less healthy. The population is over 50,000, with only a few hundred Europeans, but a large area is planted by the settlers, and copra, maize, coffee, timber and bananas are exported. The native villages are extremely neat and clean, and often adorned with flowering plants and shrubs, which, however, do not conceal the stone walls that fortify them. The inhospitable and savage character attributed to these natives is largely the result of gross abuses and cruelty committed by traders and labour recruiters. However, the men are constantly fighting, among their weapons being beautifully made clubs, and bows with poisoned arrows. The status of women is very low, and at times widows are buried with their dead husbands' bodies. There are many curious or repugnant customs among these people.

The treatment of native labour of late in the New Hebrides, by the planters and others, in some cases, has been the subject of grave complaints, and attention drawn thereto in the British Parliament has led to "conversations" with the French Government. It was alleged that natives were bought and sold in the condominium. The islands, as regards their British interests, come within the Australian sphere of influence, and dissatisfaction has been expressed by Australia in its non-representation in the Government. Port Vila, in Efaté Island, is the seat of joint administration.

The British colony of Fiji, which we now approach, is about the size of Wales, the 250 islands of the archipelago covering about 7500 square miles, with over 121,000 inhabitants. Viti Levu, the largest of the group, is nearly 100 miles long, with an area of over 4000 square miles, and Vana Levu is about half the size. The remainder are small.

These beautiful islands, covered with luxuriant vegetation, rise in some cases to hills 3000 feet high, but have wide tracts of undulating country, and, in the larger ones, navigable streams. They are generally surrounded by coral barriers.

The climate is often cool and dry, but there is a wet, warm season, and hurricanes at times destroy the crops. The volcanic mountains are no longer active, but there are hot springs in their vicinity.

The capital of the colony is Suva, standing upon an excellent harbour in Viti Levu, which is the most important of the islands, whether in size, population, fertility or general resources. Here are the Government buildings and other institutions, established since Great Britain accepted the "unconditional cession" of the archipelago, offered by its principal people in 1874.

The inhabitants of Fiji are Melanesians crossed with Polynesians, and are superior physically and more skilful than other offshoots of the same race. They are handsome in feature and well proportioned in frame, tall, dark in colour—the chiefs of a fairer complexion—with an elaborate mop of crisp hair and a bushy beard. They combine the more acute intellect of the fairer Pacific people with the savagery of the negroid type. Little clothing is worn, but decency and morality are more marked than in some others of the Pacific Islands.

However, the most brutal and repugnant customs formerly prevailed among these people, including cannibalism and human sacrifice. The "long pig," or cooked body of a human being, was the chief luxury of the savage Fijian potentate. When a chief built a house a slave was buried alive under each corner post, and when a chief died his wives were buried alive with him. War canoes were launched over a slip-way of living human beings. Every day human sacrifices were made.

Yet, despite these repugnant and heathen customs, the Fijians are described as courteous and hospitable, possessed of humour, conversational power and tact, with an elaborate code of social etiquette and strongly marked system of class distinctions. The chiefs are a "natural aristocracy," being physically and mentally superior to the common bulk. The economic and land systems were, and still are, peculiar. Everything belonged to the chiefs, but the chiefs' property all belonged to the people, and was given to the common weal in times of famine. Taboo is a sacred institution, and a savage priesthood and sacred groves—though not idols—entered into the native religion, which, nevertheless, embodied a belief in a future life. It was indeed to this belief that popular acquiescence in human sacrifice was due.

The population of Fiji is decreasing. The native woman does not mate with the large Indian coolie immigrant element. Most of the natives now are Wesleyan Methodists, and there are Roman Catholics, Mohammedans and Hindus, the Church of England being most active among the urban Kanakas and the Europeans. Due largely to the missionary schools, in every village, most of the natives can read and write their own language, but they are slow to adopt European methods generally. This last may or may not be advantageous.

When Indian coolies were introduced into the islands, in 1878, the cultivation of sugar fell into the hands of large companies, with modern appliances, and decadence of the Fijians' activity in the development of their own country set in. But it had been largely with the purpose of putting an end to Polynesian labour abuses that Britain acquired control of the islands, although the regulations enacted for the protection of the natives gave rise to great dissatisfaction on the part of the colonists—a not uncommon circumstance in tropical lands.

The Fijians are skilful agriculturists, boat-makers, carpenters, and pottery and mat makers, and many of their handicrafts show great taste. Native cloth is made, and plants are cultivated, not only for their use, but for their beauty, for the native is alive to this. The kava root furnishes the national beverage. Coco-nuts, bananas, bread-fruit, sugar-cane and many fruits are raised. Bread is made from roots and bananas, not grain, by a peculiar but simple native process. The main exports are sugar, fruits and copra. Coco-nut oil is used and soap made therefrom, and good tobacco is grown.

Continuing our journey for 600 miles or more to the north-east we reach the islands of Samoa, beautiful in their scenery, hot but not unpleasant in their climate, traversed by high mountains, among which are lovely vales and marvellously fertile plains of igneous soil. Here the forests are remarkable for their size and the variety of their timber, and the tree-fern stands unrivalled for its beauty and luxuriance. The coco-nut palm and the bread-fruit, of a score of varieties, are Nature's most valued gifts to the inhabitants of these tropic isles.

Nature has, however, tempered her generosity here by the

occurrence of disastrous storms, and the fierce hurricanes which sweep Samoa at times destroy houses and ships. Notably did this occur in 1889, when two United States ships were beaten to pieces on the coral reefs and H.M.S. *Calliope* barely escaped in the teeth of the gale. Eight of the principal Samoan islands were under German control,[1] their areas ranging from the 660 square miles of Savaii, the largest, downwards, and six belong to the United States. The total population is about 40,000, with a few hundred whites. The people are pure Polynesians, and from Savaii this race dispersed in prehistoric times over the Pacific, interesting traditions being interwoven with their migration, of heroic voyages and sunken continents. They are the best type of their race, of light brown colour, handsome and of fine physique, the average height being five feet ten inches. Tattooing is carried out all over the body. The Samoans are famous boat-builders and mariners, and are simple, generous, hospitable and honest, but are also brave fighters. There was no human sacrifice in their native religion, nor idolatry, and to-day they are mainly Protestants, Catholics or Mormons. Their soft and expressive language has been termed the " Italian of the Pacific." Agriculture has extended under American and German rule, and the chief exports are copra and cocoa. Great Britain withdrew her claims to what was an unsatisfactory joint control of the islands in 1900. The home of the famous Stevenson and his burial place overlooks Apia, one of the principal seaports of Samoa.

Southwards lie the little Tonga or Friendly Islands, as Captain Cook termed them, remarkable in many respects, picturesque and mild in their scenery, and rich and beautiful in their vegetation, scourged by active volcanoes and earthquakes, and in some cases liable to disappearance beneath the Pacific waves. Here again we find some remarkable monuments of great stone blocks, the remains of some unknown builders.

The islands fell under British protection in 1900. They number about 150, with an aggregate area of less than 400 square miles, the largest being Tongatabu, or Sacred Tonga, on which the relics exist. Many of the small islands are extremely fertile, veritable gems of the ocean. The total population is

[1] Taken by Britain in 1914.

about 20,000, people of Polynesian race, coming next in point of intellectual development to the Hawaiians.

The soft climate and abundant natural food render hard toil unnecessary to these islanders, though they are not idle, and are brave, kind to their women and value their children. They are good cultivators, skilful sailors, fishermen, boat-builders and so forth, many occupations being hereditary. The grounds around their houses are laid out with much taste and care. The nation desires to rank as a civilised state, but the native rulers, although capable enough with their own people, become an easy prey to more or less unscrupulous Europeans. The native king is assisted by his hereditary nobles and elected councillors. This king wisely refused to alienate the land except on lease, and Tonga long retained its independence. Copra, bananas and oranges are among the principal exports.

Striking now north and east from Samoa, we pass the three groups of British atolls of the Tokelau Islands, whose few hundred natives are all Christians; and the small Phœnix Islands, also British, with above them some still smaller American islets.

Crossing the Equator we reach the American Islands, just beyond the line, belonging to Britain, which include Fanning, Christmas and Palmyra Islands. These were annexed in 1888, largely for purposes of a cable station. In the war of 1914 Germany took Fanning Island and held it for a few weeks. The population numbers about 300, and guano and mother-of-pearl are exported. This group of islands is surrounded by vast stretches of ocean on every hand, and we voyage to the north almost to the tropic of Cancer before reaching other land —that of Hawaii.

The Hawaiian or Sandwich Islands occupy an important position in the trade routes of the Pacific Ocean, the capital, Honolulu, being a little over 2000 miles from San Francisco in California, on the route to the Philippines and Asia, and at the " cross-roads " to Australia and New Zealand from North America. The area of the inhabited islands—eight in number —is 6650 square miles, of which Hawaii, the principal, occupies nearly three-quarters, with Maui and Oahu about 700 and 600 square miles respectively.

Hawaii contains the largest volcano in the world, Mauna Loa, rising to 13,675 feet, and Kilauea, its companion, has the largest active crater in the world, eight miles in circumference. Kilauea in eruption is a grand spectacle. Pele's Hair, the delicate, glassy fibres spun by the wind from the falling lava, is a curious phenomenon here. Pele was the dreaded volcano-goddess of idolatrous times.

The islands generally enjoy a relatively cool and healthy climate, with a cloudless sky. The scenery varies ; there are mountains, bare or tree-covered, beautiful valleys, whether of picturesque and rocky ravines, whether of broader area, Kauai being known as the garden isle. Rolling tablelands in some cases afford pasturage for sheep, and high elevations are reached, of over 4000 feet in some of the islands, with numerous craters. Much of the forest has been destroyed by the domestic animals introduced since the time of discovery by Captain Cook in 1778.

Cook was received by the natives with " demonstrations of delight and astonishment," and in the native temples priests offered up prayers for him, and although he was killed here by a native, his bones were reverenced and offerings made thereto until Christian missions succeeded in abolishing idolatry.

There are many valuable timber-trees in the islands, and rubber has been somewhat extensively planted of late. A wide range of native and exotic fruits is produced, and the islands are famous for the brilliance and variety of their flowering plants.

In earlier times land tenure in Hawaii was of a feudal character, but from impoverished chiefs much of it passed into the possession of foreigners, although under the Americans the " crown " lands were subjected to laws designed to promote the establishment of small holdings. Of the cultivated area by far the major portion is planted to sugar-cane, perhaps 250,000 acres ; the yield of sugar—thirty to forty tons of cane and ten tons of sugar per acre under irrigation—being the largest in the world. Some co-operation is practised on the plantations. The sugar is exported to the United States.

Rice and coffee are also cultivated, and the pine-apple is canned for export. Tobacco of good quality is grown, and the large cattle ranches supply the home consumption of meat. Wool and hides are exported. The most important manufacture is that of sugar, in which much American capital is invested. Hawaiian sugar has produced some American millionaires and powerful " trusts."

The native population of Hawaii is decreasing, in spite of progress, and this would seem to be a result of modern civilisation, under which this island race is dying out. From Cook's estimate of 400,000, or the American missionaries' in 1823 of about 142,000, the natives numbered at the beginning of the century less than 30,000. It is said that foreign diseases have poisoned the native blood. Immigration of various nationalities has been systematically encouraged, and there are over 72,000 Japanese, 18,000 Chinese, 23,000 Portuguese, with other Asiatics and Europeans in their thousands, forming a total population for the islands of under 200,000.

The native Kanakas—the Malayo-Polynesian race—are reddish brown, and have been described as among the finest of the Pacific peoples. The chiefs were generally of large stature, which may have been induced by the *lomi-lomi* or system of massage. Polygamy and human sacrifice, barbarous and sanguinary idolatry with cruel and bloody rites, were practised, but there were sacred enclosures to which refugee prisoners and others might flee, and were there inviolable. Taboo was practised, and under it women suffered much, and cannibalism was connected with religion.

The Hawaiians love sports and games, especially the famous surf-riding, in which, swimming out to sea with a small polished board, which they then mount, they are hurled ashore on the breakers. Beautiful hats—rivalling those of Panama—and mats are woven from grass. Women are less numerous than men, and it is said that the maternal instinct is very low. Under the Americans the spread of education has advanced considerably. The United States assumed control of these islands in 1898, after revolution and disorders, and the fall of the native sovereign, poor Queen Liluokalani, who journeyed to the United States to cast herself upon American clemency, but lost her throne.

Honolulu has a population of over 40,000, of whom only one-third are females. Another third are Asiatics, and there are about 5000 Portuguese. The newer part of the town, in which are many pleasing residences, whose gardens are planted with tropical shrubs, rises to the high, adjacent hills, whence commanding views are obtained. The wide streets are paved with crushed lava, the houses are generally of brick, with broad verandahs, the Government buildings and hotels are good, and the parks and other institutions of a fine character. There are a number of clubs and various scientific societies, and the public lighting, street cars and other utilities are modern and well equipped. The poorer quarters of the town are less extensive and prominent than in many places. In brief, Honolulu is a town of pleasing appearance and progressive. character, which speaks well for its American regimen.

Passing south of the Equator again we reach the Manihitu Islands, rarely heard of and seldom visited—a scattered archipelago in the Central Pacific, annexed by Britain in 1889. Here the coco-palms produce copra and the sea-birds guano, which, with some pearl-fishings, form the principal resources of these remote land spots. The area is not much over twelve square miles, and the population only about 1000. Upon Malden Island exist some ancient stone structures of earlier peoples.

Still farther south are the Society Islands and Cook Islands. The first-named, belonging to France, have an area of 640 square miles, and the principal of them is Tahiti, beautiful, but suffering the reproach of decadence as regards its people.

Some of the Society Islands—there is a Leeward and a Windward group—are among the most remarkable examples of volcanic formation in the world, with upraised coral beds high in the mountains, and travellers have generally agreed that in their varied scenic form and beauty they are the principal gems of the Pacific. The heavy rains give rise to a myriad rills, which descend the slopes and unite as lovely streams, plunging over high cascades down the faces of great precipices, surrounded by the green foliage of the dense forest. Chains of verdant islets often surround the larger isles, affording often peaceful havens and excellent anchorage. The total population of these islands is about 19,000.

Tahiti covers over 400 square miles, with about 11,000 inhabitants. The tower-like peaks rise, in the double mountain of Orohena, to 7350 feet above the sea. Along the flat coast, to which the hills form a background, lie in pleasing order the numerous villages, palm-belts and fertile gardens, traversed by streams which descend the fertile vales, and on the beautiful harbour of Matavi stands Papeete, the chief town and centre of French administration.

For a tropical climate that of Tahiti is healthy, although hot, with a heavy rainfall, and hurricanes blow at times. The banana thickets are remarkable, and there are splendid orchards where many fruit-trees have been established. From Papeete pearls and mother-of-pearl, copra, vanilla and oranges are exported, and sugar and rum are produced.

The Tahitians are typical Polynesians, with the softest language in Oceania and the most beautiful native women of the Pacific. It has been said that the physique of the people is unsurpassed elsewhere in the world, although possibly these matters have been somewhat exaggerated. Generations of European vice and disease, and of inebriety, have not entirely banished these characteristics, and people six feet or more in height are frequently encountered. The complexion is a light olive to dark brown, and the eyes are expressive and the teeth fine. Beautiful native costumes were made, and the tattoo marks bore strange meanings. The upper class are physically distinguished by their fairness and physique. Cannibalism was not practised, but infanticide was common.

Steering by the stars, many of which they had named, the Tahitians, skilful navigators, made long voyages in their great canoes, often double-bodied and fitted with sails and outrigger, the curved prows rising high from the water, carved with the images of their gods. In the fields, which were well cultivated, irrigation was carried out. The power of the king was absolute, and the land system feudal. Great temples were built, and that at Athura, of hewn blocks of coral stone and basalt, was 270 feet long and 50 feet high, reached by a flight of steps. Here human sacrifices were offered.

The people have been described as courteous, generous, light-hearted, frivolous, cruel and deceitful, extremely immoral, with a system of exchange of wives. Their favourite sport is surf-swimming. The working classes prefer handicrafts to crop-raising, and many are employed as artisans by European masters. The importation of labourers for the plantations,

x

which lent itself to abuses, is now under Government control. One of the early kings—Pomare—who had proclaimed a "sensual paradise" for his followers, died of drink, and after various international complications and national decadence the French established a protectorate in 1880. The character of Tahiti has been a byword throughout the Pacific lands as far as San Francisco, which trades therewith.

Cook Islands cover about 110 square miles, the largest of the group being Raratongo, with its majestic volcano reaching nearly 3000 feet. Fertile, well watered, with luxuriant palm-groves and a healthy climate, some of these islands are well suited to white settlement, but access to others is difficult, by reason of the coral reefs and lack of good harbours.

The people, who number less than 7000, are mostly Polynesians, whose ancestors, according to tradition, were voyagers from Samoa, and upon arriving they found black people on the island—Melanesian, whose type still exists. The inhabitants of Cook Islands now wear European dress, live in stone houses and are christianised, a wide departure from their earlier cannibal times. The local "kings" and "queens" form a governing council under the British governor and protectorate, which was established in 1888, but is now within the sphere of New Zealand. The islands bear in revered memory the name of John Williams, of the London Missionary Society. Coconuts, fruits, coffee, woven hats and lime-juice are produced here.

To the south-east are the Tubuai Islands, belonging to France, and covering about 100 square miles, scattered across the tropic of Capricorn. Some are volcanic and their mountains reach an elevation of over 2000 feet above the sea. The land is fertile and well watered, and coco-nuts, bananas and arrowroot flourish. The Polynesian natives, much more numerous when Cook visited them, number about 2000. On the summits in Rapa Island we encounter some of those remarkable massive stone platforms and buildings which are so curious and mysterious a feature of the Pacific, resembling the extraordinary structures of Easter Island.

Northwards we penetrate the extensive Paumotu Archipelago, belonging to France, stretched out for 1300 miles, clusters of small islets rather than islands, some of them covered with rich vegetation after the character of Tahiti, and enjoying a healthy climate. Coco-nut palms flourish, and many varieties of tropical fruit have been introduced, and pearl oysters are abundant, giving the name of Pearl Islands to some of the group. Devastating hurricanes and heavy rollers sweep upon the coral reefs at times. A scanty but fine race, numbering about 6000, of Polynesians inhabit these islands, whose total area is about 330 square miles.

Near the eastern end of the archipelago is the small but famous Pitcairn Island, which with Henderson and Ducie Islands belong to Britain. Pitcairn was the landing-place of the mutineers of the *Bounty*, a British Government vessel which was carrying bread-fruit trees from Tahiti to the West Indies, and this beautiful island was, in its first years, "filled with treachery and debauchery," under Englishmen and Polynesian women. Remains of carved stone "giants," like those of Easter Island, are found here. The few islanders form among themselves a governing body.

The Marquesas, to the north, belong to France, and cover nearly 500 square miles in the aggregate. They are inhabited by a pure Polynesian people, who have been described as the finest, physically, of all the South Sea Islanders. These people, by their traditions, came long ago from Samoa. Their skin is bronze in hue, and their houses are curiously built on platforms raised from the ground. Friendly, hospitable and industrious, they have at times been bloodthirsty, and like their cousins of Tahiti have suffered from debauchery and infanticide.[1]

When France entered upon possession in 1842 the people numbered 20,000, but now are reduced to a fifth of that figure, although they have adopted Christian beliefs, at least outwardly. The vegetation of the islands is luxuriant, and bananas, sugar-cane, bread-fruit, coco-nuts and other tropical products are raised or grow wild. The climate is hot and damp, but not unhealthy. There is a small trade in copra and cotton, and the people are keen traders. There are some ancient remains and giant stone statues of a mysterious past here.

Last of the Pacific Isles is Easter Island, a mysterious land standing alone, separated by thousands of miles of open ocean

[1] R. L. Stevenson draws a mournful picture of their decadence.

from its companions of Oceania and from the Pacific coast of Chile, to which country it belongs. Here we encounter those astonishing remains of stone " giants," some of them monoliths seventy-five feet high, which one tradition says were set up " before the Flood," set in rows along the shore, with houses and platforms of hewn stone, the whole being a veritable enigma of the ocean, as no key to the period of construction or to the builders of these monuments has been discovered. Easter Island, however, is not in the torrid zone, but lies some 250 miles south of the Capricorn tropic.

In our survey of the Pacific Islands explorers, whalers, traders and missionaries have followed each other's paths. Copra, pearls, trepang, tortoise-shell, debauchery, kidnapping, " labour " recruiting, robbery, murder, deserters, beach-combers and finally missionaries and reputable settlers and administrations have all been elements in their " development." It is an undoubted fact that the natives have often suffered at the hands of Europeans and Americans, and unfortunately in some cases they are dying out. Even the endeavour to civilise, and to enforce the observance of Christianity and modern customs, have been conducive to this deterioration. Doubtless the future holds better things for these simple people, or those who may succeed them, than their past history has too often recorded.

Here we leave the vast Pacific, with its beautiful archipelagos, for the shores of America which lie before us.

CHAPTER XXX

TROPICAL AMERICA

A REGION of infinite variety unfolds before us in the tropical lands of the New World. Their history carries us back to the romantic days of the Spanish Conquest and the " ocean chivalry " of the Elizabethan Age, and beyond that period to the old empires of the Aztec and the Inca, of Mexico and Peru. Here we traverse the boundless forests and waterways of the Amazon, the wild interior of South America, with its still unexploited resources ; the snowy Andes, with the stupendous scenery of the Great Cordilleras, where, as the traveller ascends :

> " Hills peep o'er hills
> And Alps on Alps arise."[1]

We cross the " American Mediterranean," drop anchor in the seaports of the beautiful West Indian Islands, and voyage to or from the Pacific Ocean through the Panama Canal.

We have in this survey approached America from the east, although the twin-continent was first reached by sailing west, and the streams of humanity which peopled it have set that way, whether emigrant or slave—the one from Europe, the other from Africa. However, we may remark in passing that if America was first peopled by the Mongolian race, as ethnologists hold—long prior to the Columbian era—such Asiatic immigrants would have come eastwardly across the Pacific or by Behring Strait.

The aboriginal, tropical brown race of Mexico and South America, whatever its origin, possesses a marked advantage or superiority over the aboriginal black race of Africa in that it assimilates or intermixes with the white race. In Africa the negro and the white dispute the occupancy of the soil and are divided by an eternal and impassable barrier of colour. In Latin America, on the contrary, the brown (or red) man and the white, intermingling, have formed strong, self-governing nations, of which much may be expected in the future. This is perhaps the main outstanding difference between the two greatest tropical regions of the earth.

A further advantage enjoyed by the Latin American tropics —as regards Mexico and Central and South America—is the great zone of elevated territory which sweeps through them, with its temperate or cool climate, suitable in great part for occupancy by people of European race, with a wide variety of natural and cultivated products, in which the staple food-stuffs and fauna of Europe flourish almost as well as in the land of their origin, yet removed often but a brief march from the most delightful fruits of tropical kind.

Thus, the portion of the Western hemisphere lying within the torrid zone has much to offer, despite the defects which its governance has revealed throughout the 400 years which have elapsed since the Europeans first set foot upon its shores.

As the northern tropic cuts Mexico midway and the southern the tapering portion of South America—the Equator passing through almost the broadest portion of that continent—by far the greatest area of Latin America lies within the tropics.

The region broadly described as Latin America embodies seventeen republics. The areas of some of these states can only be stated approximately, due to boundary questions which involve large portions of debatable territory in the republics occupying the northern and north-western part of South America, whose possessions stretch across the Andes into the Amazon Valley. Similarly, the populations given are in general more or less approximate estimates, as an exact census cannot be made, due to the vast and scattered areas and the aversion from enumeration shown by the native people or Indians, fearing taxation or enforced military service.

To the foregoing eighteen independent republics must be added the various possessions of foreign powers lying within the American tropics, as British, French and Dutch Guiana and British Honduras (in which Spanish or Portuguese is not the official language), although geographically a part of Latin America.

However, the political homogeneity of the continent is very marked in comparison with the African continent, and Spanish is spoken over a zone 7000 miles long. The homogeneity

THE SNOWS OF THE TROPICS: MOUNT HUASCARAN IN THE ANDES OF PERU, 22,180 FEET ELEVATION

in some respects is remarkable. Spain stamped upon her enormous extent of colonies in the New World, throughout Mexico and Central and South America, one system of laws, literature, language, type of social life, character of city and house-building, ecclesiastical and municipal edifices and administration, and moulded the character of the people upon the same lines from one extremity of the region to the other. The power, endurance and extent of Spanish civilisation in the New World is one of the most remarkable features in the whole history of colonisation.

Thus, in whatever Spanish American country we travel, we encounter almost the same type of individual, house, city and institution, from Mexico to Peru and Chile, and Brazil is also Ibero-American in its character. There are, of course, local variations, due to the extent of aboriginal population and immigration. It might have been supposed that the great bulk of the Latin American people, after more than four centuries of civilised European control and the influence of Christianity, would have become well advanced in the enjoyment of stable, social and economic conditions and profitable occupation. But unfortunately this is not the case. The bulk of the people are, as a class, poor and ignorant, their dwellings are of the most humble character, their food is insufficient to sustain them in hard work or to produce a healthy and vigorous frame, their clothing insufficient to protect them from the elements, whilst the conditions of social life, education and hygiene are almost hopelessly backward, with comparatively little exception. As the exercise of the suffrage is generally conceded only to those who can read and write, or are property owners, it follows that the greater bulk of the people are unrepresented in their governance.

Throughout the whole of Latin America we may generally distinguish but two classes : the upper, which almost monopolises the wealth, education and opportunity of the country, and the lower, so far condemned to be mere servitors, labourers in field or mine, under the most inadequate conditions of pay, and indeed in great part subject to peonage and semi-serfdom. It is true that in the most advanced communities, such as parts of Mexico, Brazil and Argentina contain, a middle class is slowly coming to being, but this is small in comparison with the overwhelming bulk of poor of the mixed race and native Indian.

The traveller in these lands notes first the attractive and intelligent class of which " society " consists, of statesmen, leisured persons, lawyers, and doctors in all branches of science, for to be a doctor, whether in medicine, law, science or divinity, is often the favourite aim of the well-born—and the often beautiful and vivacious woman of the leisured class, in her Parisian costume. He remarks the palms and the music, the glitter of uniforms and sumptuous palaces, the delightful courtesy, the ostentatious public buildings. But if he penetrates ever so lightly below this atmosphere the miserable figure of the untaught, underpaid labourer, peon or Indian, is revealed, in sharp contrast.

The term " republic " is indeed almost a misnomer in many respects in Latin American lands. They are often republics only in the absence of a monarch, whose place is supplied by some arbitrary dictator-president, who may have obtained office by means of the illegal pressure of his particular adherents and the misuse or suppression of the ballot, frequently accompanied by crime and bloodshed. There are, of course, well-meaning and capable presidents and officials in the Latin American states, in many instances, but self-government in these communities but slowly outgrows the extravagance of its youth, and in none of them can it be said that the true principles of a democracy and a commonwealth have yet been put into practice. The man of Latin American race has the peculiar trait of enacting excellent laws for the governance of the community, whilst contravening them himself on slight provocation. The machinery of Government is good, and if the laws were fully and impartially administered, civic growth would more rapidly result, but in this lies the defect.

The upper-class Spanish American or Brazilian, whilst he does not set up a hard and fast colour-line, instinctively looks down upon the " Indian," which name becomes almost a term of reproach. Yet the mark of the Indian is upon his own skin and features generally, from President and cabinet minister downwards. This pretension is partly Iberian arrogance, partly a natural " race nostalgia," or yearning for association with the superior race.

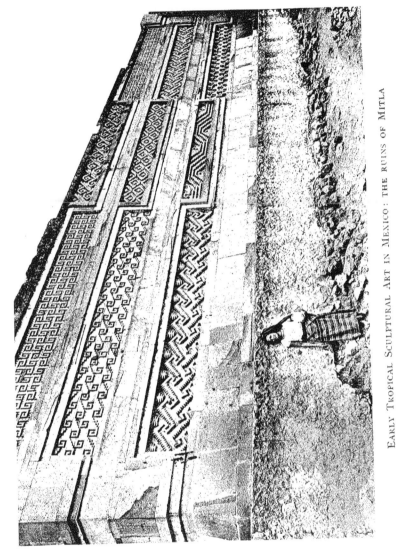

EARLY TROPICAL SCULPTURAL ART IN MEXICO: THE RUINS OF MITLA

The greatest vice of the Latin American working classes is drunkenness. The manufacture and sale of drink is a source of great wealth to the upper class and large landowners and sugar-growers, who make raw spirits and in Mexico grow maguey for *pulque*.

On the other hand there must be conceded to the Latin American a keen desire for advancement and a belief, at least in theory, in liberty and fraternity. Every state produces its class of able statesmen, administrators and professional men, and a class which strongly upholds the regard for arts and sciences. Education is highly esteemed, although at present it is almost confined to those classes which can afford to acquire it.

Among the most difficult economic conditions here displayed are the absence of manufacturing industries and the financial dependence upon European countries. The talent of the land does not so far reveal itself in the growth of that economic independence which alone can form the basis of a people's permanent security. The soil is dowered with everything necessary for the life of man, but, largely due to the absence of mechanical knowledge and initiative, the Latin American states remain, when not stagnant, largely producers of raw material for other nations whose manufactures they take in exchange, articles which it is time they should now create within their own shores.

The economic difficulties are increased by the failure to provide adequate means of transport in roads and railways. Indeed, in the majority of Latin American countries the traveller will often be undecided as to whether he shall expend his epithets in cursing a region whose roads and general accommodation leave so much to be desired, or whether, by exercising greater patience he shall rather endeavour to build up a picture of a land and people in the making, full of undoubted resources, whose better development, whenever a true *mañana* shall dawn, might make of them truly prosperous and pleasing communities. The failure to build roads is a Spanish defect, and the independent construction of railways is an art the Ibero-American people have not yet been able to compass. However, in the more advanced republics, railway construction has been active, and Brazil and Mexico especially possess large net-

works of lines. In the Andine countries railways have often
been constructed at enormous cost, and rarely yield dividends
on their capital outlay.

As regards the higher side of social and economic life among
the Spanish American people, the traveller will early find himself in a land over which, vast as is its extent, Don Quijote de la
Mancha might have waved some magic wand to endow it from
Mexico to Peru with his own characteristic. traits. The
recollections we retain of them are in general pleasing, and we
do not lose faith in their future.

The natural resources of the Latin American countries are
extremely rich and varied, and in some respects they form
veritable storehouses, upon which the world is beginning to
draw largely. However, enormous areas are undeveloped and
indeed almost unknown, at least as regards the heart of South
America. Almost every variety of food and fruit is produced
in these lands, whether of the native or imported varieties, and
so varied are the zones of temperature and cultivation that the
inhabitant, by going up or down the slopes or valleys of his
country, can reap any crop and enjoy any degree of temperature. Each Latin American republic contains practically
everything for the luxury and comfort of its inhabitants, and,
under a more advanced economic system, might become almost
self-sufficing and independent of the outside.

However, the more efficient development of these lands is a
necessary matter for European and North American commerce.
How considerable is British commercial influence in Latin
America is indicated by the fact that more than a thousand
million sterling is invested therein, in plantations, railways,
mines and public works, yielding a steady dividend in prosperous
times of five to thirty per cent. American and German influence
are both growing. To the Americans of the United States
Latin America, lying as it does in their own hemisphere, will
be of growing importance in the future, whether as outlets for
trade or as sources of raw material. But the Anglo-Americans do
not yet well understand the Spanish and Portuguese Americans.
They are poor linguists and their code of social etiquette—
matters which bulk largely in the southern republics—is very
different.

One of the principal political factors controlling the relations

TYPES OF WEST INDIAN NEGRO WOMEN, BRITISH WEST INDIES

of the three Americas with Europe is the somewhat nebulous Monroe Doctrine, which establishes as a principle that no territorial conquests by the latter upon the former shall be attempted. It is undoubtedly founded on a concept wise and just. European nations might have attempted to obtain control of territory in South America but for the promulgation of the Doctrine and the belief that the United States would act up to it. Whether the Americans are in a position effectively to take up arms in South America if such were necessary is perhaps open to question. The Latin American people do not necessarily regard the Doctrine with enthusiasm, suspecting, or pretending to suspect, that the apophthegm of " America for the Americans " may be translated as *America para los Norteamericanos*. This idea is probably a mistaken one. The people of the United States, however, have arrogated to themselves the exclusive title of " Americans."

The West Indies form a world of their own, beautiful and prosperous in many respects, but presenting special problems for the future. The sociological conditions therein concern not the brown race but the black, whose presence in tropical America is a legacy of the crime of slave traffic, which earlier disfigured its natural amenities.

CHAPTER XXXI

MEXICO

AMONG all the countries of the New World Mexico is perhaps the most peculiar and romantic, and the traveller who sojourns therein retains impressions of its wild and sunny landscape and characteristic people which time does not readily efface.

Yet Mexico, from one point of view, presents conditions perhaps the most regrettable of any tropical land. Here a people from whom much has been expected, who, for a generation, had painfully striven to build up a name for civic order and economic progress, have chosen to sweep away their reputation and fall back to that state of political anarchy which for fifty years of their earlier republican life sowed Mexican soil with disaster and watered it with blood.

It is, however, our purpose here to contemplate the more normal picture which the old land of New Spain presents.

The traveller to whom Mexico is familiar will depict first a wide, semi-arid plain, broken by the dry beds of *arroyos* or watercourses and rugged, outcropping rocks interspersed with the cactus and the prickly-pear, bounded by a shimmering horizon of steely blue hills, whose distant canyons stand out marked by the sharp shadows flung by a midday sun, or tinged by the glories of a tropic sunset. Across the dusty track come a white-clad, sandalled peon, followed perhaps by his woman, bearing in a shawl slung at her back a brown-faced baby, and as they pass the Hispanic-Egyptian—for such he might be termed—raises his wide sombrero and in musical Spanish, with native courtesy, wishes us *Buenos dias, Señor*, and God-speed. Or perhaps some brown-complexioned native maiden, with her *olla* or earthen water-pitcher gracefully poised upon her head or shoulder, stands for a moment at the spring which issues from the rock coyly to observe the passing stranger. Hard by stands a rude cross—one of a myriad such, found on every hill—which may mark some pious motive for its position there, or

MEXICAN PEONS GATHERING THE JUICE OF THE MAGUEY FOR MAKING THE NATIONAL DRINK OF PULQUE

signal the spot where human life has fallen in an accident or assassination.

Mexico is the Arabia or the Egypt of the New World. Its flat-topped, white-walled houses, its landscape, alternating between " the desert and the sown," its simple, primitive people and the steep, narrow streets of its ancient towns all bear out the analogy.

The foregoing picture, however, must be amplified. Mexico is by no means all a semi-arid land, for as we descend from the plateau on either side of the country, whether to the Atlantic or Pacific slope, or to the plains of Yucatan and other southern states of the republic, we enter a region of tangled forest and profuse flora, with a humid, tropical climate and those conditions of life which we more familiarly associate with equatorial lands.

Mexico forms a cornucopia-shaped land, extending for 2000 miles from the United States border southwards to where, beyond the Tehuantepec Isthmus, it merges into Central America, which in its turn terminates at Panama in the South American mainland. The area of the republic is 767,000 square miles—as large as Great Britain, France, Germany and Austria-Hungary. It is crossed midway by the tropic of Cancer, its southern half consequently lying in the torrid zone.

We may enter Mexico either from the north by rail from the United States and along the great semi-arid plateau, or from the Gulf of Mexico, through the port of Vera Cruz or other. The one route carries us through the drier and non-tropical highlands : the other traverses the tropical lowlands, with their characteristic scenery and vegetation. Otherwise we may approach the country from its Pacific seaports.

As the steamer approaches Vera Crúz, across the blue waves of the gulf, roughened perhaps·by the crisp white foam of the *norte* or characteristic gale of this sea, we remark, as if hanging in mid-sky, a single snowy mountain peak piercing the heavens, its base wrapped in clouds. It is Orizaba, Mexico's highest mountain, reaching an elevation of 18,250 feet above the level of the sea. The seaport and city of Vera Cruz presents from the sea a picturesque aspect, its buildings and towers lying low along the shore, and jutting seaward is the promontory upon which stands the ancient castle of San Juan de Ulloa, which was Spain's last stronghold when Mexican independence asserted itself. Vera Cruz was founded by the intrepid Cortes and his companions, who landed on the sandy beach on Good Friday, 1520, the first white men to tread the strand of the ancient empire of the Aztecs. The startled Indians drew rapid sketches of the white men, their ships, or "water houses," their guns and horses—after the manner of the characteristic picture-writing of the early Mexicans—and these they sent by swift messengers inland over the rocky Cordillera to the Aztec emperor Montezuma, in his capital of Tenochtitlan, the place where Mexico city now stands, on the plateau of Anahuac.

The story of the Spaniards' march, against the exhortation of this monarch, to Tenochtitlan, and the achievement of the famous Conquest of Mexico forms one of the most stirring and romantic pages of early American history, but we cannot dwell upon it here.[1] The Spaniards stamped out almost everything pertaining to native Mexican rule, in lieu of preserving what was useful and enduring therein, and their acts have been severely censured by historians.

As we ascend the eastern or western versant of Mexico we shall soon remark three vertical zones of climate : the *tierra caliente*, or hot tropical lands ; the *tierra templada*, or temperate lands, and the *tierra fria*, or cold lands, and the vegetation and native products are correspondingly marked in their variety. They range from coffee, bananas and a host of tropical fruits and food-stuffs in the lowlands, to wheat and coniferous forests in the uplands. The heat in the coastal and Yucatan lowlands is excessive and the atmosphere humid, but on the plateau the diurnal range of temperature is very marked, with often intense cold at nights. Here indeed we realise the truth of the assertion that "night is the winter of the tropics." In the temperate or intermediate zone are districts whose climate has well been termed that of a "perpetual spring."

The scenery of Mexico is extremely varied. We may contemplate with pleasure the wide-spreading valley of Mexico, with the picturesque capital lying surrounded by its verdant, irrigated lands, backed by the mountains from which arise the beautiful forms of the snow-capped Popocatepetl and

[1] In his *Mexico* (third edition) the author has described this journey.

In the Mexican Tropics

Ixtaccihuatl, the one nearly 18,000 feet high, and the other only 1000 feet less. These two mountains are among the few Mexican peaks which pass the perpetual snow-line, and are landmarks over a wide area. Pine-woods crown the rim of the valley of Mexico in places, and in the gardens of the palaces in the residential towns adjacent a profusion of foliage and flower is displayed. The capital offers numerous points of attraction, both by reason of its chequered history and from its newer developments, and it is a source of regret that so pleasing and well situated a city should be the scene of constant political strife and bloodshed.

The magnificent old cathedral is the finest structure of its kind in the American continents, and there are many institutions which reveal the scientific, literary and administrative aptitude of the Mexican people in normal periods.[1] The population of Mexico city approaches 400,000, and the high situation of nearly 8000 feet above sea-level ensures a generally cool and healthy climate, where, during the period before the recent revolution, American tourists wintered in considerable numbers, escaping the harsh climate of New York and Chicago. The shops in many cases are handsome and well stocked, and the upper-class people are fashionably attired. In the street life of the city we remark the admixture of this well-clad Europeanised class with the blanketted and humble Indian, a contrast which is perhaps the outstanding feature of Latin American life.

The republic of Mexico presents a marked difference from the Latin American states of South America, by reason of the better distribution of its population. It is a land of innumerable cities and villages, scattered freely almost all over its area, in distinction with the southern republic, whose inhabitants and towns are largely congregated near the seaboard. Many of the Mexican state capitals are cities of considerable beauty and merit, whose architectural and other features and general organisation are a revelation to the traveller, who may vaguely have pictured the republic as a semi-savage land. Especially striking are the splendid churches and other ecclesiastical buildings—a heritage from the Spaniard and the Roman

---

[1] The capital and people are fully described in the author's book, *The Republics of Central and South America*. Dent & Sons, London.

Catholic religion, and in part the result of Jesuit activities. Some of the most lavishly decorated interiors in the world are found in the churches of Southern Mexico.

Among the foremost of Mexican cities are Guadalajara, in the west, and Puebla, south-east of Mexico city. Both have fine public buildings and institutions, and are the centres of thickly populated agricultural districts. Puebla forms the foremost cotton-manufacturing district of the republic.

The textile industry of Mexico has acquired great importance, supplying a large part of the home demand for cheaper cotton goods. Native capital is largely employed in this, as in the sugar industry, which is also carried on to supply the native demand. Cotton is grown in various parts of the republic, the Nazas Valley—a miniature Nile in its regimen—being specially worthy of remark. Some of the districts—such as the state of Morelos —are practically vast sugar estates. It is a noteworthy condition of Mexican agriculture that the produce of the soil is almost entirely consumed at home rather than being raised for export. The same is the case with the cattle industry.

The population of the republic is a large one, numbering perhaps 17,000,000. This is perhaps the most considerable bulk of homogeneous brown labour in the world. But in Mexico we encounter, very strongly marked, the division of classes common to the Spanish American countries, a relatively small upper class—and the beginning of a middle class—mono-polising the education and wealth of the country. The great bulk of Mexico's millions are the *peones*, the illiterate labour which lives and works upon the great *haciendas* or landed estates under a system generally of debt-bondage to their masters, the wealthy landowners, themselves of white or Mestizo race. The peones are not necessarily pure Indians but there are numerous Indian tribes, some in a semi-savage condition, in the wilder parts of the country.

The Mexican peon class has many good qualities, which have been kept back by the generally disastrous social and political regimen of the republic. Under Porfirio Diaz perhaps some advancement in their condition was made, but it was far from being commensurate with the growing wealth and stability of the country as a whole at that period. The Diaz policy greatly enriched the upper classes, including foreign concession-holders,

THE SNOWS OF THE TROPICS POPOCATEPTL, MEXICO

but the ill-paid labour and primitive conditions of life of the labouring classes remained practically unaltered. An adobe or dried-mud hovel, bare of furniture, and food consisting mainly of beans and maize—meat being rarely forthcoming—forms their diet, upon which the peon performs what must be regarded on the whole as efficient labour. His great failing is drink, the failing of Latin American labour from Mexico to Peru. This drink habit is by no means all his fault, for the manufacture and sale of potent liquors—the native *pulque*, made from maguey and the sugar-cane rum—is in the hands of the wealthy landowners, who, being also legislators, are not likely to take steps towards a decrease in consumption. The peonage system has been carried to brutal extremes in certain parts of Mexico, notably in Yucatan, where the millionaire Mexican growers of sisal-hemp have been accused of practising a degraded system of slavery with their dependents.

In the upper class of Mexico we encounter a people pleasing in their courtesy and general refinement and in their aptitude for arts, sciences and commerce. The wealthy are generally well-informed men of affairs, and the women of the leisured class are handsome and refined. Indeed, we marvel that such constant political strife and bloodshed takes place among a people of such considerable pretension to civilisation, with beneath them a docile serving class. It is, however, not among these two classes that the revolutionary habit develops, but with the professional politician and an idle and ambitious military element, upstarts and ruffians whose intrigues are sufficient to throw the whole body politic into a ferment, and who, without hesitation, sacrifice their country to personal desires.

In general terms, however, it may be asserted that political unrest in Mexico, as in all Latin American states, stands on an economic basis. As long as the bulk of a people are kept poor and ignorant, with no stake in their country, carrying on no occupation permanently profitable to themselves, as is generally the case in Mexico and many of the Central and South American republics, so long will there be a field for the political renegade to secure an easy following, under promise of pay and plunder, and to cause his deluded adherents of the peon class to become mere " food for cannon."

The Mexican towns are almost invariably laid out on the

Y

customary Spanish-American plan of rectangular blocks and cross-streets, and the central *plaza*, which is an invariable feature of each, forms a centre of public recreation. Here in the evenings and upon the Sabbath there is music, and the *Serenata*, as this function of the open-air concert is termed, is a pleasing feature of life here, yielding an opportunity for neighbourly intercourse and distraction. The Mexicans are passionately fond of music, and all classes, from the fashionably attired *señorita* or " correct " *caballero*, down to the peon or the *indio*, clothed in his inseparable poncho, throng the public square on such occasions. The plaza is generally well planted with shade-trees and flowers, among which the orange and banana are found. Apart from such mild diversions the pastime of the Mexican people is in bull-fighting and cock-fights, the former cruel and retrograde sport being the favourite diversion of all classes. Its effects upon the peon population are evil.

If agriculture is the industrial mainstay of the country, mining is almost equally important, and the mines of Mexico are interwoven with its history and often surrounded by elements of romance. Many of the great towns of the plateau owe their origin to the rich mines upon which they are built, such places as Zacatecas, Guanajuato, Pachuca, San Luis, Potosi, Durango and others having been famous centres of mining for centuries. Some of them possessed royal charters as cities before the time of the *Mayflower*. Silver, gold, copper and other minerals are largely worked, and vast fortunes have been reaped. In earlier times the greater part of the wealth was drained away to Spain, although some fine churches and other institutions were built therewith.

British and other foreign enterprises reap a rich harvest from some of the leading Mexican mines, especially in gold. Upon the Gulf Coast lies a petroleum-bearing district which has become of late years one of the most famous in the world, and this is largely worked by a London company.

The native miner of Mexico is generally hard-working and dexterous, carrying on his profession under conditions and rates of pay that no European or American miner would tolerate. There are practically no laws regulating the condition of Mexican labour, whether in the mine, the field or the factory,

MEXICAN INDIAN GIRLS OF TEHUANTEPEC

but of late years the operative has begun to awaken to his low state and to enforce his protests against it, by means of the strike, or *huelga*.

We may reach all the principal towns and nearly all the main seaports of the country by rail, and the total length of line in the republic reaches over 15,000 miles. Great trunk lines run from the United States border to the city of Mexico, the distance between the two points upon the Mexican Central Railway being 1225 miles. There are various lines from the Gulf Coast to the capital, or other parts of the interior, in some cases crossing the continent to the Pacific coast.

The railway from Vera Cruz to Mexico city is one of the foremost scenic lines in the world, and here we are carried in a few hours from the tangled tropical vegetation of the lowlands to the bare, broad plateau, through profound gorges and across torrential rivers, and beneath the shadow of the snowy Orizaba, passing on our way many beautiful places of satisfying fertility, with interesting towns. It was across this zone that the intrepid Cortes and his companions struggled so painfully, beset by savage foes on every hand.

One of the most noteworthy railways in Mexico is the Tehuantepec line, crossing the isthmus of that name from the Atlantic to the Pacific. A branch from this system, running southwards, bears us into the neighbouring republic of Guatemala.

Into a description of the wide variety of plant life and luscious fruits of tropical Mexico we cannot enter here. All the varieties of the American tropics are included, the coco-nut palm, sugar, coffee, tobacco, rubber, maize, sweet potatoes, chocolate—the word itself is derived from the Aztec word *chocolatl*—and a host of delicious fruits little known outside the country. Indeed, the remarkable fertility of the soil, especially under irrigation, is apparent on every hand, and the traveller might well ask how it is that, in the midst of such abundance as the more tropical part of Mexico displays, the bulk of its inhabitants should dwell in such marked poverty.

The traveller will not leave Mexico without giving some attention to the remarkable archæology. The ancient pyramids, temples and other structures scattered about the republic, and extending thence into the Central American States, are of a character such as astounded European antiquarians when attention was drawn to them. They display a knowledge of building and the stone-shaping arts which reveal a marked culture among the early Mexicans. These works are variously ascribed to the Toltecs, the Mayas, the Quiches and the Aztecs.

The great " pyramid of the sun " at Teotihuacan, a little to the north of the capital, is an immense structure rivalling the pyramids of Egypt, although of somewhat different construction, and there are others elsewhere. The famous ruins of Mitla, with their beautifully carved façades, are a subject of wonder to the beholder, and their origin, like that of Teotihuacan, is still shrouded in mystery, or unsatisfactorily unexplained. Equally remarkable are the carved stone buildings buried in the forest of Yucatan, those of Chichen Itza, Uxmal and other ancient centres. Some observers have endeavoured to show evidence in the ornamentation of these ruins of Indian, Egyptian and other origin. At Quirigua in Guatemala and at Copan in Honduras are other very noteworthy ruins of profound interest to the archæologist and traveller, and a considerable literature has been devoted to all these early American remains. In the national museum of the city of Mexico is preserved the remarkable Aztec " Calendar Stone," part of the very correct system of time measurement which the early Mexicans employed. Here also is the sacrificial stone, whereon the human victims were bound, their breasts slashed open by the priests with obsidian knives ; and the still-beating heart, torn out, was flung upon a golden dish before the image of the war-god. Ancient Mexico, however, was not, as a whole, necessarily barbarous. There were equitable laws and customs in vogue.

The traveller who has remarked Mexico's interesting past, and the considerable wealth and possibilities of its natural resources, is constrained to ask what is to be the future of the country. The only intelligent reply is that it will be determined by the extent to which the republic's rulers foster the welfare of the great and generally well-meaning population beneath them, abandoning their mediæval and tyrannical ways under the dawn of some better spirit : some spirit of constructive " human geography " ; the planning of national life and industry.

ANCIENT TROPICAL STRUCTURAL ART · ONE OF THE GREAT TOLLEC PYRAMIDS OF MEXICO

CHAPTER XXXII

CENTRAL AMERICA—GUATEMALA, BRITISH HONDURAS, SALVADOR, HONDURAS, NICARAGUA, COSTA RICA, PANAMA

THE Central American states are comparatively little known to the outside world, and in the popular conception are often depicted as lands of revolution and financial double-dealing, subject to fevers and earthquake disasters, and inhabited by a people of but rudimentary civilisation. If this conception be true, the half truth contained therein serves to obscure a real knowledge of what, in some respects, is one of the most pleasing parts of the tropical world.

Central America occupies an exceedingly advantageous position geographically, and as regards its resources, soil and climatic conditions, Nature has worked here upon a very diversified scale. It is true that the humid lowlands suffer the scourge of malaria and other serious disorders, due as much to the negligence of their inhabitants as to natural disabilities, but the principal centres of population are generally situated amid an upland environment among the most pleasing and healthful in the world. The people of these states, although they suffer grievously by reason of the comparatively slow evolution of their powers of self-government, are capable of a high civilisation, and the traveller who visits their capital cities cannot fail to remark the evidences of a strong ideal towards such.

The scenery of these lands is often exceedingly beautiful, and the vertical zones of temperature which we encounter, due to the high elevation the land attains by reason of the mountain backbone or cordillera which traverses the isthmus, afford the traveller any range of climate he may desire, from that of the hot lowlands to the bracing atmosphere of the *altos*, or high plateaux, or of those midway valleys whose environment enthusiastic writers love to depict as a " perpetual spring." It is, however, the case that among the mountains of Central

America are some of the most destructive volcanoes on the earth's surface, and the frequency and severity of earthquake shocks—earth unrest, which is as marked as the political unrest—is teaching the need of a special type of construction in the dwellings of the people.

In these, the Central American states, we encounter conditions which differ little from those we have contemplated in Mexico. There is the same vegetation, scenery, climate, tropical products and sociological conditions, and the people display similar virtues and defects, the legacy of their Spanish forbears. However, each state jealously conserves its own individuality and rights. The language is Spanish, and the Roman Catholic religion, whether beneficent, whether retrograde, still dominates all classes, except that among the male sex that freethought or materialism, which may be regarded as the reaction from papistry, is of gathering force, as elsewhere in Latin America.

This group of six independent republics and the colony of British Honduras occupies the great peninsula stretching southeast to the mainland of South America, severed at its lower extremity by the Panama Canal, covering an area of about 208,000 square miles, and perhaps 1200 miles long. Their geographical position is an advantageous one, because nearly all the republics have an Atlantic (Caribbean) and a Pacific coastline and harbours. The total population is about 5,000,000.

We reach Guatemala, Mexico's sister republic and southern neighbour, preferably by sea, although there is now an extension of the railway from the Mexican system, forming part of the long-projected " Pan-American " line, giving access to the capital.

Guatemala Nueva, as the capital is termed, is marked by an air of general prosperity, and is well laid out and dowered with a number of public buildings and institutions which indicate the progressive spirit of its inhabitants, and their desire to stand high in the civilisation of the Latin American world, notwithstanding their frequent civil wars and political quarrels. The city stands at an elevation over 5000 feet above sea-level, upon a wide tableland, the surroundings being marked by considerable scenic beauty. Lofty mountains surround it, culminating in the dreaded volcanoes of Agua and Fuego, or

" Water and Fire," whose activities in times past destroyed the capital on various occasions. The Fuego peak is nearly 13,000 feet high. The original town was founded by the Spanish *conquistador*, Pedro de Alvarado, in 1527.

We approach this pleasing city from Puerto Barrios, on the Atlantic coast, or from San José, on the Pacific, by the railway, which traverses the country from sea to sea, a line 265 miles long. The population numbers about 100,000. The streets are laid out with remarkable regularity, and the houses are low, as a precaution against the destructive effect of earthquake shocks.

The trade is principally in coffee, amid the luxuriant plantations of which we pass in ascending from the Pacific seaports. Few countries surpass Guatemala in the excellence and quantity of its coffee export, the volcanic soil of the western versant being peculiarly adapted to the cultivation of the shrub. The plantations are largely owned by wealthy Germans.

Ascending from the Atlantic side, we traverse some of the richest banana-producing lands of tropical America, plantations of which fruit, in some cases, have been established upon the site of the remarkable ruins of Quirigua, where beautifully carved stelæ and other monoliths, and sculptured walls, arise amid the dense tropical jungle, a witness to the culture of an ancient and almost unknown people, the former masters of the land.

The area of Guatemala is about 48,000 square miles, and the population numbers about 2,000,000. There are various interesting towns scattered over the surface of the republic, some of them, as Quezaltenango, centres of life and industry which have retained their native character almost unaltered since pre-Spanish times, and of much interest to the traveller and student. Guatemala was the home of the great Quiché people, who had attained to a considerable state of native culture and carried out many dexterous arts, which even the Spanish invasion and the inroads of modern commerce have not entirely destroyed.

Guatemala is in general a well-watered and fertile country, rich in the characteristic Central American vegetation and animal life. The beautiful Quetzal bird has been adopted as the national emblem. The climate is healthy except where malaria prevails upon the coast. In this respect noteworthy improvements have been made at San José in the work of sanitation, carried on by a large foreign banana-planting company,[1] whose extensive operations form an important feature in the economic life of this and certain other Central American republics. The rainfall is very heavy, especially on the Atlantic slope.

The population of the republic tends to increase with some rapidity, as the native women are prolific and mortality and emigration are both low factors. Over half the births among the Indians are illegitimate, and one-third among the whites, the Indians forming nearly three-quarters of the population—a very diversified range of tribes, speaking different languages. The remainder of the population are Ladinos, or whites and mestizos, and there are over 12,000 foreigners in the country, principally engaged in business.

The economic condition of the Indians is a low one, being almost akin to slavery, a result of the system of debt-bondage which prevails, under which the planters and others make payments in advance to their labourers, who thus contract debts which it may take a lifetime to work off. The system is a convenient one for the employers, who thus retain labour on their estates, but is deleterious to the workers in general. Their low condition is marked by the fact that scarcely ten per cent. of the population can read or write. Primary education in all these republics is compulsory—nominally—and is provided free by the state, but in practice is lamentably neglected, although from time to time efforts are made towards improvement.

The main articles of food of the people are maize, beans and bananas, with, occasionally, jerked meat. The lower classes suffer from the habit of excessive drinking, of the Latin American race generally. The principal and indeed almost the only industry is agriculture. There is some cattle and sheep keeping, and the Indian and Ladino women make coarse cloth from wool and native cotton for their garments, as well as linen. Textile manufactures on a larger scale are also carried on in the towns. There are some mineral resources.

The Indians are generally peaceable, and under more

[1] The United Fruit Company of America.

NATURAL VEGETATION IN COSTA RICA

enlightened control might be expected to advance far more rapidly on the road of civic progress. The whole country is one of extreme interest from many points of view, and of considerable resources and possibilities.

The small British Crown colony of Honduras, facing the Caribbean Sea on the east of Guatemala, has as its principal seaport Belize, with Yucatan on the north. The coast is low and swampy, with a fertile belt inland, giving place to broken ridges and broad savannas, backed by the Cockscomb Mountains, from which numerous small rivers descend. The country contains some fine pasture-land and forests of valuable timber.

The colony is about the size of Wales, covering an area of 7562 square miles, and the climate, which is sub-tropical, is not unhealthy by comparison with that of the West Indies, or Central America in general. The population is under 50,000, and increases slowly, the majority of the people being of hybrid race descended from negro slaves, Indians and white settlers, with perhaps 2000 people of European descent or origin. Among these last are the descendants of English buccaneers, mixed with Scottish and German traders. Ancient ruined cities, such as we have remarked in Yucatan and Guatemala, appear to show the existence of a far more numerous population in pre-Colombian times.

Approaching Belize from among the green islands of the Caribbean Sea, we remark the high roofs and wide verandahs of the houses, the town extending along the banks of the river, and coco-palms and cabbage-palms throw their shade thereover, swaying in the invigorating eastern breezes from the sea. This wind, and the high tides which wash the fever-haunted and malarious mangrove swamps and marshes, are elements which render the climate more healthy. The population is over 10,000.

The establishing of the colony was associated with buccaneers and logwood cutters, and many valuable timbers are exported, with sugar, coco-nuts, bananas, tortoiseshell and other tropical products. The coloured people are among the most expert woodmen in the world, and as we ascend the river we remark the rafts of timber which they bring down for export. Rubber and chewing-gum are other articles of forest exploitation, as is the cahoon or coyol palm, which, growing abundantly, yields a valuable oil. There are also extensive pine woods, cedar and dye-woods.

Whilst the colony is well administered, after customary British methods, it cannot be said that its economic advancement is very marked, but it contains many resources of value to the Empire.

In the republic of Honduras we approach what has been one of the most backward of the Central American states, yet a land dowered with many native resources, and, in the uplands, possessing a delightful climate. Like other Central American countries, the evil climatic reputation is derived from the hot lowlands of the coast, where diseases are propagated by the insanitary mode of life of the inhabitants. That such conditions are capable of remedy is now abundantly proved, and in Honduras the work of foreign fruit-growing companies [1] in this respect has been of the utmost value.

The republic has a coast-line on the Atlantic side 300 miles long, and extends across to the Pacific, where it enjoys an outlet in its seaport of Amapala, upon the Bay of Fonseca, one of the finest natural harbours in the world. As we approach the country from the Pacific a vast mountain wall arises beyond the littoral, culminating in volcanic peaks, the summits of the cordillera. A wide valley extends across the country, the plain of Comayagua, and large rivers descend to the Atlantic and the Pacific. The republic covers an area of over 46,000 square miles, and its population numbers somewhat over 500,000.

The capital, Tegucigalpa, stands in the uplands at over 3000 feet above the sea, and is the largest city in the republic. It possesses a fine cathedral and the customary governmental offices and educational and other institutions, and has some picturesque features. The population is about 40,000.

Railway development is very backward, and for long the roads, although well laid out to the capital, were much neglected. The ox-cart roads and bridle-paths are often execrably bad. In conjunction with the foreign fruit plantations on the eastern versant, railway-building has of late received more attention. Nearly half the export trade of the country is in bananas, for

[1] The United Fruit Company.

whose cultivation the Atlantic coast is eminently suited, as it is indeed for sugar, coffee, tobacco, cocoa and other tropical products. Puerto Cortes is the principal eastern seaport, but at Tela a large fruit business is carried on and sanitary im. provements have been made there by the foreign interests.

Honduras possesses unsurpassed resources in timber of mahogany and cedar, but there has been reckless destruction of the forests. The exhaustion of the wild rubber has, to some extent, been remedied by planting, and coco-nuts are also valuable products of the seaboard. Wheat is grown in the uplands, and cattle are exported. On the Atlantic side fish and turtles abound.

The mineral resources of Honduras are the most important in Central America, silver and gold mines having long been worked successfully. But these resources are still developed on but a relatively small scale. Some coal and copper and other metals are found.

The European element among the population is small, although nearly all the Indians are Christianised, except for some mountain tribes. The Caribs are an industrious and vigorous race, spread along the east coast. The Spanish element has principally developed upon the Pacific side of the republic. The population in general tends to increase rapidly. Honduras is notorious in financial circles in Europe, by reason of its great foreign debt and continual arrear of interest thereon.

The little republic of Salvador, which we approach from the Pacific, is the smallest in area of the whole sisterhood of the Latin American states, but not for that reason is it least worthy of attention, being the most thickly populated and among the most prosperous and advanced.

Beheld from the Pacific we remark a low alluvial littoral, backed by the Cordillera, with its volcanic cones. Through a magnificent valley flows the Lempa river, skirting the mountains and the base of the volcanoes in its course to the sea. The area of the republic is somewhat over 7000 square miles, and its population numbers about 1,250,000.

To reach the capital, San Salvador, lying in its valley at an altitude of over 2000 feet above sea-level, we land either at the port of La Libertad or Acajutla, and ascend thence by rail.

The city is well laid out and possesses numerous governmental and other institutions which attest its importance as a seat of Central American civilisation, and it compares favourably with any European or North American city of its extent. The population is about 65,000. The climate, surroundings and general social atmosphere are very pleasing, and a considerable business is carried on by the foreign element residing in the city.

The republic, occupying a strip of land facing the Pacific, has no outlet upon the Atlantic coast, and in this respect it is unlike its sister republics. The seaports are in general unattractive, like many such upon the Pacific coast of Latin America, where landing is generally difficult by reason of the heavy surf. Salvador, however, abuts upon the fine Bay of Fonseca. Formidable volcanoes threaten the security of the towns in some instances, San Miguel being one of the most terrible burning mountains in Central America, and a constant menace to the capital.

The Indians of Salvador form a minority of the population, the greater part of the inhabitants being Ladinos, of white and mixed race, and in this respect the community differs from some others of the Central American states. The considerable prosperity of the country is in part due to this circumstance. The population increases rapidly.

Agriculture occupies the energies of the bulk of the people, the principal product and export, and indeed the mainstay of the country, being coffee. Like all Central American states, Salvador is greatly dependent upon European markets and financial centres for its well-being, and is elevated or depressed in accordance with the state of these. Sugar, indigo and " Peruvian " balsam are also produced, and tobacco and rubber, with cocoa, rice, fruits and cereals. The Government Agriculture College endeavours to encourage cotton-growing. Various small mining establishments, generally controlled by foreigners, produce some silver and gold, and other metals are found in addition.

There are a number of pleasing towns throughout the republic, The state prides itself upon the freedom of its institutions, but dictators control it from time to time, with consequent political strife.

EARLY TROPICAL SCULPTURAL ART STONE STELA AT QUIRIGUA IN
GUATEMALA, TWENTY FEET HIGH, OF UNKNOWN AGE, PROBABLY WORK
OF THE MAYAS

Nicaragua, the largest in area of the Central American states, is one of the most backward economically and in the march of civic order. The republic has a considerable coast-line on both sides, and shares with Salvador and Honduras the Bay of Fonseca on the Pacific. The western coast is bold and rocky, whilst the Atlantic or Caribbean shore is low, swampy and monotonous, with numerous lagoons. From the Pacific we remark the wall of the Cordillera, with its volcanic summits, as in Honduras and Guatemala. The area covered is somewhat under 50,000 square miles. The climate varies greatly according to the district, being unhealthy on the coast.

The most thickly populated part of the republic is that of the lake depression and its adjacent highlands, and the development of the country has been from west to east, due to the earlier Spanish settlements in the healthier lake region and the Pacific slope, and to the rich soil and waterways. The population of the republic, however, is small, numbering perhaps 600,000 or more.

From the great lake of Nicaragua the San Juan river flows into Costa Rica and to the Atlantic, and through this hydrographic depression it was that the one-time projected Nicaragua Canal would have been constructed.

The capital of Nicaragua, the city of Managua, is linked by rail with the Pacific seaport of Corinto, and various large towns upon lakes Nicaragua and Managua are in railway communication therewith. Managua has a population of 30,000, and the capital was formed here in order to put an end to the rivalry for that honour between the adjacent cities of Leon and Granada, both more important, architecturally and in point of commerce and the number of their inhabitants.

The present capital is, however, extending, and has various flourishing industries. Its position on the lake is important and its rail and steamer communications are conducive to considerable traffic. Coffee, sugar, cocoa and cotton are exported.

The public buildings of the city of Leon are among the finest in Central America, and there is a massive and highly ornate cathedral, dating from 1746, and other noteworthy and pleasing examples of Spanish Colonial architecture. The place was described as a " splendid city " by travellers in 1665, and a few years later the buccaneer Dampier obtained rich booty

therefrom. Leon has been the centre of the Nicaraguan demo-
cratic interests, whilst Granada was that of the aristocratic and
clerical parties, a rivalry which led to frequent bloodshed. The
last-named city, after its foundation by Cordova, in 1523,
became one of the richest places in Central America, and was
on various occasions sacked by buccaneers. It is surrounded
by large cocoa plantations, and coffee, cotton, tobacco, indigo
and other characteristic products are exported therefrom.

The people of Nicaragua are descended from the native
Indians and their Spanish conquerors, and from the negro
slaves of the colonial period. British, Dutch and French have
also intermarried with Caribs and creoles, and a type with fair
hair and blue eyes, but of Indian caste or negro coloration, has
been produced. All Spanish-speaking people are classed as
Ladinos, and there are comparatively few pure Indians in the
country. The remains of the ancient pre-Colombian civilisation, in monuments and ruins, are of interest to the antiquarian,
as elsewhere in Central America.

Here also agriculture, mainly coffee growing and export,
forms the mainstay of the population. The coffee plantations
are largely owned by Germans. Great quantities of bananas
are also grown and exported, and some sugar and cocoa. The
scarcity of labour, however, keeps all industries backward.
Gold, rubber, cattle and dye-woods are other exports.
Nicaragua has long been notorious for its political unrest and
financial difficulties, and it has suffered much from arbitrary
dictators. Its resources and the inhabitants are in general
worthy of a better and more prosperous regimen than so far has
been theirs.

We are now approaching the narrow southern portion of the
Central American isthmus, occupied by the republics of Costa
Rica and Panama, the isthmus here serpentining towards the
east.

The derivation of the name of Costa Rica, or " Rich Coast,"
lies in some obscurity. The early Indians wore earrings of
gold when their Spanish conquerors arrived, but the latter
found disappointingly little of the yellow metal in the soil of
the country, which in this sense was not rich. It may be that
the fertile soil and the bright sunshine, with a sufficiency of

moisture, forming an environment where Nature brings her choicest vegetation to fruition, with flower and luscious fruit and virgin cedar and mahogany forests, may have given rise to the pleasing designation of this part of the tropical world.

The republic covers an area of about 18,500 square miles, embodying the coastal lowlands, and an elevated tableland cut by lofty mountains studded with volcanic cones, from some of which constant streams of vapour, and in some cases outbursts of subterranean fire escape.

From Irazu, which reaches 11,200 feet above sea-level, the view extends over both the Atlantic and Pacific Oceans, and the whole area of the republics unfolds between. We may reach this summit—higher than the highest point of the Pyrenees—on horseback, as the slopes are gradual. Turialba, often a hot and angry volcano, adjoins.

The Pacific coast displays bold headlands and broad bays, but the Atlantic shore is monotonous and unbroken. Between the seaports of Punta Arenas on the west and Limon on the east side a transcontinental railway runs, 117 miles long, serving the capital—San José—standing on a plateau between the Cordilleras at an elevation of 3870 feet above the sea.

The city of San José is of modern aspect, although the houses are built low to lessen the danger from the frequent earthquake shocks. Its well-paved streets are electrically lighted, the cathedral is noteworthy, and the various governmental, educational and scientific institutions are those of an advanced community. The population is about 25,000. The climate is temperate and pleasing, the water supply pure and abundant, and the customary amenities of Latin American town life are here well displayed. The suburbs shade off into the mean wattle huts of the Indians.

The population of the republic is about 400,000, and increases rapidly. The proportion of Spanish blood is higher than in any of the Latin American states, a result of the considerable settlement of Spanish colonists, largely from Galicia, in earlier times. The bulk of the people, however, are the Ladinos or mestizos. The Indians have in large part been exterminated or civilised, but there are tribes still in complete savagery. The European element numbers perhaps 10,000, and immigration is freely encouraged, land being purchasable on relatively easy terms.

Costa Rica is noteworthy for its comparative political stability, and this arises from the circumstance that over two-thirds of the population are small peasant farmers, proprietors of their own lands. This class, occupied in a stable industry, which depends for its prosperity upon political and economic quietude, sternly discountenances revolution and limits the activities of dictatorial presidents. In this connection the republic might well be an example to other Latin American states, where the possession of the soil is generally the monopoly of a few and the labourer is more or less of a chattel.

The principal product and article of export is coffee ; the well-known qualities of the Costa Rica berry are due to the nature of the volcanic soil. The ash from the volcanoes, during the ages, has become deposited or intermixed with the earth for a depth of several yards, and contains the constituents necessary for the best coffee berry. We have remarked this circumstance in Guatemala and elsewhere, and thus the scourge of eruption from the Central American mountains is not without some measure of compensation.

An increasing industry is of the production of bananas for export. This flourishing trade has its main outlet at Port Limon, a place which, as a result of activity, largely of foreign enterprise,[1] has been converted from what was a pestilential and dangerous place, little more than a swamp village, where yellow fever prevailed, into a modern town of marked civic beauty and cleanliness. Under this influence scientifically conducted hospitals, after the Panama model, have been established at Port Limon and the other Caribbean fruit-shipping seaports in Central American states, as before described. Many millions of bunches of bananas are exported yearly to Europe and the United States from Costa Rica by various lines of steamers, and the wages paid the native labourer in this industry are higher than those earned in coffee-growing.

Other products which, however, scarcely meet the local demand so far, are cocoa, vanilla, sugar, cotton and indigo. Stock-farming is but little developed. Some of the gold-fields

[1] The United Fruit Company.

VIEW AT PORT LIMON, COSTA RICA, BEFORE AND AFTER AMERICAN SANITATION METHODS

are successfully worked by American capital, and various kinds of valuable woods are among the export.   Cedar, mahogany, rubber and ebony abound in the vast tracts of virgin forest, and tortoise-shell and mother-of-pearl are exported.

As regards its climate, Costa Rica is one of the healthiest lands within the tropics, despite the fevers which insanitary methods still permit to prevail in places.

In the republic of Panama we approach a country made famous by reason of its topographical structure, the narrow isthmus where severance was possible for the building of an interoceanic ship canal, now accomplished after many vicissitudes.

The beautiful Bay of Panama, studded with verdant isles, and if by chance our steamer enters it at evening, overarched by the flashing colours of the tropic sunset, beyond the towers of the old Spanish colonial city near its shore,[1] is one of the most romantic havens of the great Pacific coast of tropical America. Its traditions breathe memories of the Spanish adventurers, of Vasco Nuñez de Balboa, who, on Michaelmas Day of 1513, from the "peak in Darien," first beheld the Pacific, and of Pizarro and his companions.   Drake also beheld the great sea from the mountains, and prayed that he might sail it in an English ship, which afterwards he did, in his famous voyage. In the murderous exploit of the British buccaneer, Henry Morgan, in 1671, the original city was destroyed by fire, and its ruins are a source of interest to the tourist.   Colon, the Atlantic terminus of the great canal, is more modern and prosaic, but it has been the point of entry for the marvellous American activity which realised the construction of the waterway.

The republic of Panama is a long, narrow isthmus strip, covering an area of about 32,000 square miles, 430 miles long and averaging less than 100 broad.   Low, rounded mountains covered by dense forest form its interior, hills which in earlier geological times were separated by natural waterways between the Atlantic and the Pacific, until the land was upheaved by subterranean forces.

The city of Panama does not come within the American juris-

---

[1] Due to the curvature of the isthmus the sun does not set in the Pacific, but rises there, viewed from Panama.

diction which controls the " canal zone," and is a typical centre
of Spanish-American life, picturesque, and until late unhealthy,
and before the American advent the fearful traveller hurried in
his sojourn therein, or in his journey over the isthmus, almost
loath to leave his steamer for the fever-haunted land.

It would be beyond the scope of this work to describe in
detail the Panama Canal, undoubtedly the greatest piece of
work ever carried out in tropical lands. The problems of its
construction were of the utmost interest, not only from the
engineer's but from the sociological and the medical points of
view. The labour was very largely carried out by West Indian
negroes. An American commission early decided that the
work of these people was inefficient, until the primitive and
somewhat pathetic fact was recognised that their lack of
powers was due to lack of sufficient nourishment. When this
circumstance was remedied the negroes readily fulfilled their
important part, and thus the formerly despised West India
negroes largely built the canal.

The canal is fortified and controlled by the United States,
and so is not an international waterway, as was at first in-
tended, a condition which may bring about complications in
the future.

Incidental, although necessary to the colossal piece of
engineering work, was the knowledge gained in the science of
tropical medicine, knowledge which opens up almost a new
aspect in the development of the lands of the torrid zone, and
is of incalculable value to their fuller enjoyment, both for the
native and for the white man. What has been picturesquely
termed the " conquest of the mosquito," which was the prelude
to the successful conflict with malaria and yellow fever, dirt
and disease, has been not unjustifiably regarded as one of the
romances of medical science and has strongly appealed to the
popular imagination. The discoveries involved therein were
not necessarily made at Panama, and owe much to the non-
tropical but malarious lands of Italy, but they have most
notably been applied here, and with perhaps the most specific
result. We shall also recollect that the tropics furnished their

[1] In his book, *The Panama Canal* (Collins, Edinburgh), the author has
fully described the various phases attending the construction of this
great enterprise.

PANAMA CANAL: EXCAVATING LANDSLIDE IN THE CULEBRA CUT

own prophylactic against this most troublesome tropical disease, in quinine, which first came from the forests of Peru.

The permanent population of the republic is about 400,000, the people exhibiting different degrees of admixture of Indian, Spaniard and negro. The upper class displays the customary Spanish-American traits, and there is an aristocratic element which regards the North American influence askance or with only veiled antagonism. However, the people readily forswore allegiance to their parent Columbia, in order to control their own wealth and destiny, which the building of the canal rendered important. Both races, the Anglo-Saxon and the Latin American, have something to learn from each other, here as elsewhere in the republics of the New World.

We encounter in Panama the customary flora and tropical products of Central America, bananas, coco-nuts, rubber and hides being the principal articles of export. The fertile lands offer, as elsewhere in this part of America, ample room for tropical plantations. Coffee, cocoa, cereals and sugar are also produced, and cabinet-woods, fish and pearl and other oysters, with medicinal plants and herbs, and tobacco. There are excellent stock-raising grounds. Coal, salt, gold and iron are all found, and some are worked. The republic, in fact, is richly dowered with natural resources. In the dry season the north-east wind blows keenly from the sea and refreshes the eastern coast.

CHAPTER XXXIII

THE WEST INDIES—CUBA, PORTO RICO, JAMAICA, BARBADOS, TRINIDAD, ETC.

THE isles of the West Indies stretch in a vast curve 2000 miles long, from Yucatan to the Venezuelan coast, like a necklace of emeralds flung around the Caribbean Sea from the South American continent.

Geologically, these islands are the tops of a submerged mountain chain, and they are among the most beautiful and fruitful of any tropical lands. Columbus, who hoped he had found a new way through the archipelago to India, gave them their name of the West Indies, and their second nomenclature of the Antilles was due to the supposition that the great explorer had reached the fabled land of Antilla.

Not all these numerous islands lie in the torrid zone, the tropic of Cancer cutting through the Bahamas and passing slightly to the north of Cuba. The land area of the whole archipelago is nearly 100,000 square miles, and the total population approaches 7,000,000, people of many races, in which the negro of African origin predominates, the descendants of slaves brought from the African coasts.

Cuba, the largest of the group, stretches its long, narrow form from the Yucatan Strait, near the Mexican coast, for 730 miles to the east, and the line is continued by Haiti. Jamaica lies a little over 100 miles south of Cuba, and about the same distance to the east of Haiti is Porto Rico, and thence, curving south-wards towards the South American mainland, are the Lesser Antilles, divided into the Leeward and the Windward Islands, among which are many well-known place names. Other islands lie along the Venezuelan coast, including Trinidad and Curaçoa, and the flags of various European nations are repre-sented in the control of the units of the whole group.

The West Indian Islands generally enjoy that favourable topographical condition under which the land, attaining a

THE HARBOUR, ST LUCIA, WITH ROYAL MAIL STEAMER

considerable height above the sea, is dowered with a varied climate, extending through vertical zones of temperature from hot to cool. Here also the long nights are often cool and refreshing by reason of the sea breezes, which greatly temper the heat of the torrid zone. In the cold season frosts appear, but, on the other hand, the heat becomes almost unbearable in the summer months.

We are here in the zone of tremendous and often destructive hurricanes, which occur during what are locally known as the " hurricane months," when the heavy rainfall takes place. It is also an earthquake zone and is terribly visited by volcanic eruptions.

Great beauty, variety and richness mark the tree and flower life of the Antilles, due both to the indigenous species and to those plants which have been brought from all parts of the globe. The plants of temperate countries flourish but a short distance away from those which require great heat, and by ascending or descending the slopes of the islands the traveller may witness this variety well displayed. We have remarked similar conditions upon the tropical mainland, but the elevations there are more stupendous.

A great variety of palms flourishes in these islands, and their characteristic forms soften and beautify the landscape; and the forests are plentiful and wide-spreading in places. The " locust-trees," which reach a vast height and girth, may be as old as 4000 years in some cases. However, the most famous and valuable flora of the West Indies are the cultivated varieties upon which the agricultural resources of the islands rest. The Antilles were for centuries famous for the production of sugar and tobacco, and the planters reaped great fortunes therefrom, and established a peculiar civilisation founded upon this wealth and the slave labour which underlay it.

The original people of the islands were gentle and harmless natives, especially in Cuba, but the Spaniards, who obtained such easy control over them, treated these simple children of the soil with great brutality. Some of them were sent overseas as slaves and others were driven to work as forced labour in the mines opened in the archipelago. From the very beginning of their history—as far as the white man is concerned—the beautiful West Indian Islands became the scene of that traffic in human flesh and blood which lasted for centuries. The aborigines turned upon their masters at length, maddened by their sufferings, but it was impossible that they could prevail, and many of those who were not destroyed preferred to end their days in suicide; and the Antillean native race practically ceased to exist.[1]

The Spaniards, however, were not permitted to monopolise for long the possibilities of the islands, for British, Dutch and French buccaneers and pirates, as well as more legitimate rovers, flocked to the lands and waters of this El Dorado of the Spanish main. The English settlers, with their more practical ideas, soon found that true wealth lay in the soil, not in the more or less illusory mines of gold and precious stones, and they set about forming great plantations of cotton, tobacco and indigo. In 1640 sugar-cane was widely planted, and this heralded a period of marvellous prosperity for the islands, when sugar-planting adventurers flocked to the Antilles, often taking wealth there for its investment.

Legitimate trading in these waters at this and subsequent periods was greatly hampered by the horde of pirates of all nationalities which infested them, and their doings, and those of the British navy engaged in their punishment, have formed the theme of thrilling romance and story of those days. In 1655 Britain made a sudden descent upon Jamaica, and captured it, and that beautiful and wealthy island has remained under the British flag ever since. An attempt was also made upon Haiti, but this was unsuccessful. Had it been otherwise, the history of that unfortunate island might have been differently written.

The negro slave traffic in the Antilles was begun by the Portuguese, but all the nations who had colonies in the archipelago followed their example, and the " trade " existed for 300 years. As for the pirates and buccaneers, they carried on their depredations upon the high seas into the eighteenth century, and were not finally disposed of until the beginning of the nineteenth.

Slavery in these islands has its own terrible history. The

[1] It is to be recollected that the Spanish sovereigns generally opposed the brutal treatment of the " Indians," which was the work of the colonists.

British were among the most active in the trade, and were not the first to abolish it, although afterwards it was largely due to British work that it was stamped out. In 1807 there were 800,000 slaves in the West Indies, and after their importation was rendered illegal great numbers were murdered on the high seas, in the passage from Africa, by the slavers, in their attempted escape from British cruisers, and the overworking of those already in the islands resulted in considerable diminution. Total abolition of slavery in the British Antilles was not reached until 1833, and this was against the bitter opposition of the planter and the colonist. Only a strong Imperial Government at home was able to carry out such a measure.

The severe periods of economic depression from which the Antilles have variously suffered have given place during the present century to a more general prosperity, due in large part to the revival of the sugar industry and growth of the fruit trade, as well as to a more general system of cultivation, which includes cocoa, cotton, rice and other products. It is being learned that a system of " monoculture " invites disasters.

Further, a large volume of tourist travel has grown to being, and it may be that the economic development of the " American Mediterranean," by reason of the Panama Canal, will favourably affect these rich tropical lands. The cessation of Spanish rule and the substitution therefor of American influence in Cuba and Porto Rico have been other factors in sociological and economic progress. The American intervention and general treatment of Cuba is an episode greatly redounding to the credit and honour of the United States, and is almost unique in history.

Let us cast anchor in the harbour of Havana, that tideless haven upon which stands one of the most picturesque cities of the tropics, with a population of over 300,000. The land-locked bay is spacious, and the verdure of its shores is pleasing to the eye. In the old city the streets are too narrow for wheeled vehicles ; in the new, wide streets and splendid stone houses of the planter-aristocracy are found. The smells and fevers of Spanish times have given place to the healthy condition of American sanitary methods, and the palace of the arbitrary captains-general of Spain, a large and handsome stone structure, is now the home of the republic's president and centre of city government.

However, Havana and Cuba in general owe much to their long Spanish domination, if they also have embittered recollections of it. The city is famous for its promenades and public gardens and other Spanish institutions. Its commanding position earned it the name of the " Key of the New World and the Bulwark of the West Indies," under Spanish rule.

The island of Cuba covers an area of nearly 42,000 square miles, and has a population considerably over 2,000,000. Fine cities have populations of over 30,000, among them the historic Santiago. Native white people form sixty per cent. of the inhabitants, and people of negro or mixed blood thirty per cent.

The shore of Cuba is notable for its remarkable number of good harbours. Mountains traverse the island from end to end, but the rolling, open plains are remarkably fertile. In places the coast is bold and beautiful, and a multitude of torrential streams descend from the rocky gullies of the interior, beautiful cascades falling amid dense forests of superb timber.

Especially fine is the scenery in the eastern part of the island, wild mountains and tropical forests unfolding to the view of the traveller here, whilst in the central region extensive prairies extend, merging into undulating hills and gentle valleys, where the royal palm rears its graceful form and beautiful crest for 100 feet into the air, as we approach the western portion of the island. Numerous short rivers water the land, some of them disappearing into natural underground tunnels, associated with the many caverns in the limestone strata—a formation such as we have remarked also in Yucatan.

The flora of Cuba is splendid in its wealth, embodying all the principal species of the West Indies and Central America and Mexico. Considerable areas are still covered by the primeval forest, which is so dense that passage can only be effected by cutting a pathway through it. The royal palm; every portion of which is of constructive or food value to the native ; the corojo-palm, which provides drink, sugar, oil and wood ; the coco-palm ; the beautiful Bombax Ceiba or silk-cotton tree— giant of the Cuban forest fastnesses, and often reaching 150 feet in height, with enormous breadth—are among the principal trees, and there is a host of building, cabinet and hard woods, including the cedar, so largely used for cigar boxes, familiar to the smoker of Cuban tobacco.

The economic products and fruits cover an extremely wide range. They include cocoa, coffee, cinnamon, rubber, maize, wheat, cassava and tapioca, yuccas, sweet potatoes, sago, plantains, pine-apples—giving four crops yearly—tamarinds, bananas, mangoes—the favourite fruit of the negro—oranges lemons, pomegranates, dates, figs, grapes, custard-apple, and many luscious fruits whose peculiar names are unknown to the European or North American generally.

The principal industries are agricultural, and manufacturing, as in all tropical lands, is comparatively insignificant in comparsion with the opportunities offered for such. The cultivation of sugar has always been the dominant industry, but dependence upon one principal crop has often had the effect elsewhere noted of bringing about grave economic depression at times.

By far the greater part of the cultivated lands are occupied by white men. Centralised methods in sugar production, cheap but free labour, the fertility of the soil and the adjacent United States markets, where the crop is disposed of, have enabled Cuba to outdistance the other West Indian islands in sugar production.

But if sugar is the principal, tobacco is the most famous crop of Cuba, and the industry has always been a prosperous one and is essentially a "white man's crop," calling for special intelligence and careful labour. The place names of Habana, such as Vuelta Abajo and others, are familiar throughout the civilised world, by reason of the brands of cigars.

Coffee-growing is much less important than in former years, but there is no reason why the industry should not supply the home demand. Stock-breeding in earlier times was perhaps the most important industry, and may again become of greater value. Copper-mining in Cuba for several centuries produced the greater part of the world's store of the red metal, but it has declined, and iron is now the most important mineral produced.

In Cuba there is practically no "colour line" or caste, and no antagonism between the various peoples of colour and the outlook of the black race is full of promise. Indeed the negro holds a relatively high position and is industrially as useful to the community as the white man, and he is holding his own in point of numbers. The conditions regarding "marriage" are somewhat peculiar, nearly eighty per cent. of the total population being illegitimate.

Every male Cuban over age enjoys the franchise. A little over half the white voting population can read and write, and perhaps forty per cent. of the coloured. Education is controlled by the republic, and Church and State are completely separate. The Roman Catholic religion predominates overwhelmingly.

Leaving Cuba on the north, we approach the beautiful island of Jamaica, the largest of the British West Indies. But Nature has alternately smiled and frowned upon Jamaica, for dreadful hurricanes and earthquakes have wrought havoc at times in its history, and its towns and industries have suffered and still suffer severely therefrom. The island was early the principal resort of the bloody-minded buccaneers of the Spanish main, and, later, terrible scenes were enacted when the slaves arose and murdered their white masters, and when these took vengeance upon the negroes. Jamaica was at the height of its prosperity when emancipation was proclaimed in Britain, and the negroes, angered at the delay in setting them free, revolted and killed a number of white people. The British Government accompanied emancipation by compensation, paying nineteen pounds for each slave, but abolition and the removal of protective duties on sugar ruined the planters. In 1865 the negroes again rose, against taxes imposed to cover financial deficits, and terrible excesses on both sides were committed.

The area of Jamaica is 4207 square miles and its greatest dimension is 144 miles. The form of the island, upon the map, is like a turtle upon the water, and its back rises to cloud-capped mountain peaks. We remark the abundance of its verdant vegetation, in which are included splendid forests and a great variety of neo-tropical plants, flowers and fruits. One of the principal attractions of Jamaica lies in its enjoyable climate. Upon the coast it is humid and warm, but this gives place to a delightfully equable and mild temperature in the uplands,

NEGRO HUTS IN THE WEST INDIES

where the moisture-saturated clouds induce the growth of vegetation of the character of non-tropical lands.

Kingston, the capital, and Port Royal, the older capital, stand on the shore of one of the finest land-locked harbours of its size in the world. Both towns have been ruined by earth-quakes during their history, Kingston having been practically rebuilt since its destruction in 1907. The climate is healthy and dry, and the suburbs of Kingston are famous for their beauty. The town is dowered with all modern conveniences, and has many valuable institutions. Its traditions are inter-woven with British naval exploits, as are those of Port Royal.

The population of the island is well under 1,000,000, of which only two per cent. are whites, nearly all these being British. The bulk of the inhabitants are the descendants of the negro slaves, followed by mixed British and African people, the Maroons, descendants of Spanish slaves, Indian coolies and a few Chinese. The ancient Arawak aboriginals, a peaceful and docile people whom the Spaniards practically destroyed, have left traces of their past in the island's archæology.

Agriculture and the breeding of horses, cattle and sheep are the two main industries, the chief products being in fruit, especially bananas and oranges, and in coffee, sugar, cocoa, coco-nuts, pine-apples, grape-fruit, mangoes, avocado pears, ginger, pimiento or allspice—a product peculiar to Jamaica—and limes, with rum, logwood and tobacco. Valuable timber abounds in the forest, and the Jamaica cedar and the silk-cotton tree are prominent here, with many varieties of palms. Many beautiful flowers adorn the Jamaican countryside, as well as ferns and cactuses, and orchards display their gorgeous colours in the woods, which at night are lighted up by the numerous fire-flies and lantern beetles of this idyllic tropical land.

Of more recent times the growth of a negro peasant popula-tion has been encouraged in Jamaica, with the purpose of supporting the local industries of the island, and much of the land is divided into small holdings, generally of five acres or less. The negro, however, is generally content to work for his employer for only about four days a week, with another day devoted to tilling his own plot, a measure of toil which suffices to secure his requirements. As he cares little for the future and generally has no rent to pay, nor expenditure in clothes or fuel, of any consequence, nor yet much for food, which Nature produces so abundantly ready to hand, and which he gathers or steals, he does not lay by for the morrow. However, it is shown that the Jamaica negro is improving, and that he is putting money in the banks. But he is still grossly superstitious in his *obeah* practices and illegitimate to the extent of sixty per cent. of the population.

There are but few manufactures in the island except those connected with the agricultural products. Fruit, sugar and rum are the main articles of the export trade, more than half of which is conducted with the United States, as is also much of the import trade. One of the most valuable economic features of the island is the work of the Agricultural Board and Agricultural Society, with experimental stations and branches throughout the island. Good roads and bridle-paths cross the country in all directions, and a railway traverses it.

Sixty miles to the east of Cuba lies the interesting, curious and fertile island of Haiti or Santo Domingo, known to the Spaniards as Hispaniola, a mountainous land, second in size in the Antilles, and excelled by no other country in the world in the variety and richness of its vegetable products. Sociologically, the island is of extreme interest, as being the first example of a state peopled and governed upon constitutional lines by negroes.

The area of the island is 28,000 square miles, a little less than that of Ireland, and its greatest length is somewhat over 400 miles. Two republics hold sway here, that of Haiti, occupying the western half, and that of Santo Domingo, in the east. Haiti embodies somewhat less than one-third of the total area of the land, but has by far the greater population. High mountain peaks, beautiful, well-watered plains, such as the grand expanse of the Vega Real, rich forests and fertile valleys, are the main topographical features of this favoured island, and upon the coast are magnificent bays and harbours, such as that whereon Port-au-Prince, the Haitian capital, lies, and Samana Bay, in Santo Domingo.

The soil of this island produces in perfection nearly all the tropical plants and trees of the New World, and, in addition, most of the fruits and vegetables of temperate lands may be cultivated in the uplands. To enumerate these products is but to duplicate what has been said of the Antilles in general, where we encounter so wide a variety of luscious fruits and valuable food-stuffs. Cotton, maize, rice, cocoa, ginger, tobacco, indigo, cassava or manioc, bananas, bread-fruit, mangoes and all other fruits are produced. The forests yield pine and cedar and a species of oak, and cabinet and dye woods and numerous palms.

A wide range of climate follows on the bold relief of the topography. The uplands are frequently bathed in mists, but the capital occupies an extremely hot locality, due to its sheltered position.

Port-au-Prince is well situated between two great peninsulas, the harbour protected by the beautiful island of Gonaives, and the town, which was regularly laid out by the French during their occupancy of the island, with wide main and cross streets, slopes upwards to the hills. But an air of disorder prevails, for public and private buildings and institutions are dilapidated and neglected under their negro masters. The population of the town, consisting of negroes and mulattoes almost entirely, numbers about 75,000.

Santo Domingo, the capital of the eastern republic, is the oldest settlement made by the white men in the whole of the New World, having been founded in 1496. It may be regarded as the most perfect example of a sixteenth-century Spanish colonial town, being surrounded by ancient bastioned walls, with straight, narrow streets intersecting at right angles, after the usual Spanish American town-planning model. The cathedral contains the reputed tomb of Columbus, and the old fortress, the cell in which, by order of Bobadilla, the explorer was confined. A poor harbour serves the town, but the climate is cool and healthy. The most beautiful city of the Dominican republic is La Vega, situated on a fertile savanna, which yields a famous tobacco crop. A splendid cathedral here attracts the traveller's eye, arising from the lovely plain and backed by verdant, forested hills. Near at hand is the hill whereon Columbus planted his great cross.

The population of the whole island numbers about 2,000,000, three-quarters of which form the republic of Haiti. Between the governments of the two states a strong political antipathy exists. The people of the island are mainly negroes, with, in Haiti, about ten per cent. of mulattoes, who are described as a much-hated and rapidly diminishing race. The negroes are hospitable and kindly, but lazy and ignorant in general. In Santo Domingo the people are mainly mulattoes of Spanish descent, with, however, a large number of negroes, and some whites of Spanish and creole blood, the whites having a predominating influence. Spanish is the general language, and Spanish customs prevail. The people are described as quiet but shiftless.

In Haiti French is the official language, and the upper-class children are generally educated in France. The Government of this state, though a republic in name, generally resolves itself into a military despotism, with a dictator-president. Everything that a progressive nation might require for its advancement lies in the hands of this small nation, but the Haitians have been described as being permeated with corruption and veniality, and, although Roman Catholics, are accused of adhering to and practising the repugnant Voodoo rites. However, they absolutely deny this. Weird African dances are indulged in, to the accompaniment of the tom-tom. The children of unmarried parents are regarded as legitimate, and in the interior polygamy is frequent. In this connection, however, it is to be recollected that sexual morality is often extremely loose, even among the white or mestizo Spanish American people of the mainland republics, if less openly.

The machinery of government of both these states is as well established as in any of the Latin American republics, but it has been said that the Haitians, after a century of republican evolution and independence, have not advanced, but rather, in some respects, have retrograded. Undoubtedly there are good elements in the people, which a happier spirit and circumstance might bring to fruition. Lack of capital and enterprise, backwardness in education, and political unrest are some of the difficulties which, added to the native defects of the negro character, have to be contended with.

The history of the island is noteworthy and peculiar. The native name of Haiti indicates its mountainous topography. Soon after its discovery, in 1492, the Spaniards began what was practically the work of extermination of the gentle native people, and their place was taken by slaves imported from Africa. However, mining and agriculture flourished, and towns were early established under Spain. The peculiar French and English "traders," to whom the name of "Buccaneers"— the name is connected with a term meaning "dried meat"— established themselves here in 1630, and formed a flourishing French colony.

After the French Revolution, when citizen rights were conferred upon the mulattoes, acts of revolting cruelty were visited by these and the whites on each other, the latter protesting against the citizenship of the coloured people. A British force invaded and established itself on the island, but was driven out by Toussaint l'Ouverture in 1798. This remarkable and accomplished negro restored order, but he was treacherously betrayed by the French of Bonaparte and died in prison in France, and the blacks arising to avenge him committed many excesses. Finally the French evacuated the island, and in 1804 an independent republic was established.

The first negro ruler proclaimed himself "emperor," and inaugurated his rule by massacring all the whites, but he himself was later assassinated by his own people. Other emperors and dictators followed, with a further period of Spanish dominion in Santo Domingo, again followed by revolutions, executions and assassinations. A wonderful palace—that of Sans Souci—was built by one of these "monarchs," and its ruins still remain. The island is one of the most fertile in the world. Its exports are cocoa, coffee, cotton, tobacco, sugar and various forestal products, and many steamers call at its ports.

To the north of Haiti lie scattered the Bahamas, forming part of the British West Indies, extending towards the coast of Florida. They are crossed by the tropic of Cancer, and thus in great part lie out of our survey here. These islands are of considerable interest and beauty, their picturesque scenery, rich vegetation and the green and azure of the sea and sky affording pleasing contrasts of colour. The islands of Nassau form a health resort in winter for American and Canadian tourists, and have many attractions, to which the enjoyable climate of the group in general adds markedly.

The small Watling Island, of this group, lying almost upon the northern tropic, is held to be the landfall of Columbus, that first point of land beheld by the explorer upon his first and memorable voyage to the New World. Writing to Ferdinand and Isabella, the explorer said : " These beautiful islands excel all other lands in their splendour, and here the natives love their neighbours as themselves, their faces are always smiling, their conversation is the sweetest imaginable, and they are so gentle and affectionate that I swear to your Highness there is no better people in the world."

But alas ! for the fate that befell these simple people. The Spaniards, trading on their characteristic love for their departed relatives, promised to convey them to the " heavenly shores " where these were supposed to be resting, in a better world, and so, treacherously conveying them on board ship, transported them to Hispaniola, where 40,000 perished miserably in the mines, under starvation and the lash.

The products of the islands—which are twenty-nine in number, with 660 " cays " and thousands of isolated rocks— are those in great part which we have seen elsewhere in the Antilles. The total population is about 12,000, and the total area over 5000 square miles. Cotton is indigenous, and is woven into cloth by the natives. New Providence and Nassau, in the north of the group, lie little less than 200 miles from the United States mainland of Florida, and Caicos and Turks Islands, famous for their salt industry, about a similar distance from Haiti. The group is administered by a British governor from New Providence.

Seventy miles east of Haiti lies Porto Rico, whose inhabitants n 1898, at the time of the war between Spain and the United States, gladly exchanged the ancient and oppressive dominion of the first-named country for governance by the United States, which country they hailed as an economic deliverer, and under which notable progress has been made.

A land with a beautiful and varied flora, like that of its neighbour, the island of Porto Rico—or " Rich Harbour "— covers an area of 3435 square miles, and is 100 miles long. The climate is generally regarded as superior to the Antillean climate generally. The rugged, hilly districts are clothed with

Spanish cedar, sandalwood, flowering palms and trees of beautiful foliage, whilst tree ferns and tropical fruits in profusion bear out the lavish scheme of Nature here. Perhaps a quarter of the land area is cultivated, a great deal of it as small farms, of which there are some 40,000, occupying well over half the population of the island, which numbers about 1,200,000. The sugar exports, which have enormously increased under the rule of the United States, and to which country they go, are worth many millions of dollars annually. Coffee and tobacco are also largely grown, also cotton, maize, citron fruits, coco-nuts, pine-apples and so forth.

As we approach the northern shore of Porto Rico we remark a high bluff, crowned by a battlemented castle, and we shortly enter the harbour upon which San Juan, the interesting old Spanish town and capital of the island, lies. The town is built upon a small island connected with the mainland by a bridge, and is guarded by other old fortresses and walls. The pleasing streets and plazas are lined with shade trees, the massive, flat-roofed brick and stone houses are painted exteriorly in vivid colours, reminding the traveller of many another Spanish American town. Some beautiful churches stand out prominently, and there are fine public buildings. The more modern institutions attest the American influence upon the place, wherein picturesqueness is regarded as of less importance than sanitation. The population of San Juan Bautista de Porto Rico is nearly 50,000, and the harbour is capacious and land-locked. The Roman Catholic bishopric here is one of the oldest in the New World.

The island is administered by a governor, appointed by the President of the United States, and an executive council, five of whose eleven members are required to be Porto Ricans. Education under the Americans is increasing. Of the population over half are whites and about a quarter of mixed blood, with 60,000 negroes. There are but few manufactures, and transport facilities are inadequate.

The little group of the Virgin Islands, which extend eastwardly from Porto Rico, are divided as to their possession between the United States, Denmark and Great Britain, and cover about 465 square miles. They are generally rocky or sandy, partly barren, but in places yield sugar, coffee, maize,

2 A

indigo and cotton, and the natural pastures of guinea grass afford excellent cattle-grounds. Some useful timber is also found. The climate is healthier than in the other Antillean groups. St Thomas, St Croix and St John, the largest islands, are under the Danish flag. Tortola and thirty-two other small islands are British, and Snake and Crab Islands belong to the United States.

St Thomas is the Danish seat of Government here, and its town, Charlotte Amalie, lies upon one of the finest harbours in the West Indies. Christianstad, the capital of St Croix, or Santa Cruz, is a picturesque town.

The staple product is sugar, grown by English planters, and both here and in St Thomas the predominating language is English. The fertile island of St John possesses in Coral Bay a splendid harbour of refuge. Tortola also possesses, at Roadtown, a fine harbour. The people of these islands are generally peasant proprietors. The archipelago was a favourite resort of the buccaneer in early days.

Continuing on a course curving now to the south, still in the Leeward Islands, we reach Anguilla—British; St Martin —French and Dutch; St Bartholomew—French; Saba— Dutch; St Eustatius—Dutch; St Kitts, Nevis, Barbuda, Antigua, Montserrat—all British; followed by Guadalupe— French; Dominica—British. Thence we enter among the Windward Islands, comprising Martinique—French; Santa Lucia, Barbados, St Vincent and others, and Grenada, all British.

St Kitts is a well-watered and fertile island with a cool, dry climate, producing sugar, rum, coffee and tobacco. The hill slopes afford excellent pasturage and the summits are crowned with dense forest. Basseterre, the capital, has a population of 10,000, forming about a third of the total inhabitants of the island, who are mostly negroes. The area is sixty-three square miles.

Barbuda is of about equal size, but is the home of less than 1000 people. The soil is very fertile and the flat countryside is covered with forest. Cattle and horse breeding is carried on.

Antigua is a generally flat, riverless and unforested island, with a dry, healthy climate, beneficial to invalids, and the fertile

A street in Grenada   Royal Mail steamer passing

lands in the interior yield the customary tropical food-stuffs. The island covers 108 square miles, with about 35,000 inhabitants. The capital, St John, is extremely picturesque and overlooks one of the most beautiful harbours in the Antilles. This important town has a population of 10,000. It is the residence of the governor of the Leeward Islands, and its buildings and institutions are evidences of its British civilisation. Sugar and fruits are exported from Antigua, and cotton-growing is now successfully carried on.

Guadelupe's rugged surface rises to a volcano, and the lofty uplands are extremely beautiful, their cloud-capped peaks being covered with luxuriant forest. Here, indeed, is some of the most picturesque scenery in the Antilles. The soil is rich and fertile, and the extensive sugar plantations cover more than half the cultivated area. Cocoa, cereals, cotton, rubber and other tropical products are obtained, and there are well-equipped sugar factories, and some fruit-canning is carried on.

The area of the island is 688 square miles, and its population somewhat under 200,000, consisting of negroes and mulattoes, with a few white officials and planters. The mulatto women are noted for their beauty. They predominate in numbers largely over the men, and the proportion of illegitimate births is large. The picturesque capital of Basse-Terre, with about 8000 inhabitants, is the centre of French administration and civilised life, and stands upon an excellent harbour. The French and the British have variously held this valuable island at different times during its history.

Dominica is also a rugged land, its mountains rising to over 5000 feet, and here is the curious and famous Boiling Lake. But the island is remarkable for its botanical life, and the hills are clothed with valuable timber. Numerous rivers traverse the land, which is thus exceptionally well watered, and the luxuriant growth of coffee, oranges, rubber-trees, spices, tropical fruits, sugar, cocoa and so forth attest the fertility of the soil. Fish and honey are other abundant products. The atmosphere is generally dry and invigorating. The area of the island is nearly 300 square miles, and the population is somewhat under 30,000, mainly negroes, with a remnant of the aboriginal inhabitants. The few whites are of French and British descent. Roseau, the capital and centre of British

administration, is a fortified port.   French is the language of
the peasantry, and the Roman Catholic religion predominates.

In the fruitful island of Martinique we enter a land which has
felt to the full the vicissitudes of war during its history and has
been the scene of some of Nature's most terrific outbreaks in
her subterranean forces.   Here arises the notorious Mont Pelé,
whose appalling eruption in 1902 startled the whole world and
caused the loss of 40,000 lives, leading to the idea of abandonment of a region so terribly punished.   The island has an area
of 380 square miles, and previous to the eruption had a
population of nearly 204,000, which was reduced by the
disaster.   The broken and precipitous interior is draped with
forest, and many torrents descend therefrom, whilst the
healthy uplands are famed for their sanatoria.   The low
coast-lands, however, are not healthy for Europeans.

The people are mainly negroes and half-castes, and the chief
products of the island are sugar, rum, cocoa, coffee, cotton and
tobacco.   The capital, Fort de France, overlooks a fine harbour,
defended by forts, and has a population of 18,000.   The
history of the island is not dissimilar from that of its neighbour,
with the early extermination of the aborigines, the import of
negro slaves, the establishment of sugar cultivation, the depredation of buccaneers, and the various struggles for possession
between the British, Dutch and French, which latter nation
now possess it.

The loveliest of the West Indian Islands unfolds before the
traveller as the steamer halts in the magnificent land-locked
harbour of St Lucia, upon which stands Castries, the capital.
Rising sheer from the azure sea, their summits bathed in
perpetual mists, are mountains clothed with impenetrable
forest.   Elsewhere are bays and caves, and on the plains
the luxuriant vegetation marks the fertility of the soil.
St Lucia covers an area of 233 square miles, and the population of about 50,000 is mainly of negroes, the original Caribs
having disappeared, as elsewhere in the West Indies.   The
language is generally French, but this is being superseded by
English.   The white people are mostly of French descent.
For two centuries the possession of the island was disputed by
France and Britain, and it has witnessed many struggles ; and
various economic depressions resulting upon the " mono-

culture " of sugar. There are some East Indian coolies on the plantations.

The principal products of the island are sugar, cocoa, logwood, spices, kola-nuts and vanilla, all of which are raised for export. One-third of the total area is under cultivation, the remainder being retained as crown forests, and these contain the finest kind of timber for cabinet-makers. The principal trade is with the United States. Castries is the principal coaling station of the British West India fleet, and was strongly fortified.

The island of St Vincent shares with some of its companions the attribute of scenic beauty which conceals terrible subterranean forces, and it was devasted in its most fertile portion by the eruption of the volcano Soufriere, at the time of the Mont Pelé eruption, in 1902.

The flanks of the volcanic chain are beautifully wooded, and the valleys which intersect it are picturesque and fertile, producing sugar, cocoa, spices, cotton, arrowroot and many other tropical fruits and vegetables. The area of the island is 140 square miles, and the population numbers about 45,000, the majority being black, with about 8000 coloured and perhaps 3000 white. The inhabitants have suffered greatly from the vicissitudes attending the sugar industry, and efforts have been made, under an Imperial grant, to settle the labouring population as peasant proprietors on small holdings, for which purpose estates were acquired and Crown lands laid out. Very considerable success attended the settlement of negroes as peasant proprietors here, in conjunction with agricultural instruction given by the Government; and possibly the decline of sugar " monoculture " may have proved a blessing in disguise. Cotton and many food-stuffs are now produced.

The suffering entailed by the volcanic disaster was met by subscriptions from all parts of the world, especially the United States, which country has ever shown itself ready to help the distressed overseas.

Kingstown, the capital of St Vincent, and seat of the British administration, commands a beautiful situation.

In the island of Barbados we approach one of the most densely populated places in the world, the negroes being in the proportion of nine to one of the white inhabitants, in a population of nearly 200,000, the area of the country being 166 square miles. The negroes of Barbados are strongly attached to their island, and proud of their British citizenship.

The island is encircled with coral reefs, and rises to but a comparatively low elevation, and, unlike many of its neighbours, it possesses no famous harbours, Carlisle Bay being but a shallow roadstead. But the climate is a pleasant one ; fever is unknown, and the conditions are favourable for invalids, and for arresting loss of vitality in the march of old age. Due to the extreme fertility of the soil and the abundance of cheap labour, as well as careful cultivation, sugar-planting became of great importance here, and cotton-growing has later been established on a considerable scale. The headquarters of the Imperial Department of Agriculture of the West Indies is established in Barbados, an institution which has performed extremely valuable services [1] for the British Antilles. The island was the original home of the widely used Sea Island cotton.

The pleasing capital of Bridgetown lies at the foot of the hills, and in its square stands the earliest monument erected to Nelson. Its buildings, shops, pleasure grounds, institutions, splendid beaches for sea-bathing, together form an advanced and agreeable centre of life, and splendid roads lead therefrom to other parts of the country, traversing the low hills by gentle gradients. Many steamers from Europe and the United States call here, and Barbados is an important centre of distribution, as well as of tourist travel.

Whilst the island received its baptism of bloodshed by reason of wars for its possession in earlier times, it did not suffer, like its neighbours, from the emancipation of the slaves. The traveller remarks the relatively contented appearance of the teeming black population, under British rule, although he regrets their poverty. Great numbers of them were employed on the Panama Canal works, where, after the proper organisation of their feeding was entered upon, they performed very efficient work. A severe scourge of this valuable island are the periodical hurricanes.

Grenada presents the prevailing Antillean characteristic of

[1] Under Sir Daniel Morris.

THE TEEMING NEGRO LIFE OF BARBADOS

beautifully wooded mountains, and here, over 1700 feet above sea-level, is the curious Etang lake, lying in an ancient crater, the excellence of the climate being attested by the large sanatorium upon its banks. Indeed, Grenada Island is the health resort of Trinidad and other neighbouring islands, and the sea-bathing is a further source of attraction. The soil is remarkably fertile, and in the rich valleys cocoa, the staple product, and spices, cotton, coffee, kola-nuts yield well, and live stock, timber, turtles and turtle-shells and wool are exported. But the principal article of export is cocoa, which goes almost entirely to the United States.

The area of the island is 133 square miles, and the population, only two per cent. of which are whites, numbers about 66,000. St George, the capital, stands upon a lava peninsula within a land-locked harbour, and its red-brick houses climb the slopes of its surrounding hills. The town, which has a population of about 6000, is the residence of the British governor of the Windward Isles, of which Grenada is the southernmost. The struggles between France and Britain for possession in earlier times has formed much of the island's history.

We are now approaching the Venezuelan or mainland coast of South America, and reach the small island of Tobago, under the jurisdiction of Trinidad. It consists of a single mountain, rising from the sea to 1800 feet, clothed in great part with dense forests, which are carefully preserved in the interests of the rainfall, and are known as the " Rain Preserve." The area of the island is 114 square miles, and about 19,000 people inhabit it. The fertile soil produces rubber, sugar, cotton, cocoa, tobacco—the real name of the island is Tobaco—and nutmegs, and the breeding of cattle and horses is a growing industry. Scarborough, the capital, is a small town, with a population under 1000.

The island of Trinidad, lying six miles off the Venezuelan coast and the mouth of the Orinoco, is the largest of the Lesser Antilles, covering an area of 1754 square miles, with a length of forty-eight miles. Geologically, it is part of South America ; politically, it is a portion of the British Empire, with a population, with Tobago, of about 330,000. The upper classes are of British, French and Spanish blood, and the lower of negroes, pure or mixed. One-third of the total population are East Indian coolies, imported as labourers.

The capital of this rich and fruitful isle, Port of Spain, is a revelation to the traveller, who may have forgotten that here stands one of the finest towns in the West Indies, excellently laid out and replete with all the utilities and amenities of civilisation. The streets are regular and pleasingly shaded with trees, there are two cathedrals and various fine public buildings, and the botanical garden is beautifully kept. The harbour, however, is shallow, and the steamer drops anchor half-a-mile from the quay. The population of Port of Spain is over 55,000.

The surface of Trinidad is flat or undulating, but rises to hills and to three peaks, upon sighting which topographical trinity, in 1496, Columbus piously gave the island its name. The soil is remarkably fertile, and sugar and cocoa are the staple among the many tropical products it yields, great quantities of which two articles are shipped overseas. The rubber industry has made much progress of late years, and rum, coco-nuts, bananas and bitters are among the other exports. The famous pitch lake of Trinidad yields important quantities of asphalt, mainly exported to the United States under an export duty, and the island is becoming a valuable source of petroleum production, one of the foremost within the British Empire.

There remain before us now only the three small Dutch islands of Curaçao, Buen Ayre and Aruba, lying off the Venezuelan coast. The largest, Curaçao, has an area of 212 square miles, the others being under 100. Generally flat, with some low hills, this island has several natural harbours, but the soil is semi-arid in. general, although some tobacco and sugar are produced, and salt, phosphates and cattle are exported.

Curaçao has given its name to the famous liqueur, which was originally made here from the peel of a bitter native orange. The capital of Curaçao, Willemstad, with a population of over 8000, is almost a Dutch town planted in tropical America, with a canal and houses built after the style of Amsterdam dwellings. The population is over 30,000, mainly negroes, whilst Aruba has 10,000 and Buen Ayre 5000, the total Dutch West Indies, including these islands to the north, sheltering over 52,000 people.

The political and international condition of the Antilles, the islands lying as they do under the control of such varying nationalities, is of considerable interest.

Of late years some thought has been given to the project of federation of the British West India Islands as a commonwealth, and possibly this may later take form. Widely separated as the islands are by sea, it would remain to be seen what economy in government would result therefrom. Moreover, it often happens that small units of the Empire are best served by the Imperial Home Government direct, whilst locally controlling their own affairs. Yet federation might have useful results.

As a field for investment of capital this part of the Empire offers much, for, as has been shown, the islands are capable of much greater development, and the demand for tropical products is increasing all over the world. The islands do not offer a field of work for the white labourer, " poor whites " being in an anomalous condition, for the bulk of manual work is performed by negroes, a circumstance very marked among the " mean whites " of Barbados. Efforts have been made, however, to attract small planters with capital, and this class should find remunerative occupation here.

The black inhabitants of these lands have many good qualities. If childlike, it is an undoubted fact that the assumed " stupidity " of the negro is often a reflex of that of his white masters, who have been unable to understand him, a condition which happily is passing. As soldiers, the bravery of the negroes is marked. Also their decorous manner to the white people is a matter of considerable note. The terrible examples of race antagonism, the assaults and reprisals, such as occur in the United States, are unknown in the British West Indies.

The subject of " annexation " of the British Islands to the United States has from time to time been discussed. But the negroes know that they are well off under the British regimen, and even if some economic advantage might accrue to the white trading class under such a change of administration—which at least is doubtful—this would be unlikely to filter down to the black inhabitants, who form the great bulk of the population. As it is, nearly all the cacao, coco-nuts and bananas produced go to the United States, a very moderate share reaching Britain.

Of late the increase of trade with Canada has been a valuable economic circumstance, which is likely to grow largely and to prove a valuable Imperial asset. Canada alone could consume the entire British West Indian sugar production.

A marked feature of industrial life in the islands is the lack of manufacture, but the future will undoubtedly show that greater self-support and supply in this respect will be necessary, if the black economy and civilisation is to progress. The generally miserable dwellings of the negroes are largely responsible for illegitimacy and other social evils. The negroes, whilst they are disinclined for steady and continuous labour, are by no means altogether thriftless, as elsewhere remarked, and it is worthy of note that the workers on the Panama Canal sent home considerable sums of money to their families. In the British Islands a great step forward was made in the establishment of the Imperial Department of Agriculture, whose success gave rise to a similar institution for India.

Possibly the future will bring about a far greater interest on the part of the British people at home in these fruitful islands, in the cheapening of their food supply and as centres of tourist travel. The same circumstance might apply to the Americans of the United States, with whom tropical America is so closely allied. Reference has been made elsewhere to the growth of negro peasant proprietorship in the British West Indian Islands. Land which was formerly ill cultivated or lay fallow in the hands of private owners has been bought by the State and resold to an agricultural class, upon easy terms, and now yields a rich return.[1] As regards sugar-making, the tendency everywhere is to abolish the small factory and centralise. This, of course, makes for cheaper production. Doubtless tourist traffic, to which the islands owe much already, will considerably increase.

[1] Sir Daniel Morris in *The Oxford Survey of the British Empire*.

CHAPTER XXXIV

COLOMBIA AND VENEZUELA

ACROSS the waters of the Spanish Main, as the Caribbean Sea in the days of ocean chivalry was termed, blue under the tropic sky, we approach the flat coasts of Colombia, or the forbidding granite wall which Venezuela presents to the arriving traveller. These two republics offer the nearest point of the South American mainland to Europe, and it was part of the Venezuelan coast that Columbus sighted on his third voyage.

Colombia—whose heritage was the isthmus of Panama—possesses both an Atlantic and a Pacific seaboard, and covers an area of about 466,000 square miles. Venezuela extends along the northern coast of South America eastwards to where lie the European colonies of the Guianas, and covers about 400,000 square miles. The area of both republics, however, are susceptible to recalculation whenever boundary disputes shall have been settled.

Both these countries are traversed or broken by the northern extension of the mighty Andes, and Venezuela lies upon the great Orinoco river, some of whose affluents flow through the sister republic. Almost within the mouth of the Orinoco lies the fertile British island of Trinidad. Considerable portions of Colombia and Venezuela are unexplored, especially in those wild regions which extend beyond the mountains into the Amazon Valley, or upon the Orinoco towards the Guianas, and are of much interest to the adventurous traveller.

The Atlantic steamer bears us past the islands and into the broad channel which forms the approach to the ancient and picturesque Colombian seaport and town of Cartagena, the finest harbour on the whole of this northern coast, a smooth, land-locked bay, with fishing villages and feathery coco-palms upon its shores. Beyond arise the walls and towers of the city, backed by verdure-clad hills.

Cartagena was first established by the Spaniard Heredia, in

379

1533, and he named it after the Spanish city, founded by the Phœnicians of Carthage. There is indeed a Mediterranean aspect about this South American city, and it is pervaded by a certain air of massiveness and antiquity and the old-world dignity of a place which was a former vice-regal seat and a centre of the notorious Inquisition. A long, massive stone wall faces the sea, one of the strongest and most ancient Spanish fortifications in the New World, built to withstand the attacks of Drake and other rovers. Cartagena was a rich place, the outlet of the productive land of New Granada, and a storehouse of treasure in those bygone days, forming a tempting bait for the buccaneers of the Spanish Main.

The chief highway into the interior of Colombia lies upon the great Magdalena river, which enters the Caribbean Sea ; Baranquilla, the principal seaport, lying at its mouth. Light draft steamers, of a type such as are used on the Mississippi, carry us for 560 miles inland to near Honda, 100 miles above which town the railway runs to Bogotá, the capital of the republic. The upper reaches of the river, and its affluent the Cauca, serve beautiful and fertile districts, but navigation suffers severely by reason of frequent low-water periods. It is infested by crocodiles, some of them remarkable for their size, colour and ferocity.

The ancient city of Bogotá, on its high Andine plateau, 8600 feet above sea-level and remote from the outside world, is a typical Spanish-American capital, of picturesque aspect, with its characteristic churches, houses and public buildings, and contrast or mixture of Indian and Europeanised life. It lies but 4° north of the Equator, but its upland environment is temperate, or cold and healthy. The city has the pleasing qualities, and defects, which we remark in all the Spanish-American centres of life. In Bogotá there are handsome plazas ; streams of fresh water flow through its sloping streets from the mountains, and the surrounding plain is very fertile. The plateau is indeed remarkable in this connection for one lying at so considerable an elevation.

Bogotá has always held a reputation for being a centre of literary learning and polite life, and the educated Colombian displays an affinity for the doctorate to the full as strongly as elsewhere in Latin America. The population of the city is estimated at about 125,000, formed of the customary small white or mestizo ruling and wealthier class, shading off into the great bulk of poncho-clad, poverty-stricken and ignorant Indians, who form the majority of the people of the republic.

The population of Colombia is estimated very approximately at 400,000, an exact account being impossible in view of the extent of wild territory and the aversion displayed by the Indians from enumeration, fearing, as they do, taxation and enforced military service. The upper class have very strongly retained their Spanish character and traditions, the highlands having been settled by Spaniards in early times. There is no " colour line " between the whites and the mixed people and Indians, social distinctions being mainly a matter of wealth. In the seaports, especially at Baranquilla, there is a considerable negro element, a heritage of slave times in part, but due also to the proximity of the West Indies; and this, both in Colombia and Venezuela, has left some mark upon the population.

The upper-class Colombians entertain a high opinion of their country and culture, but are somewhat blind to its defects and slow to remedy them. They suffer from frequent outbreaks of the terrible political struggles which are so marked a feature of Spanish-American life. However, the Colombian upper class are a pleasing, courteous and intellectual people, and the lower class have many good qualities, which, so far, have not been able sufficiently to manifest themselves. The women of the upper class are generally handsome and refined.

There are several important and often pleasing towns scattered about the Colombian uplands, in which part of the country modern life and activity most freely flourishes. Chief among these is Medellin, whose population numbers about 60,000. Like Bogotá, Medellin is a university town and has other advanced institutions.

Means of communication are rendered difficult in Colombia by reason of the mountainous nature of the country in great part, two-fifths of the area being of this character. Swampy and often unhealthy lowlands cut off the interior from the Atlantic, whilst the Pacific coast is rugged and steep, and to the south lie the almost untrodden Amazon forests. The high, inclement *paramos* or uplands, over which fierce gales sweep and a piercingly cold mist often prevails, are at times almost impassable. Between the Magdalena and its tributary the Cauca the Cordillera summits reach a height of over 18,000 feet, and, viewed from Bogota, we remark the white drapery of the Mesa de Herveo, which rises 3000 feet above the perpetual snow-line. From Cali, in the beautiful Cauca Valley, a line of railway has been built to Buenaventura on the Pacific coast.

Agriculture forms the principal industry of Colombia, and coffee is the chief article of export. Sugar, cocoa and cotton are produced for home consumption, but not yet in sufficient quantities. Bananas and coco-nuts are now being grown for export, partly under the same auspices that influence this industry in Central America. Potatoes are a staple food of the upland people. Tobacco of excellent quality, equal to the Havana variety, is grown, and its cultivation might be largely increased.

The increased production of all these staple articles, for home consumption and for export, awaits improved facilities of labour. The agricultural classes, both here and in Venezuela, have been greatly depleted by the fratricidal civil wars of past years, and the population is almost stationary as regards its increase, and in general is extremely poor and backward. The land is mainly in the hands of large proprietors, with a consequent absence of a thriving peasant class. Wild tribes inhabit the more remote parts of the country, and in some cases attack travellers.

Education is free and compulsory, but only a small proportion of Colombia's population is able to read and write. Bitter struggles in the past have resulted in freedom in religious matters, but the community is strongly dominated by Roman Catholicism, whose influence is not generally progressive.

There are fine pasture lands in the interior of the country, but only hides and other easily handled animal products are exported, due to the difficulty of transport, but these form an important source of revenue. Colombia is rich in tropical flora and fauna, and from its almost unexplored woods, and those of Venezuela, the horticultural world obtains some of its most beautiful orchids, whilst bird, insect and reptile life is second in its profusion only to that of Brazil.

The republic is richly endowed with mineral wealth, and great quantities of gold have been obtained from its soil. In earlier times this wealth was obtained under the brutal system of forced and slave labour, which prevailed under the Spanish dominion of the New World. At the present time a number of foreign mining enterprises are at work upon the rich reefs and alluvial deposits, and undoubtedly a promising future offers in the exploitation of the precious metals. Coal, iron, copper and lead also exist in abundance, but their exploitation generally awaits improved means of transport. Petroleum beds are being worked with some success, but some jealousy has been displayed in this industry between British and American capitalists.

Venezuela offers natural and political conditions generally analagous with those of its sister state, in its high mountain ranges and broad lowlands and forested areas, and in the life of its people.

To reach the capital of Venezuela we land at the hot and unattractive seaport of La Guayra, and from here one of the most remarkable railways in the world carries us over the mountain rampart which faces the Caribbean Sea—the line running along the face of the precipice like a creeper against a wall—to the beautiful Venezuelan capital of Caracas, a journey amid stupendous scenery, in which we observe the seaport and the ocean mapped out far below.

Caracas lies in an environment whose climate is likened to that of a "perpetual spring," the city standing at 3000 feet above the sea, and is therefore free from the sweltering heat of the littoral. It was founded by the Spaniards in 1567, and the modern place may be regarded as one of the most attractive cities in South America. Much thought and work have been expended upon its planning and construction, and the national and private buildings and institutions are in harmony with the fine surroundings. The more showy structures are, however, in strong contrast with the poor dwellings of the working classes, as ever.

The greater part of the Venezuelan republic lies within the basin of the Orinoco, which is very thinly populated. This vast river has always appealed to the imagination of explorers and

travellers, and its navigable waterways afford means of transport for thousands of miles throughout the country, upon the
main stream and its numerous affluents. The Orinoco, however, does not lend itself to the passage of ocean-going steamers
such as ply upon the Amazon and the River Plate, vessels of
the stern-wheel type being employed, as on the Magdalena.
These ascend to Ciudad Bolivar, nearly 400 miles up-stream,
and beyond this point, up the Apure, Arauco, Meta and
Guaviare, the principal affluents, smaller craft provide means
of transport. The rise and fall of the river, according to the
seasons, is very considerable, but the gradient of its bed is
slight, and doubtless in the future hydraulic works of canalisation will greatly improve the conditions of navigability.
Indeed, the Orinoco, in this connection, offers an interesting
subject for the work of the engineer. A singular hydrographic
feature of the Orinoco is in its junction with the Amazon
fluvial system, where, by means of the natural Casiquiare
"canal," vessels may obtain access to the Rio Negro, the
great tributary of the Amazon, which enters that river near
Manaos.

One of the distinguishing topographical features is the
enormous grassy flood-plains or *llanos* of the Orinoco, which,
in earlier times, were the seat of a thriving cattle industry, and
produced the famous *llaneros*, a race of hardy horsemen cattle-
keepers, which furnished some of the best fighting material in
South America. But these resources, both of cattle and men,
have been heavily drawn upon in civil wars, and what might be
an extremely important centre of meat production and food
supply for the world, in these grassy plains, lies half abandoned.

When the Orinoco overflows, all living creatures are driven
far back to higher ground, and the tangled jungles become
inundated over vast areas. The endless green walls of vegetation, as our craft pursues its way, enclose a silent and deserted
waterway, in which not a living thing is seen, except perhaps
it be a miserable Indian in his frail canoe, seeking to barter
his produce. Even alligators are rare. The great cattle
pastures are seriously affected by the annual inundation, which
is generally followed by prolonged drought, and agriculture
is similarly prejudiced thereby. Such service as has been
performed upon the Nile might possibly be undertaken here,

in order to control the vast volumes of water from the Andes and the Guiana highlands, which sweep down the Orinoco, so bringing about a more useful regimen of the stream.

The population of Venezuela numbers less than 3,000,000, three-quarters of which are of mixed white and Spanish race—the largest mestizo proportion in South America. The conditions of the governing and governed elements are similar to those we have remarked elsewhere. The laws are well codified, but, in practice, are badly carried out. It has been recorded of Venezuela that no president has ever attained office through a legally conducted election, or left it without the accompaniment of a revolution. Of late years, however, there has been some improvement in the country's governance.

The great bulk of the people are illiterate, and the Church and the civil authorities clash in the provisions of instruction. But education is a matter in which unfortunately the people as a whole take little interest. The inhabitants of the country are sparsely scattered over what in reality is a region of great resource, and their generally miserable condition attests the lack of economic organisation of the class which pretends to rule and succeeds in exploiting them. Thus labour is scarce and industries flourish very slowly.

Like many of the Latin American countries, Venezuela is a producer of coffee, the growing of which is the principal industry, there being tens of thousands of plantations. The yearly value of the export of the berry reaches several millions sterling. Cocoa-growing and export is also important, as is sugar cultivation. Cotton-growing might be very largely carried out in the excellent lands of Maracaibo. Maize and beans are the staple food of the working class, but wheat is grown in the highlands. Venezuela is capable of producing a very wide range of food-stuffs, tropical, sub-tropical and those of temperate lands, due to the influence of vertical zones of temperature and the fertility of the soil. The forestal products are of considerable importance, including rubber from the Orinoco forests, also coco-nuts, fibres and timber.

Mining in copper, iron and gold is a source of some wealth and industry, the iron ores of Imataca, worked by a Canadian company, being plentiful. Coal, asphalt and petroleum are also worked, the Maracaibo and other oil deposits being of

2 B

considerable importance. Venezuela indeed promises to become a useful producer of petroleum, under the efforts of foreign capitalists ; largely British.

Maracaibo and its district form one of the leading industrial centres of Venezuela. The town, which is situated upon the lake of the same name—an arm of the sea—has a population of 60,000, and its institutions and general amenities attest the growing importance of the place. Manufacture is making some headway here, and water-power is employed to some extent therein—an asset of great value in nearly all South American states, but in general so far scantily employed.

Both Colombia and Venezuela must be regarded as tropical lands capable of a much wider development, both for their own enrichment and for the benefit of oversea commerce, but this awaits an economic organisation superior to that which so far has controlled them.

Our way now lies through the Andine republics, upon the Pacific coast.

CHAPTER XXXV

THE ANDINE REPUBLICS—ECUADOR

A VAST stretch of coast unfolds before the traveller who has essayed the journey into western South America, a coast beaten by the tearing surf of the boundless Pacific Ocean between the comparatively few havens which exist upon the singularly uniform shore-line on this side of the continent.

North of the Equator the shore is clothed with green vegetation, in Colombia and Ecuador, but soon after having crossed the line we remark a region of generally barren appearance, inhospitable and uninviting, inhabited by innumerable sea-birds which fly at times like darkening clouds upon the face of the waters, and by seals, whose strange bleatings arouse the echoes of the arid, rocky islets ; the few hamlets which cluster here and there, with an occasional larger town, serving rather to mark the general absence of man upon the seaboard than to proclaim it a centre of population.

On the one hand stretches the broad surface of the Pacific, its bosom uninterrupted by any object, unless it be the disc of the setting sun, flaunting its crimson and yellow clouds across the west over a sea blue or grey, according to the season, throbbing in long swells and bearing the steamer on from day to day in almost mournful solitude. On the other hand is the sandy or rocky coast, interspersed here and there by green oases, which mark the courses of scanty rivers, and on the far horizon arises, like an endless wall grey-blue in the distance, the Cordillera of the Andes, exhibiting upon rare occasions throughout its course within the tropics a snow-clad peak.

Yet this long Pacific coast, extending from the Equator to far below the tropic of Capricorn, is in reality a region of considerable interest to the traveller, upon whatever errand he may be bent. Here stretches the classic land of Peru, along whose coast and through whose rugged mountain defiles Pizarro and his companions pressed, marking the beginning of

387

history in this strange land, scattered throughout which are the fortresses and temples of the ancient and mysterious Incas. And from among those rocky terraces and half-inaccessible mountain valleys of the Andes, all blended in one grey distance, as we view it from the steamer's deck, a stream of rich minerals pours to the seaboard, with many tropical products, both from the Cis-Andine plantations and the Amazon forests of tropical Peru, Ecuador, Bolivia and Chile. Moreover, arid as is the Pacific coast of South America, for 2000 miles or more, perhaps the absence of vegetation and the bare appearance of the rocky headlands and sandy deserts render it less monotonous than the interminable mangrove fringe of the West African coast in the same. latitude.

The states whose description we here enter upon have much in common as regards their topography, people and natural resources. Ecuador and Peru occupy the seaboard and extend across the Andes into the Amazon Valley ; Bolivia is cut off from the coast by the northern portion of Chile. Dry desert areas, extremely rugged mountainous hinterlands and dense tropical forests beyond make up the greater part of these territories, the highlands of the Andes rising to very high elevations, subject to the extremes of a cold and snowy climate. All these contain, however, magnificent fertile valleys, healthy upland plains, and in places are well watered, the scenery often being stupendous in its beauty and variety.

The republic of Ecuador takes its name from its position upon the Equator, and is one of the smallest of the South American states. Its area, apart from the considerable territories whose ownership is disputed with its neighbours of Peru and Colombia, is estimated between 114,000 square miles and nearly twice that amount, according to the claim allowed.

As the steamer upon which we embarked at Panama or Valparaiso drops anchor in the Guayas river, in front of the Ecuadorian seaport of Guayaquil, we may descern, if the equatorial atmosphere at that moment happily be clear, a far-off snow-clad mountain, looming above the horizon of the tree-clothed landscape which slopes upwards from the water's verge. It is Chimborazo, which shares with its neighbour, Cotopaxi, the monarchy of the Northern Andes, rising 20,700 feet above the level of the sea.

THE SNOWS OF THE EQUATOR. COTOPAXI, THE WORLD'S HIGHEST ACTIVE VOLCANO

However, this great mountain of Ecuador is rarely seen from the seaboard, nor yet is the devastating and terrible Cotopaxi, still more beautiful in form. Chimborazo is generally wrapped in the clouds which hang upon the equatorial belt of Ecuador, and the traveller's eye will rather be attracted to the emerald-green of the coast and islands which line the entrance to the Guayas river and adjacent part of the Ecuadorian coast.

Ecuador, topographically, is one of the most interesting of the South American states. Nature might seem to have chosen- to establish upon this part of the earth's surface an orographical and climatic exhibition of her powers, a model upon which are displayed, within a measurable compass, every variety of South American plant and animal life, with every form of landscape and variation of temperature, from tropic heat to Arctic cold. From the sweltering heat of the lowlands, amid the waters of whose streams the alligator lurks, we may ascend to regions of perennial spring, where delicious fruits tempt the eye and palate, regions overhung by snow-fields and glaciers, yet but a day's march therefrom, the home of the condor, in those high elevations where the vegetation ceases and where perennial winter reigns.

Nor is Ecuador interesting by reason of its topography alone, for the land was the theatre of an early culture, a strange people having dwelt there long prior to the advent of the Spanish conquerors, a race whose origin antedates even that of the famous Incas of Peru. These people were the Shiris and others, and they left curious monuments, walls and implements, notably the carved stone arm-chairs which are found among the ruins upon the hills.

The republic has an important economic claim on our attention in that the soil of the coastal lowlands, being extremely fertile, produces a large proportion of the world's supply of cocoa, its staple article of export, which has its outlet at Guayaquil.

From the steamer's deck the tropic seaport of Guayaquil presents a pleasing picture. Across the broad, placid bosom of the river, upon which float a wide variety of craft, from ocean liner to the curious canoes of gaily clad Indians, we remark the clustered lines and terraces of its white buildings, sloping upwards to picturesque hills, verdure-clad to their summits.

At night the numerous lights extending far along the front and up the terraced slopes are reflected in the river, giving the impression of a large city.

Landing is effected by boats, a horde of apparently semi-savage, but in reality eager and civil, native watermen invading the vessel's deck and cabins to offer their services to the passenger—at times carrying off his baggage despite his protests or before the bargain is completed, after the customary South American fashion. The conditions of the city—pleasing from afar—are not altogether dispelled on close acquaintance, although there are features which repel and even endanger the traveller during his sojourn therein. The commercial position is advantageous, although the passage up the river from the coast at Puná is sometimes attended with difficulty. The fluvial system of the Guayas river, serving a rich district, is navigated by numerous craft, which discharge their produce at the port.

Guayaquil has been captured or burnt at various times in its history since its foundation by Benalcazar in 1535, French and British buccaneers having reduced it, notably Dampier, in 1707. The "new town" forms the business and residential quarter of the population, which numbers about 60,000, of which several thousands are foreigners. The main streets are wide and straight, and the buildings almost entirely of wood; and although these are in general of little architectural pretension, they are in some cases admirable examples of carpentry work. The overhanging upper storeys and balconies form arcades over the footpaths, and the many first-class shops contain almost every modern commodity. Among the principal buildings are the cathedral, churches, governor's and bishop's palace, town hall, colleges and so forth, and there are electric trams and other modern conveniences.

The old town or poorer quarter is of very different aspect, with ill-paved, undrained streets and miserable houses, and to the unsanitary state of the city as a whole has been due its terrible reputation for epidemics of yellow fever. From this cause Guayaquil is frequently neglected or avoided by steamers which ply upon the Pacific coast, and troublesome quarantine restrictions are necessary in other seaports after leaving it. However, something is being done to improve the condition, an impetus having been given to the effectuance of sanitary improvements by the American occupation of Panama, and the traffic from the canal, Guayaquil being the first place of importance south of Panama. Yellow fever is not necessarily endemic here, and at times years pass without a single case occurring. A large export trade is carried on in cocoa, rubber, hides, "Panama" hats, cinchona or quinine bark and *tagua* or ivory nuts, amounting to three-quarters of the whole foreign trade of the republic.

An extensive view of the city and river is obtained from the Santa Anna fortress and its hills. The roads leading out to the suburbs are dusty and bad, after the usual Spanish American fashion, but the plantations and vegetation reveal a marvellous wealth of tropical colour and variety, with groves of coco-nut palms, pine-apple fields and banana and orange groves, and the flaming scarlet of the flowering trees and creepers is remarkable.

The coastal zone of Ecuador differs considerably from that of Peru, to the south, in being clothed in parts with dense vegetation, among which the sombre mangrove is the principal tree. This difference is due to the effects of the climate, following upon the absence of the Antarctic current, upon whose agency the aridity of the Peruvian coast partly results. Thus we encounter here conditions more normal to the equatorial zone, with a heavy rainfall and rain forests. Further, the Ecuadorian coast is markedly different from the Peruvian seaboard, in being penetrated by a large river system, the Guayas, the only navigable river of any considerable importance upon the whole coast-line of western South America.

The Guayas river traverses a very fertile coastal plain, and upon its banks are situated the numerous cocoa-producing haciendas which form the chief basis of Ecuador's economic wealth. To ascend the river and its affluents by the steamers which ply thereon is to survey one of the most pleasing regions of any part of the tropical world, and the traveller who has failed to accomplish this will be unable to form an adequate idea of the true character of Ecuador.

The system embodies a number of parallel rivers, traversing the rich coastal plain-like natural canals, with an aggregate of over 200 miles of navigable waterway. Alligators of large size haunt the waters, sometimes floating down on masses of tangled vegetation, and are at times shot at from the steamboats. We pass among extensive *cacaolates* and *cafetates*, or coco and coffee plantations, with groups of feathery coco-palms and bamboos, the last-named, the useful. *caña de Guayaquil*, used for building purposes. There are also broad cattle pastures, extending away from the banks.

Especially beautiful is the scenery upon the Daule, a large affluent of this river system. In addition to steamers, the native *chatas*, or flat boats, and the great rafts of *balsa* wood ply freely on the river, being borne up and down by the tide or current without motive-power, and carrying large quantities of merchandise. Many canoes are also seen. The great rafts are often covered over with a roof. It was a native *balsa*, but with a mat sail, that the first European voyager, Ruiz, the pilot of Pizarro, encountered upon the sea-coast, and the white men were startled at beholding a sailing craft in those unknown waters.

From Guayaquil the traveller is ferried across the broad river to Duran, whence starts the Guayaquil-Quito Railway. This line, nearly 300 miles long, traverses the Guayas Valley and ascends through exceedingly steep and difficult country up the slopes of the Andes, to the valley plateau, whereon stands Quito, the capital of the republic, surrounded by its stupendous snowy volcanoes.

We reach, on this line, an elevation of nearly 12,000 feet above sea-level, and short trains and powerful locomotives are necessary to overcome the steep gradients. The pretty hamlet of Huigra, at 4000 feet elevation, is above the " yellow fever line," and the upland region is in general healthy, although typhoid fever is a scourge, and the absence of sanitary methods is very marked in the native villages. Upon the Alausi " loop," where the railway surmounts a terrific rocky promontory, known as the " Devil's Nose," a magnificent view is obtained, and the terrace cultivation common to the Andine countries is here well displayed. There are fertile valleys, producing grain and fruits, butter, cheese and alfalfa, and potatoes of exceptional quality are raised. It is to be recollected that Ecuador was the original home of the potato, which the natives evolved from the wild, bitter variety, and which the Spaniards introduced into Europe.

As the train ascends upon this remarkable railway and reaches the higher elevation the magnificent, snow-clad Chimborazo suddenly bursts upon the view, overlooking the fertile plain of Riobamba and its important town of the same name. Beyond, the Latacungo valley is traversed, and here the great Cotopaxi, the beautiful volcano from whose crater smoke curls up unceasingly, appears, rising to 19,613 feet above sea-level. The railway bears us across the very base of the famous mountain.

Other great snow-covered volcanoes come into view as the train proceeds, and the line, descending somewhat, reaches Quito. The Guayaquil-Quito railway has had a troubled financial history and barely covers its working costs.

The great "avenue" of volcanoes which form the approach to the city of Quito have been described as the most remarkable assemblage of cyclopean snow-clad peaks in the world. There are a score of such over 14,000 feet high, all within sight of each other, some of them being active volcanoes which from time to time have devastated the land with lava and mud outbreaks. Some of them are of weird, picturesque appearance, with sunk and jagged cones, their slopes covered with glaciers. One of the volcanic group is Tunguragua, one of the most beautiful, but its volcanic activity has been terrible.

The region has always been subject to terrific earthquake shocks, which have constantly wrecked the towns of the plateau, Quito having suffered severely throughout its history, since or before the time of the Incas. It is not shown, however, that the eruptions and earthquakes are necessarily connected with each other.

The city of Quito lies upon a somewhat bleak plateau, at 9400 feet above sea-level, overlooked by the great assemblage of snowy volcanoes, and it is a place of considerable topographical and historic interest. The city was originally founded by the more or less cultured natives of the pre-Spanish era, and was the capital of the ancient kingdom of Quito and the Shiris, who were overcome by the Incas of Peru under Huayna Capac. The Incas built roads between Quito and their Peruvian capital, more than 1000 miles to the south, roads which have been famous in antiquarian lore. In 1534 Pizarro's associates, Almagro and Benalcazar, fought with and overcame the Incas, and Quito became a Spanish city.

The aspect of Quito is picturesque, the white walls and red-tiled roofs of the houses standing out well against the mountain background. The houses are of the old Spanish or Moorish type, with barred windows and inner courts, built of the customary adobe or sun-dried bricks, plastered, whitened and often decorated outside with vivid colours. This use of colour upon the walls of Spanish-American houses gives a pleasing aspect even to the poorest village.

The shops of Quito are generally well stocked with wearing apparel and other articles from London, Paris, New York and other parts of the world. We shall immediately remark the sharp division of the classes here, the humble Indian clad in his cotton garments and *poncho*, almost rubbing shoulders with the correctly attired upper class in their European clothes. Officers in uniform, men in silk hats and frock-coats, priests, traders, and Indians almost without garments are all encountered among the crowd, whilst mules and llamas throng the streets.

The population of Quito is about 80,000. The climate is often very cold, but pleasing in certain seasons. The death-rate is high, due to lack of sanitary appliances and methods, which here is very marked. The chief streets, however, are wide and paved, and the situation is naturally healthy. Quito has always been a stronghold of Roman Catholicism, and quarrels between the clerical and reform elements have often been prolonged and bloody.

The eastern part of Ecuador, as before remarked, extends to the upper part of the Amazon basin, and descending the slopes of the Andes from beyond Quito we reach the head of navigation of the great Napo river, one of the principal north-western affluents of the Amazon. The Napo rises amid the snows of Cotopaxi and other Andine volcanoes, traversing a wild region, to where its waters are navigable.

This portion of Ecuador, the Oriente, consists of broken and precipitous valleys, wooded in their lower portions, and the path we take is that over which Gonzalo Pizarro and Orellana and their followers performed their terrible journey in the discovery of the Amazon, in 1540. The Spaniards had heard from the natives of a great El Dorado, which lay in that direction, but which really existed only in the Indians' imagination. The adventurers set out well equipped, with a large body of

HUITOTO INDIANS OF THE PUTUMAYO, PERU, VICTIMS OF THE
RUBBER TRADE, WITH NEGRO OVERSEER FROM BARBADOS

natives. But their provisions gave out as they reached the Napo river, and it was decided to construct a " brigantine " to follow their course down-stream. The vessel was built of wood hewed painfully in the forest, with nails made from the horses' shoes, but it held only a portion of the company and was captained by Orellana, who was instructed to explore and return. Borne down by the swift current, the vessel continued onwards, but finding no land of gold, Orellana and his men sailed down the main stream of the Amazon, emerging after a perilous voyage upon the coast of Brazil, the first white man to navigate the river.

The population of the republic, numbered at about 1,250,000, is composed of the few white people and the mixed race, with a great bulk of pure Indians. The Indians of the uplands and the coast are generally christianised, and their language is Spanish, although the native Quechua is largely used. The tribes of the Oriente, or eastern forest and river region of the Amazon Valley, are in most cases still savage.

The upper class, composed of the white and wealthier mestizos—between whom there is no barrier of colour in Ecuador—form a governing caste or oligarchy, and in their hands lies the entire wealth and education of the republic, and the greater part of the cultivable land. This class displays the customary qualities of the educated Latin American, and is courteous, sensitive and pretends to a high civilisation. The men, when wealthy, have often received a European or North American education and are generally well informed and strong advocates of " progress." The ladies are vivacious and pleasing, but kept back by the habitual social restriction of the Spanish race. The greatest drawbacks to advancement in Ecuador is the constant tendency towards revolution ; and the terrible political murders which so frequently take place in the country form a dark stain on its character.

The great bulk of the labouring class and Indians are kept back by lack of education and the system of practical serfdom or peonage which prevails, and by their extreme poverty. Only a small proportion of the population can read and write. Wages are excessively low, and food and housing of the most primitive nature. The chief industry is agricultural, and vast haciendas or estates belong to single owners, and under these feudal conditions the labourers live much at the mercy of their employers. The rich cocoa estates of the coast are in the hands often of absentee proprietors, who, drawing income from their possessions, love to pass their days in Paris. There are not wanting, however, some elements of progress, and labour in the towns is learning to organise in its own protection.

The principal products of Ecuador are cocoa, cotton, sugar, rice, maize, yucca, indigo, rubber, potatoes and a large variety of fruits, whilst valuable timber is exported from the coastal forests. In the uplands wheat, maize and other cereals are grown. Above 10,500 feet wheat does not ripen, and below 4500 feet it does not form into the ear. There are no forests on the high slopes and plateaux. The fertility of the soil in the lowlands is very great, and with more plentiful labour and better conditions governing it unlimited quantities of food-stuffs and raw material could be produced.

Of forestal products the ivory-nut or *tagua* is exceedingly important, being exported for button-making and other uses. Ecuador produces the finest " Panama " hats in the world, made from the peculiar *toquilla* grass, growing upon the coastal plains, and the industry has reached considerable proportions. The gathering of the grass and the making of the hats requires special knowledge and skill.

The native Ecuadorians are deft and clever in their handicrafts, and they produce their own blankets and other woven textiles, upon hand-looms. They still preserve, to some degree, the exquisite taste and skill in colour and material in textile fabrics of the old Incas, but these are now giving place to cheap and gaudy imported wares.

CHAPTER XXXVI

PERU

THE Peru of the ancients, the land along whose coast-line we continue our journey southwards from Panama and Guayaquil, was a country with a civilisation and organised social system, ruled by the famous potentates of the Inca dynasty, an empire where gold was used for household utensils, and where beautiful stone-built palaces and temples looked down from the high region of the Andes. In this strange empire none lived in poverty and men were made by law to love and regard their neighbour and share with him the common heritage of the land's natural resources. Hand industries were specialised and intensified; agriculture was regarded as a noble occupation, and the season for cultivation was opened by the turning of a furrow with a golden plough in the hand of the emperor. Such, the chroniclers tell us, was the life that flourished up to four centuries ago beyond that grey and distant wall of the Andine Cordilleras, which rises faintly upon our eastern horizon, when the Spanish conquistadores broke suddenly in upon it.

The Peru of to-day cannot be said to have so beneficent a social system, and that which was planted by the Spaniards and taken over by the republic, although inheriting much that is useful and fine, is still awaiting an energetic and prosperous development.

Peru is a land of many natural resources and covers a very large territory of 440,000 square miles—or half as much again if disputed territories are included; extending from the Pacific seaboard far across the double or treble chain of the Andes into the dense and almost unknown Amazon forests. The coast is over 1400 miles long, and on the south lies the great nitrate deserts of Tarapacá, part of Northern Chile, but formerly belonging to Peru, before the terrible struggle in which the former country triumphed.

The seaboard of Peru, the coastal plain, extends back towards

397

the Andes for nearly 100 miles and is crossed at wide intervals by rivers of small size, not navigable, but of great utility for purposes of irrigation of the fertile soil of their valleys, where large and valuable crops of cotton and sugar-cane are raised. The seaports, with the exception of Chimbote, Payta and Callao, are open roadsteads upon a surf-beaten shore. Vast stretches of sandy or stony desert, or semi-desert, intervene between the seaports. But upon the littoral some of the principal towns exist, notably Lima, the capital, and Trujillo, with Arequipa farther inland.

Upon the Andine uplands are various important towns, where ordinary life and occupations are carried on, situated at from 10,000 to 13,000 feet or more above sea-level. Some of these upland towns are the most important centres of population, but whilst in some cases they are reached by rail, in general they are difficult of access, being served only by the execrable mule-roads which traverse the Cordillera, over passes which at times lie near or above the perpetual snow-line.

Callao, in whose harbour the steamer drops anchor, is a well-known seaport, the haven of vessels from many lands, and its position, central to the enormous Pacific seaboard of South America, is of great importance. Callao is the starting-point of the famous Oroya railway, which crosses the coastal plains and ascends the Andes, reaching an elevation of over 15,000 feet, carrying the traveller within a few hours from the tropic lowlands to the perpetual snows, and serving rich mining districts.

Far more pleasing than its seaport of Callao is Lima, the beautiful capital of Peru, distant from the coast less than ten miles. The "City of the Kings," as its founder, Francisco Pizarro—whose body lies interred in the cathedral—named Lima, is surrounded by irrigated plantations, forming, when seen from afar, a green oasis amidst the brown land, its towers rising from the flat-topped buildings under the frequently blue sky, with the faint range of the Andes on the distant horizon.

Lima, with its broad plaza and rectangular street plan, is a typical Spanish American capital. The fine old cathedral, the quaint architecture of many of the older houses and public buildings, with their carved balconies, barred windows and interior *patios*, or courtyards, serve to retain for Lima a semi-mediæval aspect, heightened by the numerous and handsome

RUINED INCA FORTRESS OF SACSAIHUAMAN, AT CUZCO, PERU, IN THE HIGH ANDES, WITH A GROUP OF LLAMAS. THE WALLS ARE OF LARGE MONOLITHS

old churches, whose towers arise in every quarter of the city, attesting the strength of the Roman Catholic cult which dominates its inhabitants. It is not a particularly wealthy city, for Peru squandered its wealth, derived largely from guano, and lost the incalculably valuable nitrate fields in the war with Chile. But Lima is a centre of some considerable literary, social and scientific culture. The population numbers about 150,000, among which are many foreigners, principally engaged in trade or mining.

We enter the Peruvian interior by rail, if our road happens to be in those districts so served, or on horse or mule back away from such. The Oroya or central railway, with its branches, embodies about 300 miles of line, and is one of the most daring pieces of engineering work of its nature in the world. It traverses the fertile Rimac Valley and ascends the steeper mountain slopes by a series of switchbacks, and reaches Oroya, with branches to Cerro de Pasco and Huancayo. In travelling upon this line the effect of the high elevation reached frequently induces an attack of *soroche*, or mountain sickness, among the passengers.

The other principal railway system of Peru is the southern line, which, starting from Mollendo—one of the worst landing-places in the world perhaps—ascends through terrible volcanic wastes to the town of Arequipa, a pleasing and healthy city standing nearly 8000 feet above sea-level. The line then crosses the Western Cordillera, at a height of 14,660 feet, and descends to Lake Titicaca, upon which steamer communication is maintained with Bolivia. To the west it runs to Cuzco, the old Inca capital. The railway, with its branches, is about 450 miles long, and it is a very notable engineering work. These railway systems are under the control of a British company.[1]

We encounter in Peru three main zones of territory and climate ; the coast, which is temperate and relatively healthy, due largely to the peculiar conditions brought about by the absence of rain, which never falls on the Peruvian littoral ; the mountainous region, consisting of vast plateaux, upland lakes and rugged ranges and peaks, culminating above the perpetual snow-line and subject often to a cold and inclement climate, and, third, the hot forestal foothills and plains lying in the

[1] The Peruvian Corporation.

basin of the Upper Amazon, much of it an exceedingly wild and savage territory, and at present very difficult of access.

The traveller who chooses to overcome the difficulties incident to journeying in these remote places will find some measure of compensation in the magnificent scenery of the snow-clad Cordilleras and the profound gorges which the rivers have carved out in descending therefrom. Many of the high Peruvian peaks are almost unknown. Some of them, as Coropuna, the Huascaran [1] and others rise to over 22,000 feet, and the White Cordillera, in the department of Ancash, and that upon the borders of Bolivia display extensive snow-fields and are exceedingly striking and beautiful.

All the great rivers of Peru flow eastwardly from the Andes to swell the Amazon, the principal being the Marañon, the Huallaga, the Ucayali and the Madre de Dios, all coming from the south, with the Putumayo and others from the north. The possession of the latter river is, however, in dispute with Colombia and Ecuador. These rivers, throughout the lower part of their courses, form navigable waterways of great value and importance, penetrating vast regions of forest and plain full of undeveloped resources, but almost without inhabitants beyond the savage forest tribes, or the rubber merchants who exploit them.

This eastern portion of Peru may be entered more readily from Brazil, by ascending the Amazon in the ocean steamers which reach Iquitos, the Peruvian river port, lying over 2500 miles from the Atlantic. In its great navigable rivers on this side Peru possesses what is almost equivalent to an Atlantic seaboard.

An interesting hydrographic feature of the Peruvian and Bolivian uplands is Lake Titicaca, shared by both republics. Upon this great upland sea, the most remarkable in certain aspects of any lake in the world, we lose sight of land as the steamer crosses it. The lake lies nearly 12,500 feet above sea-level, and is 165 miles long. At its southern end it is connected by the Desaguadero river with the smaller Lake Poopo, the whole forming a separate system with no outlet either to the Pacific or the Amazon, its surplus waters being absorbed by evaporation.

[1] Partly ascended by the author.

QUECHUA INDIANS OF THE ANDINE UPLANDS OF PERU, AND PART OF THE ANCIENT INCA FORTIFICATIONS OF OLLANTAYTAMBO

To turn to the people who inhabit this varied land of Peru—
a population of about 3,500,000—we encounter those marked
divisions and qualities so characteristic of the Spanish American
communities. Peru was the seat of the Viceroy in colonial
times and preserves much in its culture and tradition proceeding from that period. The people of purely white race are few,
but there is no line of distinction between this class and the
people with whom is blended in greater or less degree the blood
of the brown or native Indian race. This mixed class, the
typical Peruvian people, form perhaps a third of the population, the remainder being made up by the *Cholos* or Quechua
people, long associated with the whites and merging indefinitely
into the mestizo class, and the savage Indians of the forests.

The educated Peruvians display in marked degree the
Latin American traits of courtesy, vivacity and political
turbulence. They generally follow one or other of the polite
professions—doctors of law, science or medicine, statesmen
and politicians, or are land proprietors or wealthy business
men. In their hands lie practically all the wealth, education
and power of the community, but they form a very small proportion of the whole. This wealthy class has in its possession
practically all the large areas of fertile land in Peru, excepting
those which lie in inaccessible regions, or those owned by
foreign corporations. Laying claim—and in some respects
justly—to a high civilisation, they nevertheless are constantly
subject to political outbreaks, accompanied by terrible scenes
of bloodshed and reprisal. The upper-class women of Peru
are famed for their beauty and vivacity.

The great bulk of the Peruvian nation is extremely poor and
backward, generally landless and without more resource than
that of labouring for the lowest wage, and living under the most
primitive conditions of life as regards food, clothing and domi-
cile. But a small proportion of the population can read and
write. However, this backward people contain many good
qualities, which so far have had little opportunity to manifest
themselves. As regards its natural resources, Peru is a rich
land, and might be said to contain a mine or a farm for every
family. But its social and economic life is much disorganised,
with consequent backwardness.

We shall recollect that under the Inca regimen the land of

2 C

Peru was all measured and apportioned to each village and family, and that everyone lived, if primitively, at least in sufficiency. We see to-day the remains of the intensive Inca cultivation of the soil in the thousands of small holdings or terraced farms and gardens with which the slopes of the Andes are covered in many districts. Here the people raised produce for their own sustenance and to pay the nation's tax, and carried on the often beautiful and valuable native industries of textile and other work.

This system was swept away under Spanish rule, and the Indians were treated by the colonists with such inhumanity that the ill fame of their rule has stained for ever the pages of Spanish occupation of the New World. However, something of the old regimen survives in the possession by the Cholos and Indians of the small holdings or *chacaras* in the upland districts, which the modern law preserves for them. These lands are in general situated under topographical and climatic conditions so difficult and inclement that the governing or monopolising class is not disposed to encroach therein, although land is frequently and unjustly taken from these poor but not unworthy sons of the soil. There is fortunately a growing party in Peru which is desirous of fostering the Indians.

Let us draw rein a moment upon the mountain pass. Around us on every hand stretch the deep ravines and steep slopes of the treeless Andes, and from the distant snow-field torrents rush downwards among the contorted rock strata. But we remark the evidences of older human occupation in this desolate region, in the innumerable terraces, often half obliterated, which cover the lower slopes, extending in chequered or step-like form from the plain up to the very clouds which sweep along the mighty crests. These small terraced farms are known as *andenes*. In the drier districts irrigation channels, cunningly contrived along the broken ground for miles, brought water from the streams, and gave or still give life to the maize crop, or to the potatoes and alfalfa.

Set in the midst of such scenes, our gaze rests perhaps upon some ruined temple, palace or fortress of the bygone Incas, who so beneficently ruled this vast land of old Peru. It is remarkable that a people should have spent such gigantic labour in constructing edifices of cut stone in places so elevated and

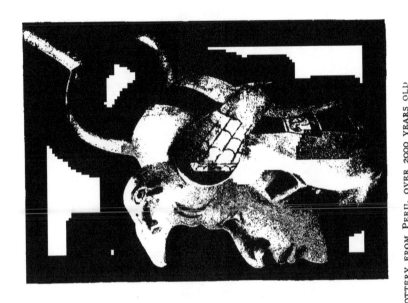

EARLY TROPICAL CERAMIC ART: PRE-INCA POTTERY FROM PERU, OVER 2000 YEARS OLD

inclement that corn will not ripen there, and where there is no timber, and often no water. Some of the famous ruins of the Incas are over 13,000 feet above sea-level, and we find their ancient dwellings crowning the summits of mountains up to 16,000 feet or more.

The stone buildings of the pre-Spanish era of Peru and Bolivia are numerous, and a detailed description of them would require many chapters. Some of them are of the Inca period, and their age may be a thousand years or even less. Others are of an earlier time, either of the Aymarás or of a race who can only be termed the "Andine People," and their age may be many thousands of years. The student of Peruvian archæology soon distinguishes the two types. The remarkable ruins of Tiahuanako, with its beautifully sculptured monolithic doorway, standing upon the Titicaca uplands, in Bolivia, are of the earlier time, and there are other examples throughout Peru, among them the stone of Chavin. The massive fortresses at Cuzco—the "Mecca" of the early Peruvians—and Ollantaytambo, the observatory where the solstices were determined at Intihuatana and Pisac, and the ruins of Huanuco Viejo are examples of Inca work.[1]

These buildings are characterised by the singular form of the stone masonry, with very fine joints, often mortarless, and often irregular-sided blocks, beautifully fitted. The arch was unknown, as was the use of the column, and there was no knowledge of iron; tools of copper and hard stone being used to cut and carve the blocks. Pottery of beautiful manufacture, principally of the early period, is common in Peru, having been unearthed from the *huacas* or tombs, and exquisite textiles are recovered and preserved, as well as objects of gold and copper, in making which the people were extremely dexterous.

In fine, it is shown that industrial arts of a high order were practised by the early Peruvians, and it is lamentable that not only should these have been practically destroyed, but that, as regards the textile industry, the handicraft of the present-day Indians of the uplands, of fine texture and artistic native dyes, should be driven out, as is happening, by the cheap products of Britain and Germany.

[1] These structures are fully described in the author's book, *The Secret of the Pacific*. London, Unwin, third edition.

The ancient ruins and native life of the simple people of the Peruvian uplands will doubtless in the future form matters of considerable interest to the traveller, and the economist will also find material worthy of study here. The native life and status is inferior now to what it was in the time of these non-European rulers, and it must be long before the modern Peruvian or Bolivian governing class, with their peculiar character, learn truly to foster and improve the hardy and well-meaning brown people, who, the true sons of the soil, are the only ones who can carry out manual labour in their particular environment of the uplands, by reason of the rarefied air. The country's greatest asset, its native population, is politically and economically neglected or oppressed, the same unfortunate story which we find throughout the greater part of tropical Latin America. It is true that schools are maintained to educate the people, but the spread of knowledge is slow.

As regards the forest Indians of Peru, these stand in a category apart. They have been hunted and enslaved by those who exploit the natural forest wealth of rubber and other matters, and the law does not even interfere in their enslavement and spoliation. It is true that the Amazon forest region of Peru is remote from the civilised part of the country, with the mighty wall of the Andes intervening, and in this circumstance the infamous occurrences of the Putumayo have sought some excuse. It is difficult to police such vast, wild areas, and the republics who own or dispute territory in such are too poor to spend much of the national resources upon them, and their officials often too corrupt to see that the law is carried out.

The forest tribes in many cases are industrious, and a valuable asset in their place. Some may be truculent, or may even attack and murder the white man on occasions, possibly from their native savagery but often as a matter of reprisal. The system of debt-bondage is widely followed in parts of Peru, with the purpose of forcing the Indians to work. A common method is that by which they are given a liberal supply of liquor at the beginning of their work, in advance of wages, a debt which is so manipulated by their masters that it is never worked off.

Protestant missionaries cannot easily reach these people, or any of the natives of South America, as the State religion is

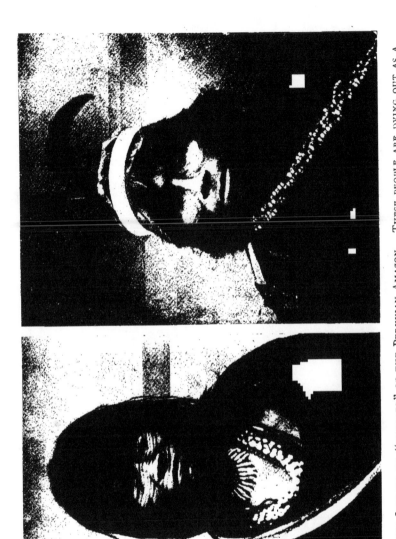

Types of Forest Indians or "savages" of the Peruvian Amazon. These people are dying out as a result of the white man's treatment.

Roman Catholic, and the jealousy and antagonism of the priests is incurred by proselytising. Every village in Peru has its local *cura* or priest—except in the savage forests—and this functionary, whilst doubtless acting as a restraining force and doing good in his way, is himself often a despoiler of the Indian, and under his regimen advance cannot be made.

In the remoter parts of Peru and elsewhere here the Indian and even the mestizo villages are practically self-supporting and cut off from the outside world. In some cases the people will not even sell food to the traveller. "We produce only enough for ourselves," they will say, and the traveller's money is of no value to them. Were it not that a sense of hospitality sometimes prevails, the traveller might almost starve.

The principal industries of Peru are agricultural. In the hot lowlands of the coast cotton and sugar, with maize, alfalfa and so forth are the main products. There is no forestal wealth or rubber here, as the whole coastal belt is treeless. In the uplands maize and, at elevations where this will not grow, potatoes are largely raised, forming the food of the people, with chillies, beans and other matters. In the Montana, as the Amazon forest district is termed, true tropical products are found: rubber, quinine, *coca*—the leaf with the sustaining properties so largely masticated by the Indians—coffee, cocoa and a wide variety of fruits.

This last-named region is of amazing fertility, and whenever a closer development of this part of South America shall be brought about it will become a veritable storehouse of food and raw material. But cultivation will only follow on the advent of roads, railways and a more enlightened administration. At present the deep, uninhabited forests and sombre rivers soon pall upon the mind of the traveller who has ventured there.

The pastoral industries of Peru are important. Whenever we journey upon the high uplands we remark the herds of graceful llamas and alpacas; and sheep, goats and cattle are moderately plentiful. The llama is a valuable possession of the native, performing the carrying trade away from the railways often, and yielding its wool for his clothes. Peru and Bolivia are the greatest producer of alpaca wool, which finds its principal outlet at Arequipa.

Mining in Peru has received some considerable impetus of late years, in the discovery and working of the huge copper deposits of Cerro de Pasco, where an American company has installed an extensive modern plant for treating the ores. Peru is one of the richest mineral-bearing countries in the world, every known metal being found, with the exception of tin, which is the peculiar product of Bolivia.

Gold, silver, copper, lead, zinc, quicksilver, iron and the rarer metals abound. There are deposits of coal so enormous as to excite surprise that they have never been made the subject of modern enterprise. Outcropping seams of splendid anthracite cross the country for miles. Peru has always been a fabled land of gold and silver, and in the white metal many fortunes have been made. Gold-mining, however, seems not often to have been attended with success. The old quicksilver mines of Huancavelica, among the largest in the world, were long worked as a possession of the Spanish Crown. In brief, the mining adventurer in Peru encounters a field of such extent and attraction as exists in few parts of the world, and the innumerable old workings in the mountains serve to show that the native for many centuries has known of the metalliferous treasures of the soil.

A famous natural product of Peru is the guano, a fertiliser from whose export great wealth has been derived. It is produced by the myriads of sea-birds which inhabit the lonely headlands and islands, but the supply has now become greatly reduced. Much of the wealth was squandered, and political and financial corruption early attended its disposal. Indeed it has been asserted that the Peruvians at one period in the life of the republic were spoilt by this easily gotten wealth.

The Chincha Islands, where some of the principal guano deposits lie, were marked by a barbarous system of exploitation under coolie labour. After the abolition of slavery Peruvian agents in China—from 1847 to 1854—dispatched shiploads of coolies to work the guano deposits and sugar plantations of Peru, and terrible treatment was meted out to them. In the guano pits the unfortunate Asiatics were driven to their toil under cow-hide lashes, and out of 4000, who were fraudulently transmitted to Peru, not one survived, the greater number committing suicide in order to escape from their sufferings.

Almost similarly brutal treatment was dealt out to the sugar plantation coolies.

A recrudescence of this treatment appeared in the infamous Putumayo rubber working, under Peruvian and Bolivian task-masters, in connection with which it is regrettable to record that the negligence of British directors was a contributing agent. We remark, therefore, that the winning of Peruvian treasure, from gold, silver and quicksilver, under the terrible slavery of the viceroys, to the guano, sugar and rubber of more modern times, has been attended with the most ruthless treatment of the unfortunate tropical labourers.

With all its defects of governance, Peru is a country of vast resources, which so far are only partially exploited, and its people contain elements of greatness which are capable of raising the nation to a much higher plane. This can only be accomplished when the republic puts into practice the principles of a true commonwealth, principles the machinery of which already exist to a large extent. Much might be expected, under more enlightened economic and administrative conditions, of this broad and interesting country. The brown race of Peruvians, like those of the Andine republics generally, are in reality intelligent, and call for a much better system of education than that which prevails at present.

CHAPTER XXXVII

BOLIVIA AND CHILE

THE republic of Bolivia lies remote from the sea, having lost its littoral provinces in the war with Chile in the middle of last century. It is largely a mountainous country, lying amid the ranges of the Andes, sharing with its neighbour, Peru, some of the highest inhabited places upon the globe, in the towns of those inclement uplands where the unaccustomed traveller with difficulty draws in the attenuated atmosphere.

The republic, however, is not entirely mountainous, as it extends eastward to the low central plains of South America, watered by the River Plate, as well as into the dense forest area of the Amazon. Bolivia covers an area of over 500,000 square miles, and its population numbers about 2,000,000. With the exception of a small white and mestizo class, who control the destiny and resources of the community, this population is formed of Indians.

We reach the principal city of Bolivia—La Paz—by rail, either from the Chilean seaports of Antofagasta, or that of Arica, or by rail from Mollendo and thence by lake steamer across Titicaca. The line from Arica—which lies in the former Peruvian province of that name, still held by Chile against the Peruvian claims—was recently built at the cost of the Chilean Government by British contractors, at an outlay of nearly £3,000,000. Here, in approaching Bolivia, we traverse an interesting region, ascending the steep western versant of the Andes, reaching nearly 14,000 feet above sea-level. The high, snow-clad peaks of the Andes arise as the train proceeds on its way, and some of them appear on either hand. The Desaguadero river is crossed, with Indian villages and ancient Inca remains interspersed through the bare landscape, and large herds of llamas and alpacas are seen.

An important copper-mining district is passed at Corocoro, and, beyond, the railway comes suddenly to the edge of an

NATIVE WOMEN OF TIAHUANAKO, BOLIVIA, IN THE HIGH ANDES

enormous chasm or valley, at the bottom of which, spread out as on a relief map, lies the city of La Paz. Beyond arises the snowy range of the White Cordillera of the Andes, extending for 150 miles between the giant peaks of Illimani and Sorata, a magnificent sight. This remarkable view is also obtained in crossing Lake Titicaca.

La Paz is not the political capital of Bolivia, that distinction having been assigned to Sucre, also a high upland town. La Paz stands at 11,000 feet above sea-level, and possesses a bracing and often cold climate. It is noteworthy that the people of these high towns—and we remark the circumstance equally in Mexico and elsewhere—do not heat their houses with stoves nor use fires except for cooking. Of the 75,000 inhabitants of La Paz, about half are Indians, with whose primitive and picturesque costume—the sandal and the coloured poncho are worn—the European dress of the upper classes contrasts strongly. Ladies in Parisian attire and black-coated professional men, lawyers, doctors, business men, politicians, landowners, wealthy miners and so forth form this class, as in Peru, Ecuador, Mexico or elsewhere.

This class is generally formed of the mestizos, with a sprinkling of white families, but the invigorating mountain climate of the *puna* or uplands of Bolivia gives a reddish tinge to the complexion, which the more pallid people of Lima and other coastal towns lack. This hue is not considered fashionable, however, among Latin American ladies, who seek to enhance their undoubted charms by the free use of face powders.

There are a number of fine buildings in the steep streets of La Paz, and in the national museum are fortunately preserved interesting archæological remains of the ancient culture of the Incas. Unfortunately great quantities of sculptured stones were taken from the Tiahuanako ruins by the European builders of the Bolivian railways to form bridges and warehouses. The houses of the poorer classes are but earthen or mud hovels. The cost of living here, as in all large Bolivian and Peruvian towns, is, in general, excessively high.

We find in Bolivia the same backward, propertyless, ill-paid and ignorant, but well-meaning native people of the uplands, such as are encountered in the other Andine republics. Education is extremely backward, only a small proportion of the population being able to read and write. Spanish is the official language, but the Indians prefer their native Quechua, Aymara or Guarani. As in Peru, public worship in any but the Roman Catholic faith is prohibited, but Protestantism is tolerated among foreigners. Roman Catholic mission work among the Indians is subsidised by the State. In these lands the Indian is of a generally melancholy disposition, or rather it should be said that his attitude seems to display melancholy. At nightfall he often ascends to some lonely hill-top, and there makes solitary and mournful music with his flute or pan-pipes, reflecting possibly on his fallen state.

Manufacture is very backward in Bolivia. However, the natives still weave most of their own garments, as cottage industries, employing primitive spindles and looms, as in early times. The Andine native woman is seen twirling her hand-spindle even when minding her sheep on the lonely hills. Very fine fabrics, as in Peru, are woven from vicuña, llama, alpaca and sheep wool, the women being especially clever in this work. Ponchos are sometimes woven so finely as to be water-proof. The beautiful native dyes which the Cholos of the Andine uplands used are unfortunately being displaced by the cheap German article, and the importation of foreign fabrics is already ousting the excellent home and cottage manufacture.

As in Peru, the llama is the most valued possession of the Indian. It carries merchandise, mineral ores and other matters, marching along at four miles an hour and feeding itself by the wayside as it proceeds. The llama is the only hoofed quadruped or beast of burden indigenous to South America, and prior to the advent of the Spaniards, who brought in the horse and the ox, was the only friend of man in the animal world upon these bleak upland steppes of the Andes. The vicuña, its wild cousin, and the huanaco, another of the sheep-camel family, cannot be tamed, but is found in great bands, traversing the highest and bleakest regions.

Although agriculture and pastoral industries form the main-stay of the Bolivian people, mining is extremely important in tin, copper and silver, and some gold is also produced. Bolivia is one of the leading tin-producing countries in the world, large quantities of the metal being exported. Among the chief centres of production of this metal is the picturesque mountain town of Oruro. There are also important petroleum deposits, but these so far are little worked.

Silver-mining at Potosi—upon what was literally a "mountain of silver," discovered in 1545—has formed one of the romances of the vice-regal days, and here Spanish hidalgos and other adventurers won or squandered huge fortunes, adventurers who had flocked to this high, bleak place from all parts of the world. We remark the picturesque costumes of the Indians in these places, and the singular customs of the wealthier Cholo women, who load themselves with petticoats as evidence of their possessions.

Leaving the Andine uplands of Bolivia or Peru, and surmounting the snowy range of the Andes—where probably we suffer the pangs of the *soroche* or *puna*, as mountain sickness here is termed—we descend towards the Amazon Valley. Here, by contrast, the climate is exceedingly agreeable. In both these countries, within a day's ride, we may travel from the coldest and bleakest districts, where vegetation is almost absent, to valleys where the temperature is that of a perpetual spring, and where oranges and grapes hang in clusters. Here the Indian has discarded his thick, home-made woollen "tweeds" for a thin, native calico, and, indeed, when we penetrate sufficiently far into the forest wilds we find Indian tribes absolutely naked.

Of late years the Bolivian rubber-bearing region upon the Acre, Madre de Dios and other rivers of the Amazon system have acquired great value, and numerous rubber stations have been established in what were unexplored wilds. These places are reached by mule-roads, or, where the jungle is dense, by forest trails, known as *trochas*, cut out with *machetes*. All these means of communication are extremely difficult.

In these tropical forests the trees are often loaded with lianas and flowers, whose petals, detached by the wind, drop like falling snow to the ground, and myriads of splendidly coloured birds and numerous varieties of monkeys have their home here. Streams run—swift or sluggish—on every hand, often over gold-bearing sands. Elsewhere the forest aisles are dim and mysterious, and along the navigable rivers the solitude and the monotony of the green walls of foliage, sometimes without any animal life whatsoever, become depressing at times, and the traveller's mind reverts to the vast, open spaces and glorious horizons and sunsets of the cold Andes, which he has left far behind, as in another world. The forest, its stillness broken perhaps only by the cry of the jaguar, now seems more lonely than the most remote Andine mountain, even if there the only living creature to be seen was the solitary condor or the vicuña.

The most interesting region, perhaps, is that where the mountains begin to give place to the forests, for in such is displayed a wider range of resources, scenery and inhabitants, from the cedar to the banana, from the llama and his shepherd to the gold-digger and the rubber-collector, or the planter of cocoa and sugar-cane, and distant valleys and mountains are not obliterated in the sea of leafage which the lower forests display.

The heavy and frequent rainfall of this vast belt of territory, extending throughout Bolivia, Peru, Ecuador and Colombia, is due to the east winds, which, laden with moisture from the Atlantic, sweep across Brazil and are intercepted by the Andes. Severe floods occur, detritus and decomposed vegetation are brought down, and as a result malarial fevers are prevalent, the most severe scourge of these regions. It is, however, possible in large measure to guard against malaria if the traveller exercises a wise and intelligent mode of life. A further scourge of the region is the mosquito. But if the days are hot the nights are often cold. Blankets, mosquito nets, quinine, good food and temperate living are matters constantly to be borne in mind in the Amazon Valley, if the traveller would journey in comfort.

The exploitation of rubber is one of the principal resources of Bolivia, and is almost as important as the mineral trade, although it is adversely affected at times by very low prices. But the economic conditions of its gathering are unsatisfactory in the extreme, and grave charges are laid against the state of peonage and semi-slavery maintained by the white masters of the natives. Debt-bondage is a common feature, drink or goods being advanced to the Indian by his " employer " with the purpose of creating an obligation, for under Bolivian law the Indian must work this off before he is free. The debt is, however, so manipulated that it increases instead of decreasing. The following is the process which largely obtains. The *serin-*

TYPES OF FOREST INDIANS OR "SAVAGES" OF THE PERUVIAN AMAZON. THESE USEFUL TRIBES ARE BEING EXTERMINATED BY COMMERCIALISTIC METHODS

guero, as the poor native rubber gatherer is termed, after spending his time often in danger and hardship in tapping the rubber forests, is " paid " for his rubber, a price greatly under market value, by the owner of the forest. This " payment " is made by crediting the gatherer's account with the sum and permitting him to draw out goods, such as farinha, dried meal, matches, sugar, clothing and so forth, which articles are charged at so exorbitant a price that the unfortunate worker rarely has a balance in his favour. This system of imposture is common throughout the whole of Latin America. It is often necessary perhaps to pay in goods, but cheating accompanies the payment. The advent of civilisation must be looked for with the building of railways and the growth of wider industries.

The social and economic life of Bolivia in general awaits a more vigorous and equitable national spirit, such as will strive to uplift the humble and ignorant classes which form the bulk of the population of the republic.

The republic of Chile occupies but a small part of our attention here, as only the northern portion of this more enterprising country lies within the tropics, north of the Capricorn line. The fertile provinces of Tacna and Arica, with the pleasing little seaport of Arica, famous for the memory of the terrible struggle between the forces of Peru and Chile for its possession, give way as we proceed southward to the sterile deserts of Tarapacá, with their enormously valuable nitrate-fields.

The portion of the Chilean coast lying in the tropics extends from the Peruvian boundary slightly north of Arica to Antofagasta, which seaport stands almost upon the tropic of Capricorn, a seaboard about 400 miles long. From Antofagasta a railway—the highest in the world—runs to La Paz in Bolivia, traversing a wild and interesting country. To the east of the nitrate-fields stretches the dreadful desert of Tamarugal, wherein no living thing exists, and beyond are seen the snow-capped peaks of the Bolivian Andes. The wild interior region has, however, its own peculiar beauty of sun-beat plain, rocky ravine and snow-clad volcano. Among the objects of natural interest is the singular borax lake of Cebollar. Iquique, the principal nitrate port, is a town pleasing in some respects, built mainly of wood, and inland extends the deserts of Tamarugal, in parts forming a veritable Sahara. The British element here is an active one, occupied mainly in the nitrate trade. The nitrate-fields lie upon the *pampa*, forming the western part of the desert, and are reached by railway from Iquique and other ports.

The nitrate industry and its environment, and the Chilean *rotos* or miners who carry it on under the supervision of their British masters, furnish elements of considerable interest to the traveller. The industry is unique among the world's mining enterprises. The natural beds of saltpetre extend over a very large area, at several thousand feet above sea-level, and so far are not nearly exhausted, although considerable inroads have been made upon them; and it is estimated that in twenty years the supply will begin to fail, at least at the present rate of production of 2,000,000 tons yearly.

Surveying these deposits, before work has been carried out upon them, we remark no indication of the wealth beneath the surface. A dry, bare, rainless desert, strewn with fragments of igneous rock, stretches to the horizon, broken by arid and rocky hills, under a steely blue sky and beating sun, whilst in the far distance arises the faint grey range of the Andes, upon the Bolivian border. Excavating a few feet below this unpromising surface, the Chilean miner lays bare a solid bed of white, glistening and compact salt, which is blasted out and treated in the *oficinas*.

These are extensive works where the mineral is boiled in huge vats and the resulting liquor evaporated in open pans, the pure nitrate crystallising in the sun like heaps of snow or salt. A valuable by-product of the operation is iodine. The *oficinas* employ a large number of hands and have European staffs. The managers of these colonies—for such in effect they are, surrounded by the desert—maintain British traditions, and evening dress for dinner is generally regarded as the rule. These establishments are interspersed over a wide tract of country and many millions of British—and some German— money are invested therein.

The Chilean Government draws a large part of the revenue from the duties upon the nitrate export. The charge has been made, however, that the money at times has been too freely diverted to merely political issues, with a consequent insufficiency of expenditure on public works or the educational system. The upper class Chileans are, however, energetic and enterprising, among the most vigorous element of Latin America. The non-tropical part of the country is fruitful, and often beautiful, with various handsome cities : the home of a progressive people. The Chilean navy, modelled after the British, is a source of pride to the republic.

The Chilean *rotos* or nitrate workers are a picturesque, hard-working and fearless race, admirable in their way, but given to turbulence, hard drinking and bloody quarrels among themselves, whilst the periodical strikes are matters of grave anxiety to the handful of European officials who control them. These are due to grievances, real at times, although not always. There is also in vogue in the nitrate *oficinas* the system, although in a milder form, of debt-bondage, involving the part payment in kind or commodities, which is not always carried out in strict honesty. These native workers, in large part, are descendants of the fierce Araucanian Indians of Chile, who in early times resisted the advance of the Inca Empire, and, later, kept off the Spaniards to the last.

It is recorded of an early Araucanian chief who had captured one of the conquistadores whose party had come in their eagerness to search for gold, that he spoke as follows :—" Gold is plentiful here ; you shall have your fill," in the act causing the unfortunate Spaniard's mouth to be stuffed with the yellow metal—boiling hot and molten.

Here we leave the Pacific coast states for the Atlantic side of the continent : the Guianas and Brazil. No railway traverses the great width of the South American tropics, and the adventurous traveller, journeying from Peru or Bolivia to Brazil, must entrust himself to the rugged mule-track of the Andes and the frail canoe of the Amazon for a considerable part of his journey.

CHAPTER XXXVIII

GUIANA—BRITISH, DUTCH, FRENCH

ALONG the north-eastern coast of South America, from Venezuela to Brazil, stretches a land of considerable interest, known geographically as Guiana, and embodying the British, Dutch and French colonies of that name, the only European possessions on the mainland of this continent. On the one hand lie the tributaries of the Orinoco, on the other those of the Amazon, cut through by the Equator.

The Guiana coast was first sighted by Columbus, when he discovered Trinidad and the Orinoco; later by Amerigo Vespucci; and again by Pinzon, who, venturing into the region south of the Equator, discovered the mighty Amazon. Guiana was the land where the golden city of Manoa or El Dorado was thought to lie, and its rivers and forests were first explored by the adventurers who vainly sought those fabled places.

The three Guianas present somewhat similar topographical conditions. The low coastal plains, upon which by means of dykes and drains the numerous sugar plantations have been established, give place as we proceed inland to hills and plateaux, with districts rich in gold-bearing soil, covered with dense jungle and tall forest. The hinterland contains undulating savannas, open and grass-covered, or clothed with thick forest and broken by high mountains. Each colony is endowed with its individual river system, which affords access—often dangerous—into the interior.

British Guiana, frequently known familiarly as Demerara, covers somewhat over 90,000 square miles of the territory, and has a population of well over 300,000, of which less than 17,000 are Europeans. Over one-third are negroes, and there is a still larger proportion of East Indians or coolies, who number 125,000. The remainder is made up of mixed races and the aboriginal Indians.

Scenery of extraordinary grandeur opens to the view among

the rugged hills and dense forests of the interior of the colony, and some of the highest waterfalls in the world leap from the rocks of the Guiana highlands. The Essequibo river is navigable for some distance, and its huge mouth contains fertile islands, where sugar is, or has been, cultivated. It is upon a tributary of this river, far inland, that the remarkable Kaieteur Falls were discovered a few years ago, until then unknown— a witness to the unexplored character of the region.

The climate is not unhealthy for a tropical land, and the malarial fevers and other diseases largely result upon the habits of the natives. The luxuriant vegetation rivals that of any part of the tropics, and the vast forests are inexhaustible in their splendid timbers. The great trees, draped with the most beautiful orchids, struggle for life in the gloomy forest aisles, covered with a network of lianas, and the monotony of the forest walls along the river banks is broken by the giant palms. Rubber, balata, oils, gum, febrifuges and other forestal products tempt the prospector in their exploitation, and in some cases yield commercial profit. Bananas, guavas, avocado pears and all kinds of luscious tropical fruits grow or are cultivated, with maize, yams, cassava, sweet potatoes and many other food-stuffs.

British Guiana is famous for the profusion of the *Victoria regia* water-lily, inevitably pointed out to the traveller, in the lagoons and ditches, and along the canals paralleling the streets of the capital. Wild animals, birds and insects of infinite variety abound—the puma, the alligator, the manatee and baboons and monkeys. Alligators swarm in the canals and lagoons, as do fishes, crabs and shrimps, and iguanas among the rocks—all articles of food for the coolie and the negro.

Georgetown, the capital of the colony, is one of the most pleasing towns in northern South America. The streets, which are straight and wide, are laid out at right angles with each other, and in some cases canals covered with the beautiful water-lilies run down the centre, and banks of green turf, with shade trees, line the sides. The gardens of the private residences are ornamented with a profusion of flowers and foliage. The public institutions and buildings—Government offices, theatres, churches, law courts, hospitals, museum, scientific societies and botanical gardens combine to form a well-

2 D

established tropical town. Georgetown stands on the Demerara river, and was established by the Dutch in 1784, but fell under British control in 1812. In contrast with the pleasing aspect of portions of the town are the miserable slum dwellings of the poor, scattered about it. There are various local manufactures, and steamer and rail service with adjacent places is well established. The population is somewhat over 50,000.

The staple industry of British Guiana is sugar cultivation, and coffee and cocoa are also grown, and rice, which rapidly increases in area planted. The coastal and riverine lands are extremely fertile, and tropical fruits and vegetables of any kind can be grown. The wide cultivation of cotton is not now carried on. Indeed agriculture in general is conducted in a somewhat haphazard way, except on the larger estates, and serious problems have to be overcome in this connection.

The question of labour in the colony is a difficult one. During last century negroes and Indian and Chinese coolies were freely imported, and the system of indentured labour has been in vogue since 1840. Economic disaster was brought about as a result of slave emancipation, prior to that date, and afterwards negro riots, due in part to alleged injustice, were serious. In 1905 the strikes of the dock labourers and cane-cutters, who demanded higher pay, brought about grave disorders and bloodshed.

The lands privately owned call for a better system of cultivation, and the state lands, which can be cheaply obtained, require capital expenditure in improvements. Gold-mining has variously suffered serious reverses and attained to some success. Precious stones are found, and forestal products are exported.

The development of British Guiana offers much for the future, but so far is regarded as tardy.

Surinam, or Dutch Guiana, has an area of about 57,000 square miles, and the population is about 85,000, exclusive of the forest Indians. A large number of the inhabitants are negroes, Chinese, Javanese, Hindus and other coloured people, brought in originally or recently as labourers. Cocoa, coffee and sugar are grown on the larger plantations, generally upon the coast and rivers, but only about 45,000 acres are under cultivation throughout the country. Rice, maize and bananas are also produced, and there is some gold mining and export.

The capital of the colony, Paramaribo, is a clean and generally healthy town, with a population of about 35,000—a large proportion of the entire inhabitants of the country, whose custom tends towards town-dwelling. In appearance Paramaribo is not unworthy of a Netherlands town, with its spacious squares, solid buildings, canals and broad, tree-planted streets.

Guyane, or French Guiana, lies between the Dutch colony and Northern Brazil, and covers an area of about 51,000 square miles, with a population of about 30,000. The vegetation of the country is very rich. Palms, hardwood-trees, rubber, coffee, sugar, manioc, rice, vanilla, cloves, cinnamon, maize, arrowroot, yams, bananas, bread-fruit are among the food and other natural resources. Only the small area of about 9000 acres is, however, under cultivation. Large herds of cattle might flourish on the savannas, and timber-working offers a profitable source of industry. Gold-mining is the most important industry at present, but it has various difficulties to contend with. The colony is in part a penal settlement.

Cayenne, the capital, is a regularly built and cleanly town, with good streets. Its population is about 13,000. Cayenne has given its name to the well-known pepper, originally brought from the Guianas by the Spaniards.

CHAPTER XXXIX

BRAZIL, PARAGUAY, ETC.

THE enormous republic of Brazil is perhaps the most typical of all the torrid lands of South America, and it is the largest single country within the entire tropical zone. Here we enter a land where Nature displays her most varied and fruitful gifts, where slavery, empire and republic have flourished in their turn, and which, although backward and even retrograde in certain of its social attributes and incoherent in its constitution, nevertheless shelters some of the most progressive elements of the western hemisphere.

As we approach Rio de Janeiro, the chief seaport and capital of the republic, the beauty of the scene impresses itself strongly upon the mind. Over a brilliant, sunlit sea, studded with verdant islands, we behold the palm-fringed shores, from which arise mountains of the most striking form and hue, some of them draped in emerald vegetation to their summits, others remarkable from their nakedness. The harbour is one of the finest in the world. The steamer traverses a channel a mile wide, so much like the mouth of a river that the first navigators who entered it supposed it to be such and called it the river of Janeiro. The vast oval bay which thence unfolds could accommodate the fleets of the whole world.

The wealth of the city of Rio is largely modern, and reflects itself in the ornamentation of its public buildings and residences, which has run riot in fanciful Italian stucco, banishing the severe old colonial structures. Theatres and pavilions, squares, parks, gardens and avenues have been built or laid out with a lavish hand, and this modern splendour and luxury might almost seem to be a material expression of satisfaction whereby a grand city whose people dwell in health has been substituted for the ancient place where, less than a quarter of a century ago, they suffered the most terrible scourges of fever.

The population of Rio now reaches over 810,000, forming

420

RIO DE JANEIRO, ONE OF THE FINEST OF TROPICAL SEAPORTS AND CITIES

the second largest city in Latin America. Rio is the outlet for a large part of the most thickly populated and productive portion of the republic, coffee, sugar, diamonds, tobacco and other products all being poured forth from its rich interior. Sugar was the earliest source of Brazilian wealth, and coffee is now the greatest, being of far more value even than the rubber of its sombre Amazonian forests.

Brazil covers an area of 3,270,000 square miles, extending over the eastern and central part of South America, a giant whose boundaries touch almost every other republic on the continent, and which, at one time or another, it has squeezed out of portions of their territory. The republic is larger than the United States (excluding Alaska), with a length of 2600 miles and breadth of 2500 miles, and an enormous coast-line, or larger than the whole of Europe without Russia. It shelters a large population of over 20,000,000.

Sketched briefly, the surface of Brazil consists of two vast river basins, those of the Amazon and the Plate—which last has its outlet in Argentina—and the enormous area of rugged highlands which fills the eastern part. Narrow coastal plains, and those of Rio Grande do Sul and the Guiana highlands, complete the general topography.

Notwithstanding its great natural wealth and resources, Brazil is an incoherent and backward nation in many respects. In the large cities the cost of living is excessively high, whilst the great bulk of the population dwells in comparative poverty and ignorance, the natives, like those throughout the whole of Latin America, being in general poorly paid agricultural labourers, or collectors of forestal produce under onerous conditions.

Brazil differs from all other Latin American republics in the considerable admixture of a negro element in its population, the people of African descent, whose forbears were brought as slaves from the Guinea coast. Slavery was abolished in Brazil in 1888, and its abolition was partly the cause of the bloodless revolution which brought about the fall of the empire of Dom Pedro II., a monarch who had at heart the welfare of the nation, and under whose reign economic advancement was very marked.

The Brazilian is an admixture of the Portuguese, the Indian

and to some extent the negro. It was an early policy of Portuguese rule to encourage this union, and no prejudice of colour exists in the republic. However, the low-class negro population in the seaports and on the plantations is a difficult and degrading element of the Brazilian social economy.

The upper classes are well educated and of old-world ideas and refinement, a feudal aristocratic element having been the outcome of the long colonial rule and sixty years of monarchy. They are marked by an affinity for art, music, poetry and literature, but at the same time are active in business and the development of their factories and estates. Science, however, and the exploration of their vast country and the building of its railways are matters which have generally been left to foreigners.

Education among the bulk of the Brazilians is very backward, less than a quarter of the population being able to read and write. This condition is in marked contrast with the passion among the men of the upper class for taking university degrees, whereby they may enjoy the distinction of being addressed as " doctor," a predilection which is characteristic of many who have neither desire nor necessity to practise a profession. Of late years, however, the university and the doctorate, with cap, gown and ceremonial, have been abolished, as undemocratic, and professional men may be produced by any federal, state, or private school. This is a singular departure in Latin American educational affairs.

In considering the natural resources of Brazil, the republic is divisible into three regions—Northern, Central and Southern. The first is dominated by the great Amazon river and its forests, and has a hot, tropical climate and heavy rainfall, and the Amazon lowlands are covered with dense vegetation. Nearly all the Brazilian rubber comes from this region, growing wild along the riverine belts, and Brazil nuts, with cocoa and cotton, are other products. The region is thinly populated.

Central Brazil is more hilly and more thickly populated, with a cooler average temperature, and it yields nearly all the tobacco exported from the republic, and more than three-quarters of both the cotton and the cocoa which Brazil produces, as well as nearly three-quarters of the diamonds.

The southern division is territorially much smaller, but forms the most important and most densely populated part of the republic. Within it stands Rio de Janeiro, the capital, and the exceedingly important coffee-growing region of Sao Paulo.

The fertile, well-watered soil, cooler climate and progressive inhabitants of this part of Brazil are of great economic value to the republic. It consists partly of an elevated plateau, covered with the famous coffee plantations, and the great railway network of Brazil is largely situated within it, and is tributary to Rio de Janeiro and the great coffee-shipping port of Santos. This, once a veritable pest-hole of fever, but now comparatively healthy, due to improved sanitary methods, lies a short distance south of the capital. Between these two places Brazil is cut by the tropic of Capricorn.

The coastal plain is everywhere exceedingly hot, but the fertility of this part of Brazil, on the border of the torrid zone, is in marked contrast with the dreadful nitrate deserts of Tarapacá, in Chile, in the same latitude on the Pacific side of the continent.

The tropical vegetation of Brazil is marvellously rich, the forest life, the wealth of tree, flower, palm and climbing plant being no less remarkable or beautiful in their way than those of tropical Africa. The wild animal life, however, although varied and teeming, does not, of course, contain the mighty carnivora or pachyderms of the Dark Continent. The insect life is extraordinarily prolific. The vegetation around Rio is very beautiful, and far to the north, in the Amazon Valley, are the largest areas of virgin forest on the surface of the globe. Enormous portions of these forested plains are annually flooded, and reptile life is exceedingly abundant therein.

The Amazon, the largest river on the globe, is one of the principal topographical features of South America. It rises amid the snows of the Andes, more than 3000 miles from its mouth on the coast of Brazil, and traverses the densest and darkest of forest fastnesses. The long tributaries which flow through Peru, Bolivia, Ecuador and Colombia, in the west, traverse profound gorges, cut, in some cases, like mighty canals through the easternmost range of the Cordillera. The headwaters of these arteries are overlooked in many places by high rocky headlands, whereon stand the mysterious ruins of ancient

Inca and Aymara temples and fortresses, almost within the limit of the perpetual snows.

The Amazon is navigable for ocean steamers for 2500 miles through Brazil, into the heart of eastern Peru, Bolivia and Colombia, giving outlet to the native products of an enormous territory. The principal of these products is rubber, of whose production Manaos, a fine town 900 miles up-stream, is the centre. Manaos depends for its prosperity on the rubber trade, and during the decline in price of this commodity suffered a severe economic relapse. Its cathedral, theatres, houses and electric car lines were all built upon the proceeds of the " black gold," as rubber is there familiarly termed.

The city of Para also depends upon the Amazon traffic for its prosperity, and is an equally important place, lying about eighty miles from the mouth of the river. Para has given its name to the rubber of the Amazon valley, which, however, comes not only from Brazil, but from Peru, Bolivia and Colombia, having its outlet by the river.

Comparatively little planting of rubber has been carried out in the Amazon Valley, and the supply is dependent upon the " wild " rubber and native gatherers. Indeed the amount of cultivated land in this enormous region is almost negligible, for in an area equal to four-tenths of the whole continent which the valley covers less than 100 square miles are cultivated. Enormous quantities of cocoa and other tropical food-stuffs might be raised along the Amazon and its tributaries, under better organisation.

Brazil formerly produced all but a fraction of the world's supply of rubber, but the plantations of Malaysia have to a considerable extent superseded this. There are still enormous areas of rubber forests available, and the Government of Brazil has recently issued regulations with a view to an increase of production and to the betterment of the conditions under which rubber is obtained, a movement following to some extent upon the infamous Putumayo disclosures.

A drawback to the extended yield of rubber here is in the scarcity of labour. In the neighbourhood of Manaos the native rubber-gatherers have become specialists in the collection of the commodity, and do not readily take to other forms of labour, whilst farther afield, in Peru and Bolivia, the outrages committed upon the natives, the extermination of tribes due to raiding, brutality, alcohol, forced labour, the carrying off of their women and so forth, have greatly reduced the available labour supply, or caused the natives to retire into the more remote regions.

The inhumanity displayed towards the Indians of the Amazon valley is not merely a recent development. As early as 1537 the Pope, stirred by the cruelties inflicted by Portuguese and Spaniard upon the natives of the Amazon and the Plate, declared "that they were men like others," and slavery was sanctioned as a lesser evil and a means of avoiding the horrors carried out by the colonists. At present, in the Amazon Valley, throughout Brazil, Peru and other republics which possess territory therein, the natives have no legal rights, and they have been hunted and enslaved like animals. It is possible that under new regulations some improvement in the condition of these unhappy but indispensable children of the rivers and the forests may be wrought, but such is likely to be slow, in view of the character of the Latin American people, by whom the forest Indian has been regarded scarcely as a human being.

The virgin forest closes in near Manaos, but steamer traffic is maintained upon the great river and its affluents into the heart of the continent. The great Madeira river, one of the principal tributaries, affords access by navigation for 1000 miles from Manaos to the terminus of the new Madeira-Mamore railway at Porto Velho, and ocean steamers load here, having arrived under their own steam, at an elevation of 600 feet above sea-level, 2000 miles from the Atlantic. The railway gives outlet to the Bolivian rubber forests, and penetrates dense and practically unknown wilds, shrouded in the thick canopy of the tropical vegetation, whose only inhabitants are the harmless, naked Indians who emerge timidly therefrom, to barter with the white man. There are, however, elsewhere savage tribes who attack with poisoned arrows. The line is built around the rapids, which here form a barrier to river navigation, and is 200 miles long, giving access to thousands of miles of navigable water in Bolivia.

The coffee industry is now the most important element in Brazilian trade, and might almost be said to have produced an economy and local civilisation of its own. This industry centres about the town of Sao Paulo, in the states of Espiritu Santo, Minas Geraes, Rio de Janeiro and Sao Paulo, principally. The rise of the industry in this region was due to the character of the soil, which is a rich deep " red earth " of decomposed lava. In 1885, when the knowledge of its properties became known, a veritable mania for coffee-planting possessed itself of the Brazilians—rich and poor, townsman and countryman—who for the time almost abandoned other forms of employment or agriculture. The diabasic red earth deposits were eagerly sought for, and *fazendas*, or plantations, appeared on every hand.

The result has been in the establishment of the greatest coffee-producing centre in the world, and Sao Paulo became the wealthiest and most populous part of the republic, yielding half the world's output of coffee. Some of the largest estates are 50,000 acres in extent, and they form complete settlements, with their own schools, machinery, railways and other adjuncts, and the growing, curing and sale of coffee occupies, it might be said, the mind of the community here to the exclusion of all else.

Agriculturists of every nationality enter into the composition of the population of Sao Paulo. There was an enormous influx of Italian labour, induced by state-aided colonisation schemes and contract labour systems, but the conditions under which this labour was employed were so unsatisfactory that the Italian Government prohibited emigration until they were bettered. In addition, large numbers of the Italians left the country on account of the impossibility of obtaining small holdings. There are, however, over 1,000,000 Italians around Sao Paulo, forming the most compact oversea community of this nationality existing in any part of the world.

The Italians are highly esteemed in Brazil as colonists, as are the Germans, who number perhaps 500,000 in normal times. Immigration and colonisation by Europeans is steadily increasing in the republic. Propaganda by the Government has been made abroad; foreign colonies have been laid out, and various attractions are offered with the view of inducing Europeans to settle in the country—attractions which are often more apparent than real, but which in some cases offer positive advantage. But investigation and safeguard are very necessary in this connection for the ingoing settler.

One of the most prosperous industries in Brazil is that of cotton cultivation and manufacture. There are more than 150 mills throughout the country, with 1,000,000 spindles, and large dividends are earned by their proprietors. All the raw cotton produced is consumed locally, and the import, both of this and of cotton textiles, is steadily decreasing, showing the growth of the home industry—an advantageous economic condition for the republic. But notwithstanding high tariffs—which in some cases have been overdone—Brazil still imports large quantities of the better kinds of fabrics. Imported coal is used as fuel ; freights are high for carriage from plantation to mill ; labour is cheap, and there is a plenitude of capital in Brazil for the industry.

There is very abundant water-power in Brazil, and this has been utilised for the cotton mills in some cases. The purchase of textile machinery from Britain is very considerable. The mills are often well equipped, but the capacity of the mill hand is estimated at about half that of a Lancashire worker. Much of the cotton is grown on small plantations, and thus is a " poor man's industry," which is not the case with coffee, which requires large capital.

Brazil produces more than half of the world's cocoa supply, but more might be cultivated. A great deal of sugar is produced, especially in the important sugar-producing districts around Pernambuco, but the export is decreasing, due to growing home consumption.

Maize, *yerba maté*, or Paraguayan tea, tobacco and mandioca —farinha or cassava—are other important products, the last-named being the staple food of the masses. It has been stated that an acre of good farinha is the equivalent of six acres of corn, and requires less labour. Tapioca—from the same source—might be more largely exported. The tobacco is often of excellent quality, equal to the Cuban in flavour, and it is almost all consumed in the republic.

The pastoral industries of Brazil are of great importance, the vast plains and plateaux of the country affording a wide field for cattle-raising ; but the production of meat and other animal products fails at present even to meet the country's home requirements, and large quantities of *charqui*, or jerked beef, are imported, which might easily be produced at home. The estancieros or great cattle owners of Brazil do little more than leave the cattle to multiply on the natural pastures. Sheep-raising has received but little attention.

The famous *gaucho* of Brazil, the breed of man whose business lies exclusively with the cattle, is an extremely picturesque figure, with his gay, distinctive dress, his splendid horsemanship and peculiar habits. He is a veritable son of the pampas, devoid of fear, uncouth, generous, quarrelsome, a creature of his peculiar environment.

The principal mineral wealth of Brazil—although at present only partially developed—is in the enormous deposits of iron ore. Thousands of millions of tons of ironstone, carrying high grades of metal, exist in the highlands in various parts of the country, capable of forming the basis of a great industry. The main drawback is the absence of fuel, for Brazil is not a coal-bearing country, except, as so far known, to a limited extent, and it is probable that the industry will resolve itself into the export of ore.

Several famous gold mines, the best producing in South America, are found in Brazil, notably the St John del Rey and the Ouro Preto, the first-named being the deepest mine in the world. These have produced large dividends for many years, for their British shareholders. Numerous ancient workings exist elsewhere, but it cannot be said that gold-mining is likely to become a very important industry here, although capable of considerable extension. A great deal of capital has been lost in seeking for gold in Brazil at various times.

It is thus seen that Brazil produces a wide variety and extent of food-stuffs and material of manufacture. The earlier culti-vators developed an unsound policy of devoting their attention to the cultivation of only a few articles, and when, through condition of markets, drought, labour or other causes, prices fell, economic disaster was entailed. This system of " mono-culture " is still a failing of Brazil, but the Government has recognised the condition and strives to bring about a more general and extended cultivation of food products, and to foster home manufacture.

With its considerable population and extremely varied and

INDIAN WOMEN OF THE CHACO, PARAGUAY, IN DANCE COSTUME

abundant resources, Brazil in the coming years might become a far more important factor in the civilisation of the Western Hemisphere. To attain to this end much greater economic organisation is requisite, and a more sustained and intensive policy for upraising the vast bulk of its poor democracy.

West of Brazil and south of Bolivia, in the region watered by the tributaries of the River Plate and cut by the tropic of Capricorn, lie the northern portions of the republics of Argentina and Paraguay. This territory embodies the wild and comparatively little known region of the Gran Chaco, consisting largely of forest, jungle and swamp, and inhabited principally by uncivilised tribes of Indians, the Guaranis and others. There is, however, especially in the Argentine portion of the Chaco, a very extensive area of land well adapted for cotton cultivation, and the existing plantation of cotton upon the rich, black alluvial soil of the region has given excellent results. It is believed that the Argentine Chaco might, under wide cultivation and with an adequate labour supply, become an important element in the industry of the world's cotton production. Unfortunately the conditions of labour—such as obtain upon the sugar plantations around Mendoza, much farther south in Argentina, in some cases worked by British capital—leave much to be desired, from a sociological and economic point of view.

Paraguay is, in large part, an extremely fertile country, often well watered, but its people are poor and the republic is one that has suffered more terribly from internecine strife and revolution than perhaps any of the South American states. The admixture of European blood in the Paraguayans is less than elsewhere on the Continent, but the white element and mixed race which forms the governing class, although small, controls the life and resources of the country, and is also responsible for the political unrest which at times drenches it in blood. Some experiments in " socialistic " government and in private schemes of communism have taken place in the republic. The South American Missionary Society is doing good work among the Indians.

The principal article of cultivation in Paraguay is the native tea, or *yerba maté*, largely consumed in South America, an

ilex leaf containing excellent properties. Sugar-cane, grapes,
coffee, tobacco, cotton, oranges, ramie fibre, are all important
products, and the country is, in some respects, very favourably
adapted for cattle raising. Navigable communication is given
to the country by the river system, the Paraná, Paraguay and
Pilcomayo, all affluents of the mighty River Plate, with its
outlet at Buenos Ayres, far to the south in the temperate zone
of South America.

To the north of this region arise the escarpments of the
great tablelands of Matto Grosso, in Brazil, and on the west
northern Argentina abuts upon the vast and rugged Cordillera
of the Andes, overlooking those Chilean deserts and mineralised
territories which we have elsewhere described. Much may be
expected of this part of South America, embodied in Paraguay
and the adjacent regions, whenever the tide of economic development may flow that way.

In the above description of Brazil and neighbouring territories
we reach the close of our survey of the tropics. The Atlantic
Ocean, which rolls between South America and Africa, unlike
the Indian and Pacific oceans, contains scarcely any islands
within the torrid zone. With the exception of Cape Verde,
St Helena and Ascension, nearer Africa than South America,
the only isolated point of tropical land in the South Atlantic
is the uninhabited Trinidad island, belonging to Brazil, which
in 1700 was occupied by the astronomer Halley for Britain,
and which might have been a valuable coaling station. Even
the lost or fabled continent of Atlantis the mythical island of
Brazil, and other lands beyond the Pillars of Hercules containing those "Earthly Paradises" of which the ancients dreamed
and Plato wrote, must have lain—if at any time such territory
existed—north of the tropic of Cancer. But we are reminded
how nearly the tropic belt closes in when we remark that from
land to land, at their nearest points, South America and Africa
are but fourteen hundred miles apart.

CHAPTER XL

THE FUTURE OF THE TROPICS

IN the foregoing pages we have surveyed the entire equatorial belt and have passed in brief review its varied lands, with their people and resources. The extent, variety and frequent beauty of the tropics, together with their shortcomings, conditions of barbarism and their requirements and possibilities, form a subject of profound interest to the student of human affairs and the developing world. Their scenery, their fertility, their deserts, forests, mountains, rivers and all other natural features, the scope of adventurous travel, their animal life, beautiful or formidable, and their human races, with their varying stages of savagery or culture, are unrolled before us.

As earlier stated, it is primarily to the economic resources and possibilities and the human element of the tropics that it is the purpose of this book to point. Viewed thus, we shall have remarked two outstanding conditions in the life of these regions. The first of these is in the abundance of their natural gifts, of food-stuffs and other material resources valuable to the world, and the second is the comparative misery in which the bulk of their native inhabitants dwell.

We are well aware of the fact that the development of such regions must be slow ; that primitive peoples live in a primitive manner. But we are constrained to ask : Are the intelligent communities into whose hands the control of the backward coloured nations has been given doing all that modern intelligence would suggest and modern duty demand towards the development of these great resources and the advancement of the native people of tropical lands ?

These regions offer peculiar problems of development. The fact that they are extremely rich in natural products, yet inhabited by people of inferior intelligence, has itself invited the ruthless exploitation which in large part has formed the history of equatorial Africa, Asia and America. These people have

431

commonly been regarded as beasts of burden or mere instruments in the creation of wealth for their more intelligent conquerors. As we have seen, they have been driven to work in plantation or mine under the lash, their lands have been taken from them, they have been starved or tortured, their women outraged and their most primitive rights denied. The equatorial sun, during the past centuries, has looked down upon the cruellest treatment of the simplest children of the soil. This has especially been the case since the system of foreign trade and commerce began to grow, when the ships of the white men penetrated the bays and rivers of the darker continents, or reached the remote tropical islands.

The rapacious brutality of earlier days cannot be now carried on, thanks to the growth of public knowledge and opinion, but it is not entirely done with, as the notorious occurrences of the Congo and Putumayo show. But how far we still are from a true economic control of the backward coloured races is sufficiently evident to the disinterested observer. It might have been supposed, as earlier remarked, that even the most commonplace economic principles, apart from humanitarian considerations, would have urged the advisability of fostering and conserving the coloured labourer, who himself is generally the only instrument capable of cultivating the tropical lands upon which he dwells. But so far from this being the case, the treatment meted out in many tropical lands is such that native races have been partly destroyed, or live in conditions of such insufficiency and discomfort that their propagation is often seriously hindered.

In most tropical lands the cry of the white master is for "more labour." But, dwelling in the rudest habitations, with insufficient food and clothing and lack of hygienic conditions and general amenities, the native in many cases is not likely to increase in numbers or intelligence. It is only of very recent years that governmental control has been exercised in the betterment of these conditions, but there are still vast tropical territories within our survey whose Governments are still criminally negligent or careless in these respects.

The coloured peoples, on the other hand, have had their own serious defects, and have suffered by reason of their own sloth

TATTOOED WOMEN OF NEW GUINEA

or savagery in many instances. We find millions of them plunged in the most barbarous conditions of life. In all the remote tropical lands we find savages practising horrible rites and abominable customs. Many of them deform their bodies in a repugnant manner, placing sticks, rings or stones in their lips, noses or ears, or tattoo themselves, or go head-hunting, or indulge in cruel and degrading fetish worship and human sacrifice. Often they treat both their own people and their enemies with unspeakable brutality. From one point of view, many of the native customs are barbarous and useless, and we might almost be tempted to agree with those ethnologists who have thought there was once a time in social evolution " when all mankind was mad," or with those who would prescribe total extermination for some of the almost hopelessly brutal tribes of the tropical forests.

The growth of knowledge about the savage races, however, fortunately shows that the more we learn about the savage's point of view the better situated we are in assisting him towards civilisation. Even the sticks and stones in ears and noses, and other deformities, are practised because the savage regards them as adding to his or her beauty. Possibly we may assist our sense of proportion in this connection by referring to the fashion - plates of modern civilised life, and remark the absurdities in dress of the civilised female at times,[1] as well as noting afresh some of our own social customs.

The study of the native's point of view has, moreover, brought about a better understanding of his alleged sloth and idleness. Indeed, what may be termed the " lazy nigger " qualification has undergone considerable modification of late years. Sweepingly dubbed dirty, idle and vicious, we now find, under a less prejudiced observation, that the native does by no means always deserve these epithets. The African is frequently an active and intelligent worker after he is instructed.[2] The brown or Indian race of America, especially in such countries as Mexico, it would be extremely unjust to describe as lazy. Moreover, as we have seen, the natives of nearly all tropical lands

[1] Moreover, modern development of ' Kultur " should bid us suspend judgment somewhat upon the practices of the " native."

[2] Drummond, in *Tropical Africa*, twenty-five years ago, writes almost enthusiastically of the negro's capacities for work.

2 E

cultivate many useful and often beautiful arts. They are generally industrious in their own way. The agriculture, weaving,
pottery-making, iron-working and so forth of the American,
African or other so-called savage races, black or brown, are
often of much value and reveal great dexterity.

It is true that native peoples do not generally care to be
herded into factories to perform stated hours of work, or to toil
on land which is not their own, but they often work hard for
themselves. They have also been reproached because their
wants were few, and consequently they did not constitute themselves a sufficiently active " market " for the purchase of the
white man's manufactured goods, articles which in many cases
judicious reflection shows are unnecessary even for the white
man. The native's defect is that he does not understand the
need of labouring for the morrow, of building up profit and
wealth, of winning the fruits of the soil to an extent beyond
that which he momentarily requires. The soil for him is a
mother, not a task-master.

From one point of view it may be asked why he should over-
labour himself, when his scheme of life is simply to live, especially when, at the present stage of his " instruction," such
profits go largely to enrich his master rather than himself.
Whilst all will agree that the development of the world cannot
be blocked by the sloth or savagery of inferior peoples, nevertheless the belief early formed and still held by many that the
coloured labourer was designed as a human machine and mere
servant, who should be forced to work for his white master, is
rapidly losing ground.

It is, furthermore, interesting to consider another view of the
economics of native life. It is noteworthy that with all their
primitive ways the native tribes do not suffer from the existence
in their midst of that poor, destitute, unemployed or miserable
class which modern industrial civilisation produces in our towns
and country-side. They may suffer at times from famine, but
there are no economic outcasts among savages. In their simple
way all are provided for, all have wherewith to eat, dwell and
work. Their " political economy " does not contemplate the
spectacle of starving men, women and children, unemployed,
yet surrounded by plenty. In their primitive " human
geography " the land they live in sustains them all. In ours

CALENDARIO AZTECA O PIEDRA DEL SOL.
EN EL MES DE DICIEMBRE DEL AÑO DE 1790
AL PRACTICARSE LA NIVELACION PARA EL NUEVO
EMPEDRADO DE LA PLAZA MAYOR DE ESTA CAPITAL
FUE DESCUBIERTO ESTE MONOLITO Y COLOCADO
DESPUES AL PIE DE LA TORRE OCCIDENTAL DE LA
CATEDRAL POR EL LADO QUE VE AL PONIENTE
DE CUYO LUGAR SE TRASLADO A ESTE MUSEO
NACIONAL EN AGOSTO DE 1885.

EARLY TROPICAL SCIENCE   THE AZTEC CALENDAR STONE, MEXICO

the very richest of our lands or districts cannot do this.[1] The
native does not herd his kind into slum dwellings and congested streets where there is nothing to be obtained directly
from the soil, nor does he monopolise the soil or neglect to
make it yield sufficient for his wants, as is often the way of the
white man. In brief, he does not outrage Nature's economic
system, as do the people of manufacturing nations, and the
condition offers matter of study for civilisation.

Turning for a space from the contemplation of tropical resources as they concern the native, to consider their relation to
the teeming peoples of the civilised temperate zones and nations,
we might, from an advanced economic standpoint, marvel that
so comparatively limited organising effort is brought to bear
thereon. It is true that commercialistic development has of
late been rapid, but much of it is on lines which cannot be
permanent. The tropics are full of commodities which we
urgently require in our everyday life, for our breakfast-table
or our workshop, but for which we are forced to pay inflated
prices. When we remark the enormous areas available for the
cultivation of food-stuffs and other products in the tropical
countries, still lying fallow, and recollect that almost the only
system of their cultivation for export is that of commercialistic
adventure—which, furnishing good results in many cases, can
but skim the cream of these resources, and that often accompanied by an improper exploitation of the native—we cannot
but censure our lack of scientific organisation. We are fully
aware of the debt society owes to such individual enterprise,
but something more is now requisite.

This, of course, arouses considerations which are common to
the whole of the industrial world, in any climate. But the
tropics especially stand forth as concerned therein. The fact
is that we have not yet developed a true "science of
living." We have no constructive economic biology. The
true adaptation of mankind to its economic environment of the
earth has not yet come to being. As regards savage lands,
these have now been overrun by explorers, geographers,

[1] An example will suffice : Southern Scotland is perhaps one of the
richest lands in natural resources in the world—animal, vegetable and
mineral, but the slums of Glasgow and Edinburgh are amongst the most
appalling human dwellings on the face of the globe.

travellers and traders. The heights of mountains have been determined and the courses of rivers mapped, and innumerable books have been written about them ; and knowledge of the world's geography, thanks to these explorers, is almost complete. Scientists classify the crania of the aborigines and discover the archæological secrets of ancient kingdoms ; they classify a myriad insects and plants : and indeed every conceivable aspect of life is studied—except that primal, simple one of what we have termed constructive human geography, the science of collective living on the earth, to which we shall presently refer at greater length.

It seems almost a platitude to say that the proper exploitation of tropical resources is a matter in which the interests both of the native and the absentee white consumer are bound up together. But the interests of the former are, as we see, often sacrificed to those of the latter. Tropical produce is generally created by natives who work for very low rates of pay for planters and others. Often these coloured folk are in the position of sending away overseas the foods they produce, whilst they themselves suffer from insufficiency. The complacent British, American, German, French or other white people have become so accustomed to be fed with many things by the efforts of toiling black and brown men in distant lands, of having their breakfast-table stocked by people who themselves, it might be said, have no breakfast-table at all, that they rarely give the matter a thought.

It is true that much more attention is now given to matters of native labour, of slavery, peonage, debt-bondage, contract and forced labour, and so forth, and there is marked progress in humanitarian considerations here.[1]   But the battle against commercialism and vested interests is an uphill one, and sentiment too often is made to play the part which ought to be taken by scientific organisation. Philanthropist and missionary are carrying on their noble work in almost every tropical land, but of themselves alone they cannot furnish the constructive effort necessary to an organised economic advancement, economics which must be based upon principles scientifically ethical.

---

[1] The natives of the tropics owe an incalculable debt in this connection to the Aborigines Protection Society of London, an institution little known, however, to the general or trading public.

Perhaps the most hopeful, because at present the most practical condition affecting the tropical lands is the awakening and application of judicious governance. In this it will not be disputed that Britain has led the way, and under her control the basis of contentment and peace, with the beginnings of prosperity for many native states, has been laid. The roll of high-minded and capable administrators whom Britain has set over her tropical possessions is a long and honourable one, and is the outcome of a righteous system of conquest, such as the world has never witnessed before. When errors have crept in they have been errors of honesty and humanity, rather than of self-serving and oppression, except that at times economic selfishness has appeared.

Among the most valuable features of governmental action are the Imperial Departments of Agriculture, of which leading examples are in the West Indies, India, the African colonies and so forth. France and other European powers are not without some institutions of a similar character. Universities of tropical agriculture have been suggested also.[2]

The commercial element, which takes the field after the soldier, the governor and the official scientist have established the nucleus of civilisation and the basis of scientific industry, falls, from the very nature of things, short of the spirit of governance. There are many serious charges to be laid at the door of Commerce in the tropics, as the hastiest examination of these pages will have been sufficient to indicate. Nevertheless, without the commercial spirit, little could have been done so far. Also, it may be laid to the credit of Britain that the standard of British traders leads the way. Furthermore, if the methods of traders and commercialists have been imperfect and brutal at times, we are bound to reflect that it was often an imperfect and brutal world that had to be entered.

One of the worst vices of the tropical native, whether in Africa, America or Asia, is drunkenness. But, as before remarked, the drink is generally produced and sold by his masters. In

[1] In London the Imperial Institute carries on a work in connection with the British tropical possessions of real and growing value, but the British public knows too little of its operations. The same might be said of various of our scientific societies and institutions; geographical, botanical and so forth.

[2] By *Tropical Life*, a London periodical.

Mexico and South America the huge plantation owners manufacture crude spirits and other intoxicating drinks, and this form of commercialism is ruining the valuable brown race. In Africa British and other governments permit the import and sale of drink, to the great detriment of the native, in many instances.

The problem for the future development of the tropical lands which lies before the civilised world is, as has here been accentuated, a twin problem. It calls for a system under which the coloured native races will fully and fairly enjoy the wealth of their native lands and the fruits of their labour, whilst at the same time the oversea white inhabitants of the globe continue to draw their necessary supplies from the tropics, supplies which their own lands can, of course, not produce, and to which they are geographically entitled as inhabitants of the common earth. This problem will have to be attacked scientifically, and from a new point of view. It is the duty of the native to perform an adequate share of work, and it is the duty of his white " masters "—the term is not an adequate one, for it is a " partnership " that is involved in reality—to see that he receives his fair share of the results. The problem is not one of easy solution, but, like other economic problems, science has not yet adequately been brought to bear upon it.

In the past and the present tropical cultivation and production have mainly been carried on under a system of " monoculture "—that is, the concentration of effort on one kind of crop or product in one place. The earliest example was in sugar-growing in the West Indies, Brazil and elsewhere, and monoculture here was largely responsible for slavery. It was the outcome of the system by which one kind of product was greedily wrested from the soil under cheap, unpaid or slave labour for oversea export and the enrichment of estate owners. In less degree the same system has underlain the monoculture of coffee, cocoa, cotton, rubber, palm-oil, coco-nuts, tea and other tropical or sub-tropical products.

Monoculture has given rise, apart from slavery, to debt-bondage, forced labour, contract labour and so forth. But it also causes at times grave economic evils, one of which is in the crises brought about in tropical lands by over-production and the dependence for revenue upon a single article, subject to the fluctuating prices of foreign markets.[1] One of the worst features of the system is that the soil is very largely cultivated at the expense of its native occupants. In the growing of remunerative crops for their master the production of food for the labourer is neglected. Further, the native is kept back from the development of manufacture ; of supplying his own wants in articles of everyday necessity. The articles produced under the system of monoculture are always destined for export, for consumption overseas. The traffic, of course, returns wealth for the country of its origin, but such wealth does not filter down to the real producer in adequate measure, under present conditions.[2]

The greatest economic evil resulting from the above system is that, under it, native communities cannot become economically self-supporting, and consequently cannot attain to a high stage of freedom and civilisation. It is not to " monoculture " but to " polyculture "—that is, to varied production as contrasted with single products—that any community must look for its economic and social security. The peculiar conditions of the tropics must always of necessity call for a considerable exercise of monoculture—they must always cultivate great areas of their peculiar products for export—but it must be balanced by equitable regard for the native producer and the exercise to the utmost possible extent of polycultural principles, whereby a supply of all products and all articles necessary for life are producible locally. Until this principle be recognised economic security will never be attainable.

Reference has been made elsewhere to an undeveloped—or at present non-existent—science of " human geography." The aims of such a science, whenever it may be established, would be of world-wide scope, whether in the temperate or tropical lands of the globe, whether among peoples advanced or primi-

---

[1] There is a further economic defect in monoculture : in the exhaustion of the soil. This has resulted in the abandonment of many Brazilian coffee plantations. It remains to be seen if it will affect also the Malaysian rubber plantations.

[2] The growth of cotton-manufacturing industries in tropical cotton producing and consuming countries would, of course, lead to the diminution of the output from Lancashire and other mills, and the whole subject is one for serious economic study.

tive.[1] Its principles involve recognition of the fact that, under organisation, the earth is capable of supporting all its people in plenty, that the economic adaptation of man to his environment is the most important consideration before society now ; that agriculture and manufacture must be made supplemental and complementary to each other in given localities, as far as is physically possible, and that economic equilibrium will only be attained to under a system of development in localities of self-supply in agriculture and manufacture, as far as this be physically possible. In local industry-planning is the practical point of application of the science.

To apply such a science to tropical lands, we should have to regulate monoculture and supplement it with polyculture and " polyfacture "—to coin some new terms—that is, to multiply and foster small and varied industries. Is such a policy possible among the coloured races ? A perusal of these pages will easily show that the natives of nearly all tropical lands possess, or did possess, excellent systems of land tenure and cultivation, and that their arts and crafts were often very well established. They are, in general, perfectly capable of manufacturing what they require, and they did so to a large extent long before the white man thrust his commodities upon them, often to their detriment. The solution will lie in part in encouraging these native manufacturing industries.

We shall have remarked in nearly all the African colonies and territories, whether negro or Mohammedan, that the natives carry on small manufacturing operations, and this fact is of extreme importance. The same holds good in India, in Mexico, in South America and in almost every tropical land. We find in Africa and India clever native smelters and workers of iron, who owed nothing of this difficult art to Europe. We find excellent and often beautiful textile work in cotton and other

---

[1] The author has consistently endeavoured during the last few years to bring about the consideration of such a science. He lectured before the Economic Section of the British Association in 1913 upon it, under the name of " Human Geography and Industry-Planning : a Necessary World Science," and before the Royal Society of Arts early in 1914. Called before the Special Commission of the House of Commons as a witness in the Putumayo atrocities inquiry, he urged the need of considering and applying the science to native peoples as a protection against undue commercial exploitation,

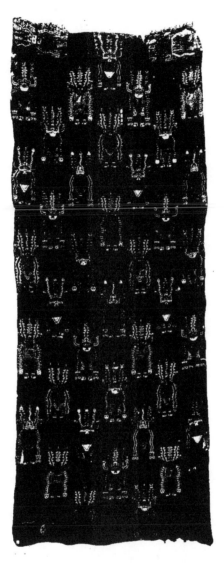

EARLY TROPICAL TEXTILE ART A PIECE OF PERUVIAN
PRE-INCA TAPESTRY, OVER 2000 YEARS OLD. ABOUT
NINE FEET LONG

fabrics throughout the whole tropical belt.   Wood-carving, gold
and silver working, stone masonry and all else are native arts in
many lands, as is boat-building and weapon and implement
making.   Irrigation and land cultivation were well understood
in the tropics by aboriginal peoples.   Native arts, dress and implements are often extremely beautiful and ingenious, and those
relics of thousands of years ago—especially in tropical America—
the pottery, sculpture, weaving and dyeing of the Aztecs, Incas
and others which we have considered in these pages, as well as
in Ceylon, India, Indo-China and elsewhere, are indications of
native culture in the tropics.

To-day the almost indefensible operations of wholesale manufacturers and traders are tending, as we have seen, to drive out
and destroy these native arts and industries.   Even travellers
and students appear to have regarded native arts and crafts as
instances of curious savagery and native cunning, rather than
as the basis or nucleus of a possible economic culture and
system, and the white man ignorantly or greedily has set himself to bring about their destruction.   In India and Burma
native textile arts are swamped by the output of Lancashire
and elsewhere.   The fine textiles of India and the Swahili
robes of East Africa have been copied by European mills, and
now unfortunately are largely supplied therefrom, and the
textiles of the clever Andine natives of South America have
also been ousted by the same thoughtless commercialism.   It
is useless to point to the fact that these garments may be
supplied " cheaper " to the native from abroad.   Nothing is
cheap that robs a people of its art and home industry.

Let us look more closely at some of these native arts.   The
paramount native manufacturing industry of the tropics has
been in textile fabrics of cotton, wool or fibre.   Often these are
of considerable beauty and of racial value.   They are, or were,
dyed from native dyes, often more attractive and enduring than
the imported aniline dyes which take their place.   To pass
from textile to metal—throughout the greater part of Africa
we remark the existence of small iron-working industries.   The
negroid peoples have worked iron from time immemorial,
whether in the east or the west of the continent, and their
blacksmiths are highly esteemed.   Other countries had their
bronze age, but this period did not exist in Africa, where the

iron age always prevailed. Why was this? Because iron is found widely distributed here, in ores of a nature such that the native could easily extract the metal, and among many tribes the chief industry was—and still is—in smelting and forging it and in making utensils and weapons.

Examples of the foregoing have frequently been noted in these pages. In the Eastern Congo and Lado Enclave, to cite one, the people smelt iron from ore, in small pits, and forge it into hoes, which they exchange for flour with neighbouring tribes. The work alternates with that of fishing, and the women take their part therein. Here is an interesting instance of home manufacture and exchange. In tropical America the natives were unacquainted with iron, but they worked and forged copper with dexterity. In parts of South America, in the Andes, their little *guayras*,[1] or lead and silver smelting furnaces, lighted up the mountains at night. The remarkable iron-working skill of India, dating from thousands of years ago, has been noted elsewhere.

Thus we see that the natives of tropical lands are capable of developing and practising their own manufacturing arts, a significant fact in "human geography which has received but scant recognition. Upon this would seem to follow the advisability of a system of native "industry-planning," under which the coloured people should be taught and encouraged to establish themselves in more or less self-supplying communities, making as far as possible their own articles of everyday use, and growing their own food, exchanging the one commodity for the other among themselves. Direct interchange of goods—the ancient system of barter—is a natural one here. Must the native necessarily refer the value of his products or labour to a European or other monetary standard, except where such products are designed for export? Why should he perforce adopt the contradictory system of low wages and high prices which are a result of this system of monetary comparison, in matters connected solely with his locality? Why should the system of the middleman and the usurer be implanted in the native political economy? The matter is capable of argument.

It is noteworthy that events in some tropical lands have laid bare the unwisdom of monocultural methods. In Brazil the

[1] Guayra is Quechua for wind, which was used as the draught.

A Mexican Cotton Factory

unsound policy of devoting too much attention to coffee or other single products is giving way, and the Government is encouraging a more varied production. Another example is in the success attending negro peasant proprietorship in the West Indies and a varied cultivation to replace the lost sugar trade.

The fostering of the self-supplying native locality and small community need not interfere with the legitimate cultivation of those products so largely required for foreign export. At present native independence is looked at askance or even discouraged, because the white master fears the depletion of his labour supply. The adjustment of these conditions is a matter for a science of human geography and industry-planning.

As before remarked, too much is left in tropical lands to the individual capitalist and commercial adventurer. The debt of gratitude owed to individual enterprise in the development of tropical resources is in no wise impugned by this statement. One of the most noteworthy agents in such developments has been the enterprising company promoter and concessionaire, with his company of absentee capitalists. These, the great explorers of oversea commerce, have invaded the soil of the most remote tropical lands, and penetrated to their darkest recesses, and they have often builded better than they knew. The joint stock company, in its hunt for dividends, has implanted flourishing enterprises in spots where philanthropic or religious influence would have been without avail. We read the reports of their labours at their annual meetings and congratulate them upon their plentiful dividends. But at the same time we point to the wretched barracks around their mills and upon their plantations, where black or brown men are herded together in conditions more savage than obtained in their native wilds, and where hordes of coloured labourers are brought overseas and condemned to live without their women.[1]

The company promoter of tropical enterprises makes attractive his prospectus with paragraphs relating to the " cheap and abundant labour " which his project may command. But how does this labour live and thrive ? The annual report refers in congratulatory terms to the reduced working costs. But at

[1] This latter condition is specially marked in the prosperous rubber plantations of Malaysia, as elsewhere remarked.

whose expense is this lower cost obtained? At the expense of the coloured labourer? By no means always, but undoubtedly often. At times when shareholders' money has been lost we obtain an inquiry into the circumstances and the veil is lifted a moment upon these oversea operations. In passing it may be noted that the company director often successfully pleads his lack of knowledge of the conditions of the business he controls, as justification for misdemeanours committed. The annual reports of such companies give little information regarding the housing, food, hygiene and instruction of the coloured workers, matters unasked for and apparently of little interest to the ordinary shareholder.

There is another side to the above picture, which must not be neglected. The multiplication of oversea joint stock concerns brings about the lowering of prices of commodities produced, which is essential to commerce and manufacture; economies resultant often upon scientific methods which are not necessarily carried out at the expense of the humble worker. Moreover, some such enterprises have already learned the primitive economic fact that, in order to get the best out of their labour, it is advisable to treat it well, and the well-being of the coloured labour in such cases is regarded as an important consideration. But such a spirit is in its infancy still.

Perhaps the future may hold possibilities for co-operative working in the production of tropical commodities, whether between the white man and the coloured, or between these and the white consumer overseas. As regards the first, it must in time become repugnant to the white planter or employer to have around him hordes of coloured dependents dwelling in filth and ignorance, with an utter lack of the amenities of life, and he himself more or less exposed to the danger that his employees will some day rise and cut his throat.[1] To give the negro or Indian a fair share in the proceeds, to arouse his sense of proprietorship instead of suppressing it, may perhaps become a feature of tropical life.

As regards co-operation between the tropical producer and the consumer overseas, when, some day, an intelligent grasp of

[1] It is well to recollect that " strikes " of coloured labour are not carried on by argument alone, as in peaceful Britain.

economics is displayed by the white rate-payer or housewife, it may be that the possibilities of such direct production and supply will begin to appear, and direct control and participation be considered. Every consumer is, in a sense, a " capitalist," in that he makes a daily outlay for food, and possibly growing economic intelligence will reveal some method of capitalising this condition. It is not inconceivable that enterprising British towns or communities at home might consider the advisability of obtaining and controlling their own areas of tropical produce, within the bounds of the tropical British Empire.[1]

Reference has been earlier made in these pages to the general apathy and lack of knowledge of the ordinary British citizen concerning the oversea possessions of the Empire, and especially of the Crown colonies and tropical possessions. The sense of proprietorship is lacking, and their business and economic possibilities often neglected. Whether a more enterprising spirit in this connection will grow to being, a greater exercise of imagination, only the future can show. There are enormous areas of fertile land lying fallow in the tropical Empire, it cannot too often be repeated, which, under a greater display of activity and a spirit of co-operation, might yield abundant products, and serve to lower the high and growing cost of living. These matters cannot be carried out by governments and companies alone, but call for national enterprise in addition.[2]

One of the most interesting and speculative conditions concerning the tropics is that of their permanent habitation by people of white race. How far Europeans may dwell in these lands in health and strength in the future will be determined by the advance of the comparatively new science of tropical medicine, which of late has made such remarkable strides.[3] The discovery of the connection between the mosquito and malarial and yellow fevers, and the remedies encountered to

[1] The author brought forward such a project in a lecture to the Royal Society of Arts in 1909.
[2] It seems scarcely necessary to say that organisation does not spell " socialism."
[3] The London and Liverpool Schools of Tropical Medicine and other kindred institutions are already famous in this respect, though of such recent establishment.

minimise or prevent those diseases are too well known to need description here. They have converted the formerly notorious Isthmus of Panama [1] into a healthy region, and other extremely dangerous places have similarly been rendered habitable and secure by the adoption of modern hygienic methods.

There are, of course, other deleterious effects of the tropic climate and soil which require further study and remedies before the white man can settle with impunity in these lands. But is it not possible that the tropical lands, as it is learned to live in them in greater security of health, might become the home of small agricultural industries of particular kinds, worked by white men with their own labour—a sort of aristocracy of tropical labour? This may be rendered possible by the demand for tropical products.[2]

A notable sociological feature among the African natives, especially in British East Africa, Nyasaland, etc., is the growing desire of the negro for education. "He is eager to learn. Young educated natives are in much request, and are found to be as competent as the Indian 'babu.'"[3] Educated natives often purchase blocks of land, for independent cultivation, and live in brick-built houses. Others become skilled craftsmen. In Nairobi the negro is becoming a skilful mechanic and taking the place of the Indian artisan here. These indications are encouraging. There is no reason why a similar movement should not take place among the brown race of tropical America under better governance.

When we come to a consideration of the part which missionary work has played in the progress of the tropics, our first consideration must be one of gratitude and appreciation of the noble work carried out, whatever "economic" criticism may be directed against it. These services to mankind can never

---

[1] *Vide* among other works the author's *The Panama Canal: Its Past, Present and Future* (Collins, Glasgow).

[2] An example is in the demand for and price of copra, or dried coco-nut, now largely used in the manufacture of food-stuffs and other articles. About ten years ago the market price of copra was about £10 per ton, but of late it has reached £30, or £40, and possibly will go beyond this. There are, of course, wide areas capable of its production, but the coco-palm thrives best on sandy seashores, and within a belt not over 500 miles north and south of the Equator.

[3] Sir Alfred Sharpe, *R.G.S. Journal*, Jan. 1912.

be sufficiently esteemed, nor can they be adjudged by the immediate results apparent therefrom. Missionaries, moreover, have carried, not only spiritual, but often economic light to the heathen of the torrid zone, and, in addition, they have done no mean exploratory and geographical work. Some of the most famous African explorers were missionaries. In British Africa the missionary has done magnificent work in regard to such matters as the "drink traffic," forced labour, the alienation of native lands and so forth, matters which even modern British (and Liberal) governments have been obliged to compromise with.[1]

It may be, however, that, in the future, the economic side of missionary work will have to be strongly developed, if the movement is to hold its own. The Christian missionary is already taking note of various factors which may direct him upon some path which will secure greater progress. One of these is in the rapid spread of Mohammedanism in Africa, even in those parts of the continent controlled by Christian nations. The Moslem traders of East and West Africa carry their faith with them, and mosques have sprung up on every side of late years where earlier they were unknown. Another factor is in the apathy of the ordinary layman in Britain or elsewhere towards missionary work, and even in open antagonism. To what is this apathy due? The layman can best understand it. Missionary affairs have been regarded as the special occupation at home of feminine effort and of children's work. Ladies' meetings, at which local clergymen officiate and movements to secure the pence of boys and girls and servant maids, for the benefit of distant and intangible negroes, are shunned by the male members of a family. It could scarcely be otherwise. The very mode of presentation of religious missionary effort is often childish rather than virile, and the layman's prejudices are aroused. Indeed, the missionary is handicapped by the too sentimental side of his work. Further, there is a class which protests against the expenditure upon the far-off heathen, when there exist so many poor and miserable in our midst

[1] A recent instance was that of Uganda, where in 1910 Bishop Tucker drew attention to what was practically forced labour, the only "volunteer" being the chief who commanded his people, at the behests of the white employers and government.

at home—a view short-sighted it may be, but not to be overlooked, and not unjustifiable.[1]

This situation is not without a natural cause. Religious thought has undoubtedly been too greatly divorced from economic work. It must be more widely recognised that, to "save souls," bodies must also be saved, a fact that religious thought too often neglects. That "men may have life and have it more abundantly" is a maxim whose economic counterpart or meaning has been grievously neglected. In reality economics and ethics are inextricably interwoven, but the lesson has not yet been fully learned in any community.

Probably the greatest success has been attained by missionary work where medical missions have formed part of the operations, and where the industrial education of natives and native children has been undertaken, in addition to the spiritual work, and these methods are growing. Some of the most successful missions ever established were those of the Jesuits in earlier times in South America, who worked upon these lines. They established what were practically self-supporting communities of Indians, who carried out all the manufacturing and agricultural work necessary for their maintenance, establishing settlements of great beauty and utility. These were ruthlessly destroyed by the white Portuguese and Spanish colonists, who did not desire to see the Indian self-supporting, but coveted him as a labourer or serf.

The same useful organising principle is at work among some modern Christian missions, and the "colonising" of the natives and their instruction in local arts and crafts is being more and more undertaken, and such lands as Uganda show excellent examples of the system. But it has to confront the commercial demand for labour at present. If missionary work is to hold its own the practice must be widely extended. These considerations conduct us once more to the need for a strong, virile, economic science of tropical industry-planning. Such a work, carried on by economists under ethical principles such as

[1] The following figures are of interest here : Annual British contributions to Foreign Missionary work amount to £1,800,000; expenditure in intoxicating liquors, £163,000,000: in tea, coffee and cocoa, £31,000,000. Christianity is apparently not an exorbitant item—food for thought in a nation of shopkeepers!

those earlier outlined, for the purpose of advancing the
native state and developing tropical resources in the double
interests of the native and the oversea consumer, of fostering
the native's arts and increasing his self-supporting powers,
would be likely to appeal to the general intelligence of the
white citizen in a way that missionary work alone cannot
attain to.[1]

In the organisation and development of tropical lands we
shall hope not to see implanted, along with greater civilisation,
the prosaic or repellent elements of such civilisation. We do
not desire to see the picturesque landscape, for example,
scarred with the purely utilitarian and generally ugly factories
and structures which the white man of the industrial nations
at the present stage of his civilisation seems to find necessary
for the making of his commodities. The outskirts of British,
American or other manufacturing towns are perhaps the ugliest
places in the world, and they embody, from the æsthetic point
of view, the most unpleasing of human dwellings, in the congested streets and slums of their whereabouts. It must be
part of a study of a true industrial science of living—almost
entirely neglected so far—to endeavour to elevate labour by the
refinement of its surroundings. There is no inherent reason
why factories should be an eyesore, or their surroundings repellent. We should indeed strive to guard against reproducing
many of the unlovely and unnecessary phases of industrial
life in the at present undeveloped lands of the tropics. There
is no reason why they should go through the fearful baptism
of slum dwellings, monopolies, strikes and so forth, or why
these simpler people should be herded into huge factories and
congested dwellings.

Again we have the problem as to whether the native should
discard his native dress for European garb. To clothe the

[1] It might be said that we now require an institution whose work
would be to spread this practical and constructive "gospel," a scientific
body, perhaps of the nature of our Royal Societies, which would send
out its "missionaries" to the economic heathens (at home and abroad),
and show them the true "science of living on the earth." There is
no such institution in any land. Every science, from astronomy to
geography, or from law to medicine, is represented by some central
authority, but "Constructive Human Geography" or "Geo-economics"
is without a habitation and a name. (See Preface.)

2 F

native in the hat and coat of the white man, in place of his often simpler and cheaper habiliments, does not appeal to the student of native life. Wherever this has been done it appears commonplace and incongruous, and sometimes has brought evils in its train. The problem is, of course, not an easy one. Many native tribes wear no clothes at all, or a mere loin-cloth. Civilisation demands that the body be covered. Also, we cannot regard simply the picturesque side of native garb. Clothes must be such as work can be performed in. But in every tropical land there exist native costumes, and the question would seem to be rather one of their adaptation and the extension of their use.

The preservation of the æsthetic values of the tropics is not a subject to be thoughtlessly brushed aside. The simple people who inhabit these lands are generally courteous and often kindly and dignified. The poorest peon of Spanish America has a courtly demeanour, and the African and other coloured man has his own qualities of intercourse. Are they inevitably to acquire the brusqueness, greed and hypocrisy of the commercial world ? [1]

And, lastly, as to the deserts and the mountains, the rivers and the plains, in their vast expanses of nature and the unspoilt beauty and solitude of their own, they are among the great assets of the tropical lands. It is true that the dreadful deserts of Africa or Western America and elsewhere, or the melancholy uplands of the high, bleak tropical mountains are, from one point of view, waste spaces, and might almost seem to embody the "abomination of desolation" spoken of by the prophet. We might say in our haste that they are of no use to man. But reflection will show that even the deserts have an indescribable æsthetic and contrast value. They are spaces apart, even now, for the refreshment of the spirit. For aught we know they may contain potentialities of unknown value for a less greedy and more advanced humanity, elemental powers which may yet be harnessed for human use. For the moment, however, we are almost tempted to wish that—like the famous mathematical formula—they may "never be of any (commercial) use to anyone." The deserts are at least a corrective for the modern pedlar spirit. The dwellings of a teeming humanity of the

---

[1] Qualities which were largely instrumental in plunging civilisation into the Great War.

future will, we trust, always have to stop at the desert's edge, to halt " between the desert and the sown."

The great undeveloped spaces of the earth open to our imagination and energy an enormous field. But our work therein must be inspired by something of a new spirit. We should regard the earth more in the light of an estate ; a common heritage, calling for a stronger sense of possession and a more planned and intensive organisation of its potentialities. Our labours and our ideals have lagged. Life and evolution call for more conscientious and scientific stewardship. As for the torrid zone, which we have here surveyed, we have much to build up. Just as the hunter now learns to exchange the rifle for the camera, so must our efforts in the tropics be constructive and not destructive. Not " you curious savage, how can I exploit you ? " but " how can we construct for you a true economic and civilised regimen ? " will have to be our " culture " in dealing with the coloured man—black or brown. The lesson of the tropics is that we have much to learn from Nature, that Nature here offers us a vast store-house of things of benefit, to be enjoyed in proportion as the mood shall come upon the world, faithfully and with method to set its house in order.

ADDENDUM

Page 213.   The railway from Jibuti to the capital of Abyssinia, Adis Ababa, has now been completed.

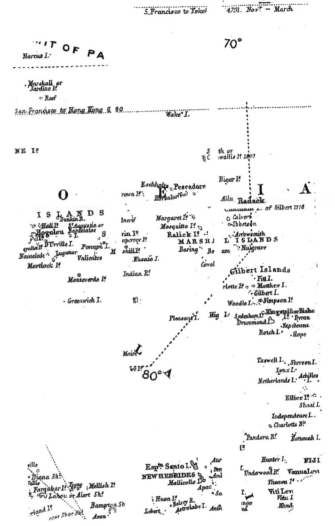

INDEX

A

453

9 781330 139851